Methods in Molecular Biology

Series Editor
John M. Walker
School of Life and Medical Sciences
University of Hertfordshire
Hatfield, Hertfordshire, UK

For further volumes:
http://www.springer.com/series/7651

For over 35 years, biological scientists have come to rely on the research protocols and methodologies in the critically acclaimed *Methods in Molecular Biology* series. The series was the first to introduce the step-by-step protocols approach that has become the standard in all biomedical protocol publishing. Each protocol is provided in readily-reproducible step-by-step fashion, opening with an introductory overview, a list of the materials and reagents needed to complete the experiment, and followed by a detailed procedure that is supported with a helpful notes section offering tips and tricks of the trade as well as troubleshooting advice. These hallmark features were introduced by series editor Dr. John Walker and constitute the key ingredient in each and every volume of the *Methods in Molecular Biology* series. Tested and trusted, comprehensive and reliable, all protocols from the series are indexed in PubMed.

DNA Cloning and Assembly

Methods and Protocols

Edited by

Sunil Chandran and Kevin W. George

Amyris, Inc., Emeryville, CA, USA

Humana Press

Editors
Sunil Chandran
Amyris, Inc.
Emeryville, CA, USA

Kevin W. George
Amyris, Inc.
Emeryville, CA, USA

ISSN 1064-3745 ISSN 1940-6029 (electronic)
Methods in Molecular Biology
ISBN 978-1-0716-0907-1 ISBN 978-1-0716-0908-8 (eBook)
https://doi.org/10.1007/978-1-0716-0908-8

This Humana imprint is published by the registered company Springer Science+Business Media, LLC, part of Springer Nature.
The registered company address is: 1 New York Plaza, New York, NY 10004, U.S.A.

Preface

Biology is nearly infinite in its molecular diversity, complex signaling cascades, and variety of organisms, each with their unique modes of adapting to their environment. Scientists have made tremendous progress in engineering these biological systems for a variety of purposes, resulting in impactful solutions across multiple disciplines. A common thread in these solutions is the requirement to manipulate DNA and engineer it in a format that elicits a response from the biological system. DNA assembly, the joining of two or more DNA parts together to create a desired construct, is perhaps the most common DNA manipulation. In its most basic form, DNA assembly consists of cloning a single DNA part into an organism-specific vector. While this type of simplistic DNA assembly has been practiced for decades, the expanding scope of biological engineering has necessitated the construction of larger, more intricate DNA assemblies encoding for complex functionalities. These constructs must be generated with high accuracy and as efficiently as possible to facilitate the rapid prototyping, testing, and iteration required in modern biological engineering. To meet these requirements, the last decade has witnessed an explosion in DNA assembly methods and standards to cover a wide range of functionalities. In this volume, we share protocols for many of the innovative DNA assembly methods developed over the last decade, with an emphasis on efficiency, quality, and complexity. The volume is divided into four parts, which cover computational tools to design and track the construction of DNA assemblies (Part I), workflows that enable high-throughput assembly of DNA constructs (Part II), standardized toolkits and characterized parts for DNA assembly (Part III), and combinatorial solutions that enable construction and optimization of entire pathways (Part IV).

Part I covers computational solutions that allow users to rapidly and efficiently design assemblies and plan laboratory workflows. In Chapter 1, Ernst Oberortner et al. describe the diverse Biological Computer-Aided Design and Manufacturing (bioCAD/CAM) tools developed by JGI. To illustrate the power of these tools, the authors share a complete workflow for the design, hierarchical assembly, and verification of a multigene biosynthetic pathway. In Chapter 2, Michael Fero and the team from TeselaGen describe j5, a sophisticated design/planning tool that provides users with robust, automatically generated protocols based on the chosen assembly method. In Chapter 3, Lena Hochrein and Fabian Machens from the University of Potsdam present the AssemblX toolkit, which integrates a software program and a novel cloning method to provide a streamlined protocol for DNA assembly compatible with any expression host. The method has been used to assemble up to 25 functional units and includes an AssemblX web tool that guides the user through each step.

Once the assembly has been designed and the protocols have been finalized, the next step is to generate the parts needed for the assembly, assemble the parts, and finally confirm the identity of the assembled construct. In Part II, Chapters 4–6, David Reif, Kristy Ip, Ron Yadin, and William Christie from Amyris describe methods to simultaneously generate thousands of DNA parts via PCR, assemble these parts into complex DNA constructs using yeast homologous recombination, and verify these DNA assemblies via a highly multiplexed NGS workflow. These protocols are applicable to common synthetic biology workflows and recommended content for readers interested in establishing high-throughput operations.

Part III is a comprehensive compilation of various DNA assembly methods, each of which offers its users unique advantages in terms of efficiency, throughput, or complexity. In Chapters 7 and 8, Sylvestre Marillonnet and colleagues share updated protocols for the now ubiquitous Golden Gate and Modular Cloning (MoClo) assembly systems. These protocols popularized the use of type IIS enzymes in DNA assembly and are the foundation for many of the toolkits shared in this volume. Ramona Grützner describes the generation of MoClo standard parts via Golden Gate in Chapter 7, while Stefan Werner shares a protocol to assemble these parts into complex multigene constructs in Chapter 8. In Chapter 9, Christopher A. O'Callaghan and Da Lin from the University of Oxford present MetClo, a flexible assembly method that uses a single type IIS enzyme for the hierarchical modular assembly of large DNA constructs. A key feature of this method is the use of "methylation-switching" whereby the type IIS enzyme recognition site is methylated so that it is no longer recognized by the type IIS restriction enzyme. This innovation eliminates many of the sequence constrains common in modular part design and provides greater flexibility in DNA assembly design. In Chapter 10, Maryia Trubitsyna et al. from the University of Edinburgh cover PaperClip, an economical and flexible assembly method that can use a variety of DNA parts stored in multiple formats in assembly reactions. Parts can be assembled in any order, and the oligonucleotides that direct the assembly can be reused for any assembly involving a given part. Chapters 11 and 12 describe methods reliant on the Phytobricks standard. In Chapter 11, Nicola Patron and the team from the Earlham Institute discuss the Phytobricks standard. Phytobricks are standardized plant-based DNA parts, which are compatible with a variety of type IIS restriction endonuclease-dependent assembly methods. In this chapter, protocols for the design and construction of new Phytobricks parts and protocols for their assembly into multigene constructs are described. The key feature of this method is the use of standardized overhangs that enable reuse of DNA parts in new assemblies with other compatible parts without modification. This chapter also contains methods for high-throughput assembly and sequencing using modern nanoscale liquid handling robotics. In Chapter 12, Andreas I. Andreou and Naomi Nakayama from the University of Edinburgh discuss Mobius Assembly, a method utilizing DNA parts based on the Phytobricks standard and a minimalistic vector toolkit. The protocol also incorporates a chromogenic *in vivo* screen to identify positive assemblies and improve efficiency. In Chapter 13, George Taylor and John Heap from Imperial College describe protocols for the design and construction of multigene libraries using the Start-Stop assembly method. A key feature of this method is the generation of "scarless" junctions between parts, thereby avoiding unwanted changes to mRNA structure or functionality. The streamlined assembly hierarchy of Start-Stop also facilitates transfer of DNA assemblies into non-model organisms. In Chapter 14, Marko Storch and the team from Imperial College present details of the BASIC assembly method, which utilizes standardized 18 bp flanking sequences and provides 99% success rates for 4-part assemblies. DNA parts formatted using the BASIC standard can be placed in any position in the assembly and in any context, thereby providing the user with modularity and flexibility in their design. Finally, in Chapter 15, Marcos Valenzuela-Ortega and Christopher French from the University of Edinburgh describe the JUMP platform. JUMP vectors adhere to the Standard European Vector Architecture and are designed to overcome several limitations in modular cloning systems. One key design feature is the presence of orthogonal secondary cloning sites that allow users to easily modify the vector backbone for their desired application.

Part IV covers methods developed to fine-tune pathway expression and optimize their flux for improving yields of any target molecule. In Chapter 16, Gita Naseri and Bernd Mueller-Roeber provide a detailed protocol for COMPASS, a sequence-independent, scarless assembly method, which utilizes a series of inducible plant artificial transcription factors to fine-tune the expression of pathway genes in yeast. The method can generate large combinatorial libraries and is also coupled with a CRISPR/Cas9 system for one-step multi-locus genomic integration. Finally, in Chapter 17, Yizhi Cai et al. present SCRaMbLE, a method that relies on thousands of LoxP sites introduced downstream of every nonessential gene across the yeast genome. This architecture enables multiple genomic rearrangements facilitated by the Cre-LoxPsym recombination system, which when applied to a strain with a heterologous expression pathway allows for generation of thousands of potentially improved production strains.

Each tool and method described in this volume addresses multiple bottlenecks that the synthetic biology community faces in solving some of the fields' most pressing challenges. These methods can be used in isolation or combined in innovative ways based on the researcher's requirements, and we encourage our readers to fine-tune these tools for their applications. Automation and software tools specific for synthetic biology have seen rapid improvement in the last decade, and biofoundries across the world can select one or more methods for deployment within their platform.

We would like to thank all the contributors for their effort in putting together these protocols and sharing their first-hand knowledge on how to apply them in any laboratory setting.

Emeryville, CA, USA *Sunil Chandran*
Kevin W. George

Contents

Contributors

ANDREAS I. ANDREOU • *SyntheSys Centre for Synthetic and Systems Biology, School of Biological Sciences, University of Edinburgh, Edinburgh, UK*
JAMIE AUXILLOS • *Manchester Institute of Biotechnology (MIB), School of Chemistry, The University of Manchester, Manchester, UK; School of Biological Sciences, University of Edinburgh, Edinburgh, UK*
GEOFF S. BALDWIN • *Department of Life Sciences, Imperial College London, London, UK; Imperial College Centre for Synthetic Biology, Imperial College London, London, UK*
YAO-MIN CAI • *Earlham Institute, Norfolk, UK*
YIZHI CAI • *Manchester Institute of Biotechnology (MIB), School of Chemistry, The University of Manchester, Manchester, UK; Center for Synthetic Genomics, Institute of Synthetic Biology, Shenzhen Institutes of Advanced Technology, Chinese Academy of Sciences, Shenzhen, China*
JOSE A. CARRASCO LOPEZ • *Earlham Institute, Norfolk, UK*
TING-FUNG CHAN • *School of Life Sciences and State Key Laboratory of Agrobiotechnology, The Chinese University of Hong Kong, Hong Kong, SAR, China*
JAN-FANG CHENG • *DOE Joint Genome Institute (JGI), Berkeley, CA, USA; Environmental Genomics & Systems Biology, Lawrence Berkeley National Laboratory, Berkeley, CA, USA*
WILLIAM CHRISTIE • *Amyris Inc., Emeryville, CA, USA*
JAMES K. CRAFT • *TeselaGen Biotechnology, Inc., San Francisco, CA, USA*
JUNBIAO DAI • *Center for Synthetic Genomics, Institute of Synthetic Biology, Shenzhen Institutes of Advanced Technology, Chinese Academy of Sciences, Shenzhen, China*
SAMUEL DEUTSCH • *DOE Joint Genome Institute (JGI), Berkeley, CA, USA; Environmental Genomics & Systems Biology, Lawrence Berkeley National Laboratory, Berkeley, CA, USA*
ARI DWIJAYANTI • *Department of Life Sciences, Imperial College London, London, UK; Imperial College Centre for Synthetic Biology, Imperial College London, London, UK*
ALISTAIR ELFICK • *School of Engineering, University of Edinburgh, Edinburgh, UK*
ROBERT EVANS • *DOE Joint Genome Institute (JGI), Berkeley, CA, USA; Environmental Genomics & Systems Biology, Lawrence Berkeley National Laboratory, Berkeley, CA, USA*
MICHAEL J. FERO • *TeselaGen Biotechnology, Inc., San Francisco, CA, USA*
CHRISTOPHER E. FRENCH • *School of Biological Sciences, University of Edinburgh, Edinburgh, UK; Zhejiang University-University of Edinburgh Joint Research Centre for Engineering Biology, Zhejiang University, Haining, China*
KEVIN W. GEORGE • *Amyris Inc., Emeryville, CA, USA*
RAMONA GRÜTZNER • *Department of Cell and Metabolic Biology, Leibniz Institute of Plant Biochemistry, Halle, Germany*
MATTHEW C. HAINES • *Department of Life Sciences, Imperial College London, London, UK; Imperial College Centre for Synthetic Biology, Imperial College London, London, UK*
JOHN T. HEAP • *Imperial College Centre for Synthetic Biology, Department of Life Sciences, Imperial College London, London, UK; School of Life Sciences, The University of Nottingham, Nottingham, UK*

NATHAN J. HILLSON • *DOE Joint Genome Institute (JGI), Berkeley, CA, USA; Biological Systems & Engineering Division, Lawrence Berkeley National Laboratory, Berkeley, CA, USA; Technology Division, DOE Joint BioEnergy Institute (JBEI), Emeryville, CA, USA; DOE Agile BioFoundry, Emeryville, CA, USA; TeselaGen Biotechnology, Inc., San Francisco, CA, USA*

LENA HOCHREIN • *Department of Molecular Biology, University of Potsdam, Potsdam, Germany*

ANNEGRET HONSBEIN • *School of Biological Sciences, University of Edinburgh, Edinburgh, UK*

KRISTY IP • *Amyris Inc., Emeryville, CA, USA*

UMA JAYACHANDRAN • *School of Biological Sciences, University of Edinburgh, Edinburgh, UK*

SALLY JONES • *Manchester Institute of Biotechnology (MIB), School of Chemistry, The University of Manchester, Manchester, UK*

DA LIN • *Nuffield Department of Medicine, Wellcome Centre for Human Genetics, University of Oxford, Oxford, UK*

WEI LIU • *MRC Laboratory of Molecular Biology, Cambridge, UK*

FABIAN MACHENS • *Department of Molecular Biology, University of Potsdam, Potsdam, Germany*

HARIS MALLICK • *Department of Life Sciences, Imperial College London, London, UK; Imperial College Centre for Synthetic Biology, Imperial College London, London, UK*

SYLVESTRE MARILLONNET • *Department of Cell and Metabolic Biology, Leibniz Institute of Plant Biochemistry, Halle, Germany*

XIANWEI MENG • *DOE Joint Genome Institute (JGI), Berkeley, CA, USA; Environmental Genomics & Systems Biology, Lawrence Berkeley National Laboratory, Berkeley, CA, USA*

BERND MUELLER-ROEBER • *Department of Molecular Biology, University of Potsdam, Potsdam, Germany; Plant Signalling Group, Max Planck Institute of Molecular Plant Physiology, Potsdam, Germany; Department of Plant Development, Center of Plant Systems Biology and Biotechnology (CPSBB), Plovdiv, Bulgaria*

NAOMI NAKAYAMA • *SyntheSys Centre for Synthetic and Systems Biology, School of Biological Sciences, University of Edinburgh, Edinburgh, UK; Department of Bioengineering, Imperial College London, London, UK*

GITA NASERI • *Department of Molecular Biology, University of Potsdam, Potsdam, Germany*

SANGEETA NATH • *DOE Joint Genome Institute (JGI), Berkeley, CA, USA; Environmental Genomics & Systems Biology, Lawrence Berkeley National Laboratory, Berkeley, CA, USA*

CHRISTOPHER A. O'CALLAGHAN • *Nuffield Department of Medicine, Wellcome Centre for Human Genetics, University of Oxford, Oxford, UK*

ERNST OBERORTNER • *DOE Joint Genome Institute (JGI), Berkeley, CA, USA; Environmental Genomics & Systems Biology, Lawrence Berkeley National Laboratory, Berkeley, CA, USA*

NICOLA J. PATRON • *Earlham Institute, Norfolk, UK*

HECTOR PLAHAR • *Biological Systems & Engineering Division, Lawrence Berkeley National Laboratory, Berkeley, CA, USA; Technology Division, DOE Joint BioEnergy Institute (JBEI), Emeryville, CA, USA*

DAVID REIF • *Amyris Inc., Emeryville, CA, USA*

LISA SIMIRENKO • *DOE Joint Genome Institute (JGI), Berkeley, CA, USA; Environmental Genomics & Systems Biology, Lawrence Berkeley National Laboratory, Berkeley, CA, USA*

MARKO STORCH • *Department of Life Sciences, Imperial College London, London, UK; Imperial College Centre for Synthetic Biology, Imperial College London, London, UK; London Biofoundry, Imperial College Translation & Innovation Hub, London, UK*

REEM SWIDAH • *Manchester Institute of Biotechnology (MIB), School of Chemistry, The University of Manchester, Manchester, UK*
ANGELA TARVER • *DOE Joint Genome Institute (JGI), Berkeley, CA, USA; Environmental Genomics & Systems Biology, Lawrence Berkeley National Laboratory, Berkeley, CA, USA*
GEORGE M. TAYLOR • *Imperial College Centre for Synthetic Biology, Department of Life Sciences, Imperial College London, London, UK*
MARYIA TRUBITSYNA • *School of Biological Sciences, University of Edinburgh, Edinburgh, UK*
MARCOS VALENZUELA-ORTEGA • *School of Biological Sciences, University of Edinburgh, Edinburgh, UK*
TRANG VU • *TeselaGen Biotechnology, Inc., San Francisco, CA, USA*
STEFAN WERNER • *Nambawan Biotech GmbH, Halle, Germany*
RON YADIN • *Amyris Inc., Emeryville, CA, USA*

Part I

In Silico Tools for DNA Design and Assembly

Chapter 1

An Integrated Computer-Aided Design and Manufacturing Workflow for Synthetic Biology

Ernst Oberortner, Robert Evans, Xianwei Meng, Sangeeta Nath, Hector Plahar, Lisa Simirenko, Angela Tarver, Samuel Deutsch, Nathan J. Hillson, and Jan-Fang Cheng

Abstract

Biological computer-aided design and manufacturing (bioCAD/CAM) tools facilitate the design and build processes of engineering biological systems using iterative design–build–test–learn (DBTL) cycles. In this book chapter, we highlight some of the bioCAD/CAM tools developed and used at the US Department of Energy (DOE) Joint Genome Institute (JGI), Joint BioEnergy Institute (JBEI), and Agile BioFoundry (ABF). We demonstrate the use of these bioCAD/CAM tools on a common workflow for designing and building a multigene pathway in a hierarchical fashion. Each tool presented in this book chapter is specifically tailored to support one or more specific steps in a workflow, can be integrated with the others into design and build workflows, and can be deployed at academic, government, or commercial entities.

Key words DNA synthesis, DNA assembly, Computer-aided design, Computer-aided manufacturing, Software, Bioengineering, Synthetic biology

1 Introduction

The JGI is a user facility providing a suite of genomics technologies capabilities including DNA and RNA sequencing, DNA synthesis, single cell technologies, and metabolomics in support of DOE mission-related scientific research. The JGI's DNA synthesis platform has established various dry- and wet-lab workflows for the design, build, and characterization tasks specific to each user project. These workflows include (a) phenotypic sequence data repository mining, (b) heterologous expression construct design, (c) screening of the sequences against biosecurity guidelines, (d) synthetic DNA requisition, (e) assembly and sequence verification of synthetic constructs or combinatorial libraries using Type-IIs/Golden Gate, Gibson chew-back, or Yeast-based assembly protocols, (f) transformation of the assembled constructs into

Sunil Chandran and Kevin W. George (eds.), *DNA Cloning and Assembly: Methods and Protocols*, Methods in Molecular Biology, vol. 2205, https://doi.org/10.1007/978-1-0716-0908-8_1, © Springer Science+Business Media, LLC, part of Springer Nature 2020

the target host organism, and (g) mass spectrometry secondary metabolite detection.

In this chapter we overview the biological computer-aided design and manufacturing (bioCAD/CAM) tools that the JGI has developed and/or uses, supporting the synthetic biologist throughout the design and build stages of an integrated, iterative design–build–test–learn (DBTL) engineering cycle. We demonstrate a common workflow for utilizing the highlighted bioCAD/CAM tools to design and build a biosynthetic pathway, using a hierarchical process. Specifically, we present a repertoire of bioCAD/CAM tools and demonstrate their applicability in a workflow of designing and building a pathway comprising multiple operons and genes. We follow a top-down design process that comprised of two levels, starting with the design of the pathway and the organization of its genes and regulatory elements into operons. Then, we define the assembly and cloning instructions for building the pathway following a Golden Gate or Yeast assembly protocol. Each part of the Golden Gate or Yeast assembly is synthesized and, in case it exceeds the maximum length synthesis, assembled using a Gibson-based assembly protocol.

At the JGI, staff use bioCAD/CAM tools to facilitate the design and build processes in service of the JGI's DNA synthesis science program (https://jgi.doe.gov/our-science/science-programs/synthetic-biology/) enabling scientific users to gain, among others, access to large-scale DNA synthesis and assembly capabilities. The bioCAD/CAM tools described in this book chapter (a) provide web-based user interfaces (UIs), facilitating access and use, (b) can be deployed outside of the JGI, JBEI, or ABF, and (c) are available at no cost to noncommercial users and entities (*see* **Note 1**).

2 Materials

In the synthetic biology community, numerous bioCAD/CAM tools have been developed to computationally aid researchers [1, 2]. At the JGI, JBEI, and ABF, we use and develop bioCAD/CAM tools that support researchers throughout iterative DBTL engineering cycles. In this section, we provide an overview of the bioCAD/CAM tools that we have developed in order to facilitate the design and build processes at the JGI, JBEI, and ABF. We also provide a brief insight into our computational biosecurity screening pipeline.

2.1 Biological Computer-Aided Design (bioCAD) of Synthetic DNA

2.1.1 Design, Implementation, and Validation Automation (DIVA)

The Design, Implementation, and Validation Automation (DIVA) platform seeks to integrate bioCAD/CAM tools in order to provide seamless and useful workflows, liberating researchers from building DNA and enabling them to focus on designing and testing their experiments of interest. DIVA's web-based UI allows researchers to graphically specify the design of synthetic DNA constructs using DeviceEditor [3] and (sub)sequences from ICE registry [4] plasmid or part entries. After finishing the design specification, researchers can submit their designs to a central construction queue, track DNA construction as it progresses, and then (in a few weeks) receive notice that their sequence-verified constructs have been completed. The DIVA platform integrates DeviceEditor, j5, ICE, BLiSS and BOOST, which we explain in more detail below. An instance of DIVA, specifically deployed for the specification of JGI's designs, is publicly accessible at https://diva.jgi.doe.gov.

2.1.2 DeviceEditor [3]

DeviceEditor mimics the intuitive visual whiteboard design process practiced in biological laboratories through a web-based graphical UI. DeviceEditor combines the visual design of combinatorial libraries, direct integration with scarless multipart DNA assembly design automation (j5 [5]), and a graphical user interface for the creation and modification of design specification rules (Eugene [6]). DeviceEditor follows a design paradigm to spatially organize abstractions of biological components. Therefore, DeviceEditor assists the aggregation and arrangement of the DNA sequences of genetic components (ribosomal-binding sites, promoters and terminators, genes, etc.) to be structurally assembled toward a desired functionality. To specify structurally composite designs in a graphical manner, DeviceEditor integrates the palette of glyphs from the SBOL Visual extension (SBOLv) [7], enabling the communication of designs in a standardized fashion. Once users have visually verified that the desired constructs are correctly designed, DeviceEditor can also direct j5 to design downstream automation processes. DeviceEditor is integrated into the DIVA platform and publicly accessible at https://diva.jgi.doe.gov.

2.1.3 j5 [5]

j5 is a bioCAD tool to design scarless, multipart combinatorial DNA assemblies using established protocols such as Gibson and Golden Gate assembly. j5 generates cost-optimal designs for these assembly protocols. The j5 DNA assembly design algorithms are generally applicable across broad classes of DNA construction methodologies and complement other DNA assembly design tools, such as DIVA/DeviceEditor.

DeviceEditor (described above) can be used to specify the inputs for j5, including (a) the desired assembly methodology, (b) the biological "parts" to be assembled, (c) the sequential order and direction (forward or reverse) of each of the parts in the final target construct(s), and (d) biological design specification rules, to constrain the combinations of parts that make up the final target constructs. j5 generates several output files, such as CSV format files describing the designed DNA assembly processes and Genbank (or FASTA or SBOL v1) format sequence files for the assembled constructs. The j5 bioCAD tool is accessible for non-commercial use through https://j5.jbei.org, as integrated functionality in DIVA (https://diva.jgi.doe.gov), and for commercial use j5 is exclusively accessible through TeselaGen (https://teselagen.com).

2.1.4 Open VectorEditor (OpenVE)

OpenVE is an open-source, web-based sequence editor for the modification and visualization of circular as well as linear sequences of DNA including their feature annotations. The source code of OpenVE, which is developed by Teselagen Biotechnology Inc., is available at https://github.com/TeselaGen/openVectorEditor. We have deployed a stand-alone instance of OpenVE (https://j5.jbei.org/VectorEditor/VectorEditor.html), enabling researchers to utilize and experience its functionalities. In addition, OpenVE is integrated into the DIVA design platform for visualizing DNA construct designs and into the ICE registry for visualizing DNA plasmids and parts.

2.1.5 Build-Optimization Software Tools (BOOST) [8]

Not every conceivable DNA sequence can be manufactured using today's state-of-the-art synthesizers. Extreme %GC regions and repeating k-mers complicate the synthesis and the quality assurance/control thereof using next-generation sequencing (NGS) technologies. To streamline the transition from in silico design to in vitro synthesis, synthetic DNA sequences need to be verified against such synthesis criteria.

Therefore, we have developed the Build-Optimization Software Tools (BOOST) that prepare designed DNA sequences for synthesis and assembly. BOOST comprises the following "apps":

- The "Juggler" reverse-translates protein sequences into DNA sequences and codon-juggles DNA sequences.
- The "Polisher" verifies DNA sequences against DNA synthesis constraints and modifies the protein coding regions to resolve eventual violations of DNA synthesis constraints. Recently, we incorporated the APIs of commercial DNA synthesis vendors to ensure consistency among the complexity screening results of the vendors' portals and BOOST.
- The "Partitioner" fragments large DNA sequences into synthesizable building blocks with assembly-specific overlap sequences.
- The "Workflow Manager" combines the execution of the Juggler, Polisher, and Partitioner functionalities in one pass.

- The "Codon Usages Tables Merger" provides the functionality of combining two codon usage tables into one table when genes need to be expressed in multiple hosts.
- The "Builder" app allows for the specification of step-by-step instructions for manufacturing (synthetic) DNA constructs, such as the steps of purchasing (synthetic DNA fragments or primers), linearizing vectors (digestion or amplification), or PCR-amplification.

Researchers use BOOST to prepare designed constructs, such as genes, operons, or pathways, for synthesis by commercial DNA synthesis vendors and for the in-house assembly of the synthetic DNA fragments. Depending on the BOOST app and the inputs of the user, BOOST generates various output files. For example, the "Workflow Manager" can output ready-to-order synthetic DNA fragments, primers for PCR-amplification of either the synthetic DNA fragments or vectors, and step-by-step instructions for carrying out the tasks of DNA synthesis, assembly and transformation into target host organism. BOOST is publicly accessible at https://boost.jgi.doe.gov.

2.2 Biosecurity Screening of Synthetic DNA

In 2010, US Department of Health and Human Services (HHS) issued the *Screening Framework Guidance for Providers of Synthetic Double-Stranded DNA*, outlining recommendations for synthetic DNA providers to ensure that existing regulations and best practices are followed in addressing biosecurity concerns. The HHS guidance defines a "*sequence of concern*" as sequences "derived from or encoding" entities on the HHS and Center for Disease Control's Select Agents or Toxins list, which are regulated under the Select Agent Regulations, and/or agents on the Bureau of Industry and Security's Commerce Control List (CCL) regulated under Export Administration Regulations. In accordance with the HHS guidance, the DNA Synthesis Science program at the US Department of Energy Joint Genome Institute (DOE JGI) has developed a computational pipeline—Black List Sequence Screening (BLiSS)—for screening all sequences that it manufactures either for internal research or for its scientific users.

As specified in the guidance, BLiSS detects "*sequences of concern*" of at least 200 nucleotides in length on either DNA strand. BLiSS follows a "Best Match" approach to determine whether a query sequence is unique to Select Agents or Toxins, or export-controlled agents, toxins or genetic elements, and to minimize false positives from closely related organisms or highly conserved "housekeeping genes" which do not pose a biosecurity threat.

At the JGI, and more recently at JBEI/ABF, researchers run BLiSS on all constructs that will be synthesized, after they are designed, but prior to building. If a construct is flagged as being a potential sequence of concern, follow-up screening is done to validate the end use of the construct (potentially resulting in contact with the US Federal Bureau of Investigation).

2.3 Biological Computer-Aided Manufacturing (bioCAM) of Synthetic DNA

2.3.1 Tracking the DNA Synthesis and Assembly Workflow

SynTrack is a web-based workflow-driven bioCAM platform that manages and tracks the DNA manufacturing processes. SynTrack imports the BOOST-specified build process information, which includes step-by-step instructions for staff (leveraging robotic platforms) to carry out all build operations.

Managing complex hierarchical, multistep DNA manufacturing processes requires (a) to define time- and cost-efficient synthesis and assembly strategies, (b) to manage the interplay between humans and automation technologies, such as liquid-handling robotics or microfluidic devices, (c) to assure the quality of the input/output components at well-defined stages in the process through adequate quality control (QA/QC), and (d) to track each step in the process including the provenance of each step's input components, such as DNA fragments/plasmids, master mixes, buffers, or enzymes.

SynTrack supports biologists in carrying out complex multistep processes that (a) begin with purchasing synthetic DNA fragments from commercial DNA synthesis vendors, (b) followed by preparing the received fragments for assembly using PCR amplification, (c) then setting up reactions based on the assembly protocol of the amplified constructs using manual labor and robotics platforms, and (d) lastly sequence verifying the assembled constructs using NGS platforms. With composed user-friendly web interfaces, SynTrack divides the routine build process into the "pre-assembly" and "assembly" stages.

The pre-assembly workflow comprises the tasks associated with purchasing DNA from commercial DNA synthesis vendors, including the tracking of the order and receiving date and the quantity of the synthetic DNA parts. As soon as all required synthetic DNA fragments are available, SynTrack can generate pipetting instruction for liquid handling robotics, such as the Biomek FX platform and Echo liquid handler to rearray received DNA fragments into the desired quadrant layout automatically. Prior to quantifying DNA fragments after cleanup, other tasks for the pre-assembly workflow include creating a set of instructions to handle liquid transfer on the condensed or uncondensed assay plates for PCR reactions.

SynTrack enables instantiation of one or more assembly workflows based on previously finished single or merged pre-assembly workflows. The assembly workflow comprises the tasks of (a) assembly and cloning using Gibson chew-back, Type-IIs/Golden Gate or Yeast-based assembly and cloning reactions, and (b) transforming the assembled and cloned plasmids into the target host and screening colonies for the presence of the transformed plasmids. To support biologists throughout the assembly process, SynTrack supports generating instructions for robotics platforms.

Throughout the entire manufacturing process, SynTrack effectively incorporates QA/QC results in order to maximize the success rate of synthetic DNA constructs that are delivered to the scientific users. For example, SynTrack enables rework on constructs that fail on synthesis by, for example, purchasing the synthetic DNA fragment from a different vendor. Integrating QA/QC outcomes from colony pickers, plate readers and NGS sequencers, the tracking system performs related measurements on targeted master plates to monitor coordinated DNA construct changes subsequently. All DNA assembly data tracking has been condensed into secure, tabular-formatted web interfaces with the ability to observe and update the status of DNA constructs with time stamp. At the core of an effective quality control assembly processes, troubleshooting needs to be managed for altering the DNA assembly workflows. The workflow task supplies a flexible method based on the feedback of the QA/QC results to allow the workflow moving into backward direction when rework is an expected occurrence for troubleshooting.

2.3.2 Sequence Verification Using Next Generation Sequencing (NGS)

For the verification of manufactured DNA constructs, we utilize DOE JGI's state-of-the-art next-generation sequencing (NGS) platforms. We developed a computational pipeline, SynBioQC, to process the raw reads of NGS sequencers in order to sequence verify the correctness of the manufactured DNA. That is, the SynBioQC pipeline ensures that the nucleotides of the in silico designed sequence matches the in vitro sequenced DNA.

To execute the SynBioQC pipeline, a user enters the sequences of the manufactured constructs and the library names of the sequenced construct pools into a web-based UI. Then, executable scripts are generated and sent to a job management system, which controls the execution of the scripts on a DOE NERSC supercomputing cluster. The pipeline aligns the reference sequences with the raw reads and calls variants, such as single nucleotide polymorphism (SNP), insertions and deletions (indels). The results of the pipeline are then visualized in a web page, enabling the researchers to analyze the results in an efficient manner.

2.4 Repositories for Storing and Retrieving Synthetic DNA Constructs

The Inventory of Composable Elements (ICE) [4] is an open-source, web-based registry platform for synthetic biological parts with support for plant seeds, microbial strains, DNA plasmids and parts, as well as protein sequences. ICE provides integrated tools for sequence visualization and editing (openVE, described above), as well as mechanisms for secure access and information sharing with other users and software tools using commonly used data exchange formats, such as FASTA, Genbank, and SBOL, as well as a RESTful API. The JGI instance of ICE is accessible at https://registry.jgi.doe.gov.

2.5 Data Exchange Using a Standardized Format

We attempt to develop and use software tools that support commonly used data exchange formats and standards. Adopting community-driven standardization efforts, such as the Synthetic Biology Open Language (SBOL) [9], enables us to provide feedback to the community and to contribute to the development of standards accepted by the scientific community.

SBOL provides capabilities to support our approach to provenance tracking for synthetic DNA requisition and synthetic construct assembly. Provenance tracking is paramount for the purpose of managing scientific intellectual property, systematically reproducing the data process, and identifying defective designs that can be replaced with nondefective alternatives. SynTrack and BOOST communicate step-by-step assembly and cloning instructions, enabling SynTrack to determine each step in the manufacturing process and to track the start- and end-time of each step, who carried out the step, as well as the input and output components of each step. SynTrack underpins the transition from the in silico design phase to the in vitro build stage while aiming to form collaborations for build elements and activities through the SBOL-based community ontology, and to further incorporate provenance tracking standards into our integrated DBTL pipeline. Tracking the provenance of the design and build tasks provides the scientific user not only with the desired synthetic DNA or strains but also with a detailed, standardized report about what steps were performed to ultimately build the designed DNA constructs. Several of the bioCAD/CAM tools developed at the JGI, JBEI, and ABF support the input and output of data encoded in the SBOL standard by utilizing the SBOL libraries [10].

3 Methods

In this section, we showcase the applicability of the bioCAD/CAM tools described above, demonstrating their use in a state-of-the-art workflow of the JGI's DNA synthesis platform. The illustrated workflow covers the stages of top-down design and bottom-up building of a synthetic biological system. The system under consideration is a 58 kb pathway consisting of five genes organized in five operons.

In Fig. 1, we outline the hierarchical bottom-up process of synthesis, assembly, and cloning. At the first level, Level-0, "intermediates" are made utilizing Gibson assembly [11] of synthetic DNA fragments with appropriate 5′ and 3′ overlap sequences for their assembly and insertion into the destination vector (Fig. 1a). In Level-1, sequence verified constructs built in the first level are digested with Type-IIs enzyme and subjected to Yeast [12, 13] or Golden Gate [14] assembly in the desired destination vector to generate the final construct (Fig. 1b).

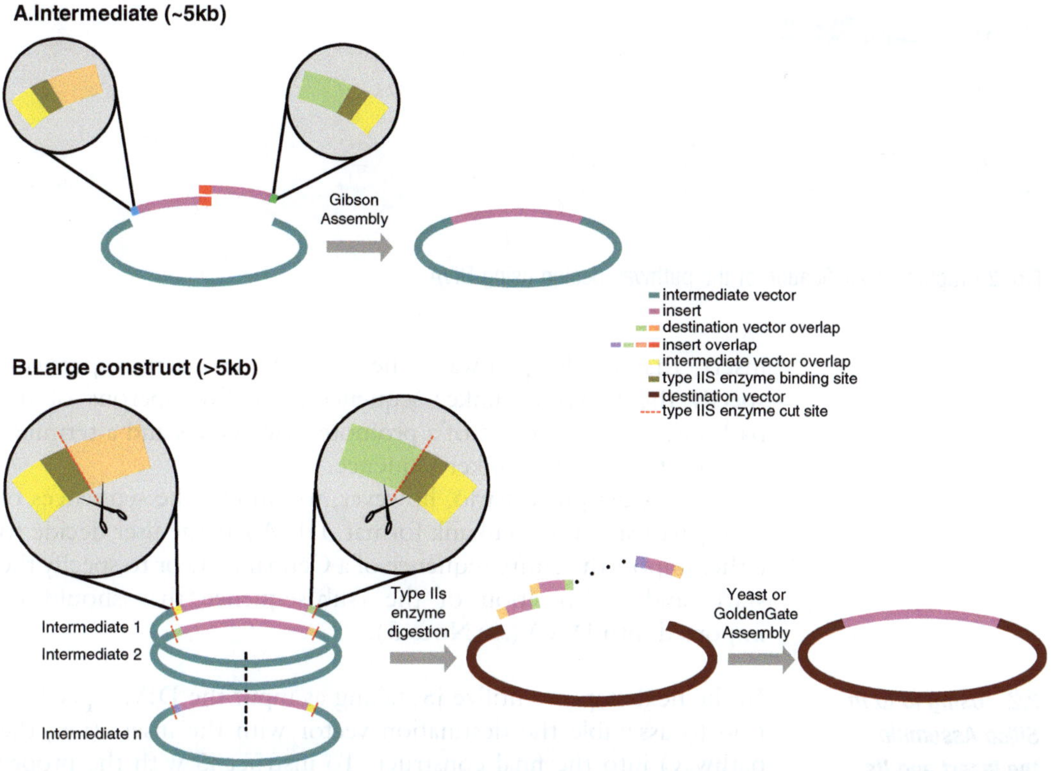

Fig. 1 Hierarchical DNA assembly of large constructs using a combination of the Gibson, Golden Gate, and Yeast assembly protocols. Bottom-up process of synthesis, assembly, and cloning of a large synthetic DNA construct (>5 kb). (**a**) Synthesis and Gibson assembly of intermediate constructs for the construction of large synthetic DNA constructs. (**b**) Assembly of intermediate constructs into large constructs

In the following sections we describe the top-down design process in a step-by-step manner and the use of bioCAD/CAM tools. The hierarchical design process focuses on the in silico design of the synthetic DNA constructs, their synthesis, assembly and cloning instructions (*see* **Note 2**). Subheadings 3.1, 3.2, and 3.3 focus on the design of the Level-1 construct, that is, the entire pathway, its design for Yeast/Golden Gate assembly, and the required destination vector. Subheadings 3.4, 3.5, and 3.6 focus on the design of the Level-0 constructs, that is, the intermediates, their sequence design for synthesis and chew-back assembly, as well as inserting the Level-0 constructs into the required destination vectors. As a result, the assembly of the intermediates ("Level-0") leads to the manufactured pathway ("Level-1"). Lastly, we briefly mention tracking the assembly and cloning steps of synthetic DNA constructs.

3.1 Design Specification of Pathway Structure and Destination Vector Using DIVA

In the first step, the designer specifies the biological system using the DIVA platform. In Fig. 2 we show a screenshot of the DIVA platform and an excerpt of the design specification of a large pathway. The first column in DIVA's design canvas represents the destination vector backbone and all other columns denote the

Fig. 2 Graphical specification of the pathway design using DIVA

insert, that is, the pathway. The bins of the insert respectively specify the 5′ vector linker sequences, the five operons of the pathway, each comprised of a promoter, two genes and a terminator, and the 3′ vector linker sequence.

In this design scenario, however, we import the sequences of each part using the Genbank format. DIVA lets the user decide to either import the entire sequence of a Genbank file or to specify the start- and end-position of the (sub)sequence that should be imported into DIVA (*see* **Note 3**).

3.2 Using j5 to In Silico Assemble the Insert and Its Destination Vector

In the next step, we utilize j5, taking as input the DIVA specification to assemble the destination vector with the insert (i.e., the pathway) into the final construct. To instruct j5 with the proper assembly strategies, we specify in DIVA that the destination vector backbone will be digested with a Type-IIs restriction enzyme and that the insert (5′ vector linker, pathway, 3′ vector linker) will be synthesized directly. In Fig. 3 we illustrate a screenshot of openVE, visualizing the final construct, that is, the pathway, which is highlighted, inserted into the destination vector.

3.3 Partition the Insert for Yeast/Golden Gate Assembly Using BOOST

Input into BOOST is a GenBank file, generated by DIVA/j5. Depending on the designs' requirements, the user inputs partitioning parameters into BOOST in order to fragment the large pathway sequence into smaller fragments. The Yeast assembly protocol requires, for example, overlaps between 85 and 110 bp with a desired optimum overlap length of 100 bp. In case the build process is based on a Golden Gate assembly protocol, the user currently needs to instruct BOOST to look for proper partitioning areas in the large sequence but not extend the sequence fragments with overlapping sequences. In this case, the user can then utilize j5 for designing the flanking sequences required for subsequent Type-IIs restriction enzyme recognition.

In addition to the partitioning, the user can instruct BOOST to verify the pathway sequence against synthesis criteria by commercial DNA synthesis vendors. In case of violations, BOOST can modify the protein-coding sequences of the pathway to reduce the cost and time of synthesis as well as to increase the synthesis success rate.

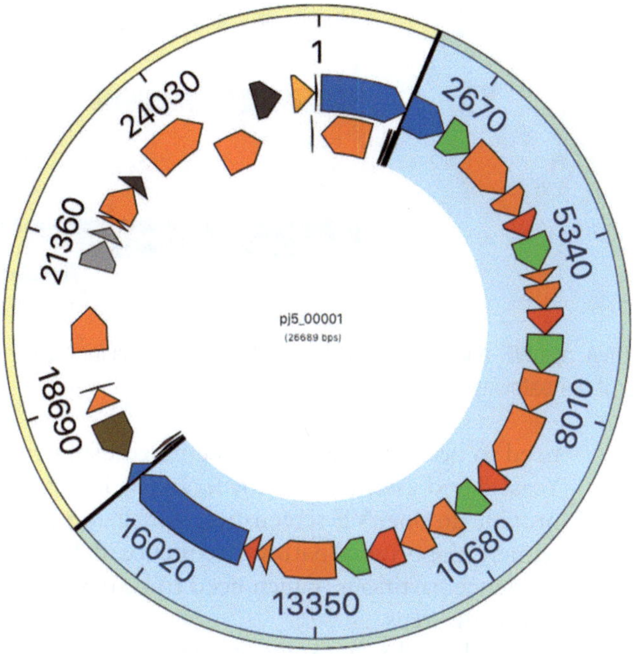

Fig. 3 j5 generated plasmid map of the final construct (destination vector and inserted pathway)

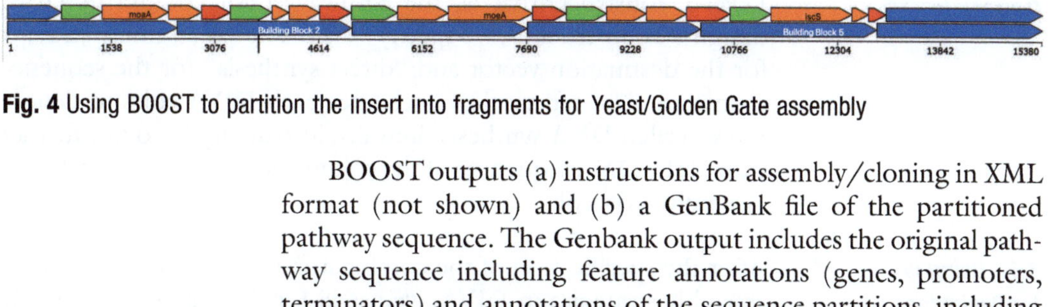

Fig. 4 Using BOOST to partition the insert into fragments for Yeast/Golden Gate assembly

BOOST outputs (a) instructions for assembly/cloning in XML format (not shown) and (b) a GenBank file of the partitioned pathway sequence. The Genbank output includes the original pathway sequence including feature annotations (genes, promoters, terminators) and annotations of the sequence partitions, including their overlap sequences.

In Fig. 4, we schematically illustrate the process of using BOOST for partitioning a large sequence into smaller fragments. The upper portion of Fig. 4 is the linear representation of the highlighted portion of the plasmid shown in Fig. 3. The user generates a Genbank file using openVE, provides the file as input to BOOST which partitions the input sequence into fragments that

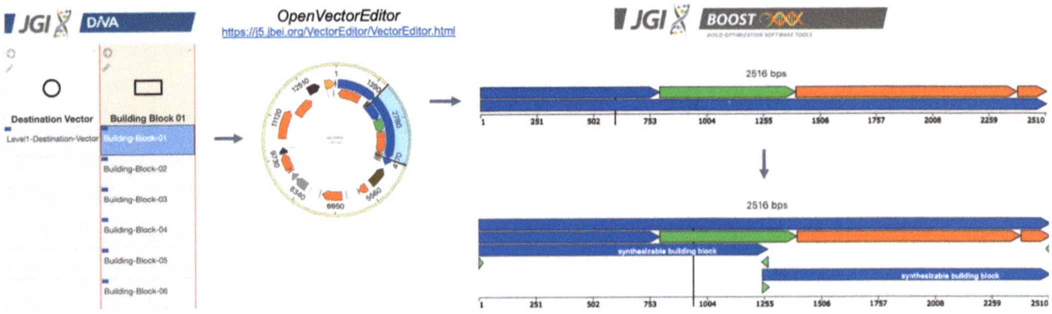

Fig. 5 Screenshots of DIVA, openVE, and the BOOST input/output sequences for the computer-aided design of the Level-0 constructs of the pathway

need to be synthesized and assembled using either Golden Gate or Yeast-based assembly protocol. In the lower portion of Fig. 4, we provide an openVE screenshot, visualizing the output of BOOST, namely, the original pathway sequence including annotations of the sequence partitions, which need to be designed for synthesis in the following steps.

In the following steps, we describe our workflow for designing the synthesis and assembly of the pathway's sequence partitions—from now on called "Level-0 constructs". In Fig. 5, we provide an overview of the design steps of the Level-0 constructs, following the same process as the design of the Level-1 construct and utilizing the same CAD tools for designing the Level-0 constructs, that is, DIVA, j5, openVE, and BOOST (*see* **Note 4**).

3.4 Using DIVA for the Specification and Design of Level-0 Constructs

The researcher starts with the specification of the Level-0 design in DIVA. Here, we build a combinatorial design of one destination (left column in DIVA design) and the six sequence partitions of the Level-1 construct, that is, the pathway (right column in DIVA design). To instruct j5, we specify "digestion" as assembly strategy for the destination vector and "direct synthesis" for the sequence partitions. The red vertical lines after each DIVA column denote the so-called DNA synthesis firewalls, instructing j5 to not further extend the DNA sequences of the column according to the selected assembly protocol.

3.5 Using j5 for Designing the Assembly of the Combinatorial Design

After the specification of the combinatorial design and the desired assembly strategies in DIVA, we use j5 for designing the assembly of the Level-0 intermediates and the destination vector. We show a screenshot of openVE, visualizing the assembled construct of the digested destination vector and one of the six synthesized building blocks, which is highlighted. The linear inserted partition into the destination vector is visualized to the right of the openVE screenshot.

3.6 Using BOOST for Synthesis and Gibson Chew-Back Assembly of the Level-0 Constructs

In the last design step, we use BOOST to verify the Level-0 inserts against DNA synthesis criteria and to partition inserts, which exceed the maximum length of synthesis, into synthesizable fragments. Similarly, for Subheading 3.3, BOOST outputs the synthesizable partitions and, if desired, the PCR amplification primers. Also, BOOST generates assembly and cloning instructions about the chew-back assembly of the synthesized building blocks and their insertion into the Level-0 destination vectors.

After the Level-1 and the Level-0 constructs are designed and the required synthesis, assembly and cloning instructions are defined, the researcher can start the build process of the in silico designs.

3.7 Using SynTrack for Tracking the Build Process of Level-0 and Level-1 Constructs

SynTrack takes as input the output of BOOST, namely, an XML file containing a description of synthesis, assembly, and cloning steps for building the desired construct. After the user uploads the XML file, then SynTrack can generate the order sheets of the synthetic fragments and PCR-amplification primers. SynTrack instructs the researchers through the assembly and cloning process, including QA/QC steps, generates plating instructions for robotics that automate certain steps along the process and also enables the user to attach the experiment data of each step.

After the assembly and cloning, the researchers perform a QC step, verifying that the build sequences matches the designed sequences. Therefore, we utilize a SynBioQC computational pipeline to perform sequence validations. Depending on the QA/QC results, the researcher either picks an error-free clone or must fix the errors, such as by reiterating the last cloning steps. The error-free clones are then stored in a master-plate, which is managed by SynTrack and shipped to the JGI user as soon as the user's constructs are ready.

4 Notes

1. The following software tools are accessible via web-based UIs at no-cost to noncommercial users: BOOST, DIVA/DeviceEditor, ICE, openVE, and j5. The SynTrack tool as well as the BLiSS and SynBioQC pipelines are currently only available for internal use at the JGI, JBEI, and ABF. Commercial licenses (where required) for the tools are available through commercial distributors or the Innovations and Partnerships Office of Berkeley Lab (https://ipo.lbl.gov).
2. The bioCAD/CAM tools described in this book chapter are developed with the intention to support researchers throughout the design process. However, the tools themselves have certain limitations that currently preclude fully automated

design workflows. In the workflow described here, we often had to use a tool as a stand-alone, requiring to export the output of each tool, for example, into a Genbank file and to import the Genbank file into the next tool in our workflow, placing a burden on the user to organize and store several input and output files. Our future work will continue toward a more integrated solution, enabling to exchange data among the tools in a more sustainable and user-friendly manner. For example, DIVA/DeviceEditor, j5, ICE, openVE, and a subset of BOOST functionality are already well integrated.

3. Here, we do not demonstrate DIVA's capabilities of importing the sequences of the design elements (vector backbone, promoters, terminators, genes) from an ICE repository. In such a scenario, the user first needs to enter all parts into an ICE repository that is connected with DIVA. For example, the designer buik-uploads all parts, their types, and their sequences into JGI's instance of ICE and specifies the design in JGI's instance of DIVA, which is configured to import parts from JGI's ICE instance.
4. The top-down design process, demonstrated in this book chapter, highlights that the design steps are repeated at each design level. For example, the steps for designing the Level-1 constructs are the same as the design steps of the Level-0 constructs.

Acknowledgments

This work was part of the DOE Joint Genome Institute (https://jgi.doe.gov) and Joint BioEnergy Institute (https://www.jbei.org) supported by the US Department of Energy, Office of Science, Office of Biological and Environmental Research, and was part of the Agile BioFoundry (https://agilebiofoundry.org) supported by the US Department of Energy, Energy Efficiency and Renewable Energy, Bioenergy Technologies Office, through contract DE-AC02-05CH11231 between Lawrence Berkeley National Laboratory and the US Department of Energy. The views and opinions of the authors expressed herein do not necessarily state or reflect those of the US Government or any agency thereof. Neither the US Government nor any agency thereof, nor any of their employees, makes any warranty, expressed or implied, or assumes any legal liability or responsibility or the accuracy, completeness, or usefulness of any information, apparatus, product, or process disclosed, or represents that its use would not infringe privately owned rights. The US Government retains and the publisher, by accepting the article for publication, acknowledges that the US Government retains a nonexclusive, paid-up, irrevocable, worldwide license to publish or reproduce the published form of this manuscript, or allow others to do so, for US Government purposes.

We thank the DOE National Energy Research Scientific Computing Center (NERSC) (https://www.nersc.gov/) for hosting the JGI's instances of the bioCAD/CAM tools described in this chapter (e.g.., DIVA/DeviceEditor/j5, BOOST, BLiSS, SynTrack, and ICE).

Author Contributions: R.E., S.N., A.T., N.J.H. and J.F.C. are end-users of the bioCAD/CAM tools. N.J.H. developed and maintains j5. X.M. developed and maintains SynTrack. E.O. developed and maintains BOOST and the SynBioQC pipelines. H.P. developed and maintains DIVA/DeviceEditor and ICE. L.S. developed and maintains BLiSS and manages the JGI's deployments of DIVA/DeviceEditor/j5, and ICE. S.N. designed and performed the pathway design workflow. All authors wrote, edited, and reviewed the manuscript.

Conflicts of Interest Statement: N.J.H. declares financial interests in the form of issued and pending patent applications related to the j5 software, and equity in TeselaGen Biotechnology, Inc.

References

1. Nowogrodzki A (2018) The automatic-design tools that are changing synthetic biology. Nature 564(7735):291–292. https://doi.org/10.1038/d41586-018-07662-w
2. Appleton E, Madsen C, Roehner N, Densmore D (2017) Design automation in synthetic biology. Cold Spring Harb Perspect Biol 9: a023978
3. Chen J, Densmore D, Ham T, Keasling J, Hillson N (2012) DeviceEditor visual biological CAD canvas. J Biol Eng 6:1. https://doi.org/10.1186/1754-1611-6-1
4. Ham TS, Dmytriv Z, Plahar H, Chen J, Hillson NJ, Keasling JD (2012) Design, implementation and practice of JBEIICE: an open source biological part registry platform and tools. Nucleic Acids Res 40:e141
5. Hillson NJ, Rosengarten RD, Keasling JD (2012) j5 DNA assembly design automation software. ACS Synth Biol 1(1):14–21. https://doi.org/10.1021/sb2000116
6. Bilitchenko L, Liu A, Cheung S, Weeding E, Xia B et al (2011) Eugene – a domain specific language for specifying and constraining synthetic biological parts, devices, and systems. PLoS One 6(4):e18882. https://doi.org/10.1371/journal.pone.0018882
7. Quinn JY, Cox RS III, Adler A, Beal J, Bhatia S et al (2015) SBOL visual: a graphical language for genetic designs. PLoS Biol 13(12): e1002310. https://doi.org/10.1371/journal.pbio.1002310
8. Oberortner E, Cheng J-F, Hillson NJ, Deutsch S (2017) Streamlining the design-to-build transition with build-optimization software tools. ACS Synth Biol 6(3):485–496. https://doi.org/10.1021/acssynbio.6b00200
9. Galdzicki M, Clancy K, Oberortner E, Pocock MR, Quinn JY, Rodriguez CA, Roehner N, Wilson ML, Adam L, Anderson JC, Bartley B, Beal J, Chandran D, Chen JY, Densmore DM, Endy D, Grünberg R, Hallinan JS, Hillson NJ, Johnson JD, Kuchinsky A, Lux MW, Misirli G, Peccoud J, Plahar HA, Sirin E, Stan G, Villalobos A, Wipat A, Gennari JH, Myers CJ, Sauro HM (2014) The synthetic biology open language (SBOL) provides a community standard for communicating designs in synthetic biology. Nat Biotechnol 32:545–550
10. Zhang Z, Nguyen T, Roehner N, Misirli G, Pocock M, Oberortner E et al (2015) libSBOLj 2.0: a Java library to support SBOL 2.0. IEEE Life Sci Lett 1:34–37. https://doi.org/10.1109/LLS.2016.2546546

11. Gibson DG, Young L, Chuang RY, Venter JC, Hutchison CA 3rd, Smith HO (2009) Enzymatic assembly of DNA molecules up to several hundred kilobases. Nat Methods 6 (5):343–345. https://doi.org/10.1038/nmeth.1318
12. Shao Z, Zhao H, Zhao H (2009) DNA assembler, an in vivo genetic method for rapid construction of biochemical pathways. Nucleic Acids Res 37(2):e16. https://doi.org/10.1093/nar/gkn991
13. Kouprina N, Larionov V (2016) Transformation-associated recombination (TAR) cloning for genomics studies and synthetic biology. Chromosoma 125(4):621–632
14. Engler C, Kandzia R, Marillonnet S (2008) A one pot, one step, precision cloning method with high throughput capability. PLoS One 3 (11):e3647. https://doi.org/10.1371/journal.pone.0003647

Chapter 2

Combinatorial-Hierarchical DNA Library Design Using the TeselaGen DESIGN Module with *j5*

Michael J. Fero, James K. Craft, Trang Vu, and Nathan J. Hillson

Abstract

Modern DNA assembly techniques are known for their potential to link multiple large DNA fragments together into even larger constructs in single pot reactions that are easier to automate and work more reliably than traditional cloning methods. The simplicity of the chemistry is in contrast to the increased work needed to design optimal reactions that maximize DNA fragment reuse, minimize cost, and organize thousands of potential chemical reactions. Here we examine available DNA assembly methods and describe through example, the construction of a complex but not atypical combinatorial and hierarchical library using protocols that are generated automatically with the assistance of modern synthetic biology software.

Key words DNA assembly, Pathway engineering, Synthetic biology, Metabolic engineering, Bioinformatics, Cloning

1 Introduction

It is common in the parlance of synthetic biology to speak in terms of forward engineering biological systems in much the same way that an electrical engineer might design and build an electronic circuit [1]. Progress has been made in translating engineering design principles to the description of biological parts and devices as well as systems [2]. Nevertheless, a biological system is not as easily characterized or simulated as an electronic or mechanical system. Progress often involves building many variants of a genetic construct or pathway, and then performing functional tests in order to find the DNA sequence that encodes the best performing biological system. DNA synthesis costs are dropping rapidly, but direct synthesis alone cannot address the combinatorial explosion that occurs when trying to optimize a biological pathway that may contain even a small number of variants tested in combinatorial fashion, leading to libraries of thousands of elements. The cost of building a large library grows linearly with library size without the benefit of part reuse. Streamlined part reuse means that library costs

Sunil Chandran and Kevin W. George (eds.), *DNA Cloning and Assembly: Methods and Protocols*, Methods in Molecular Biology, vol. 2205, https://doi.org/10.1007/978-1-0716-0908-8_2, © Springer Science+Business Media, LLC, part of Springer Nature 2020

grow only logarithmically with increasing library size, enabling large screens at low cost. Importantly, even the everyday construction of small libraries or single constructs is enabled more effectively and without error by using modern DNA assembly methods and software [3].

A common challenge is to (a) design a set of DNA constructs that span the space of potentially optimal constructs for the performance of a particular function, (b) build the constructs minimizing errors, cost and time, (c) transform into a host chassis (*E. coli*, yeast), (d) perform a functional assay to determine individual construct efficacy, (e) capture the results in a DNA parts knowledgebase, and (f) use those learnings to inform further rounds of optimization or investigation. The general problem of assembling a set of somewhat arbitrary DNA "parts" into a contiguous construct is shown in Fig. 1. Two key observations inform our response to this challenge.

1. DNA synthesis costs will track DNA sequencing costs and will continue to drop while the length of reliably synthesized constructs will increase. DNA synthesis improvements, coupled with the combinatorial DNA assembly techniques described here, will help to make building complex DNA libraries faster and easier.
2. Our knowledge of biology is incomplete. Accurate simulations of biological systems, while improving, will lag behind our ability to quickly build and test, implying that new approaches

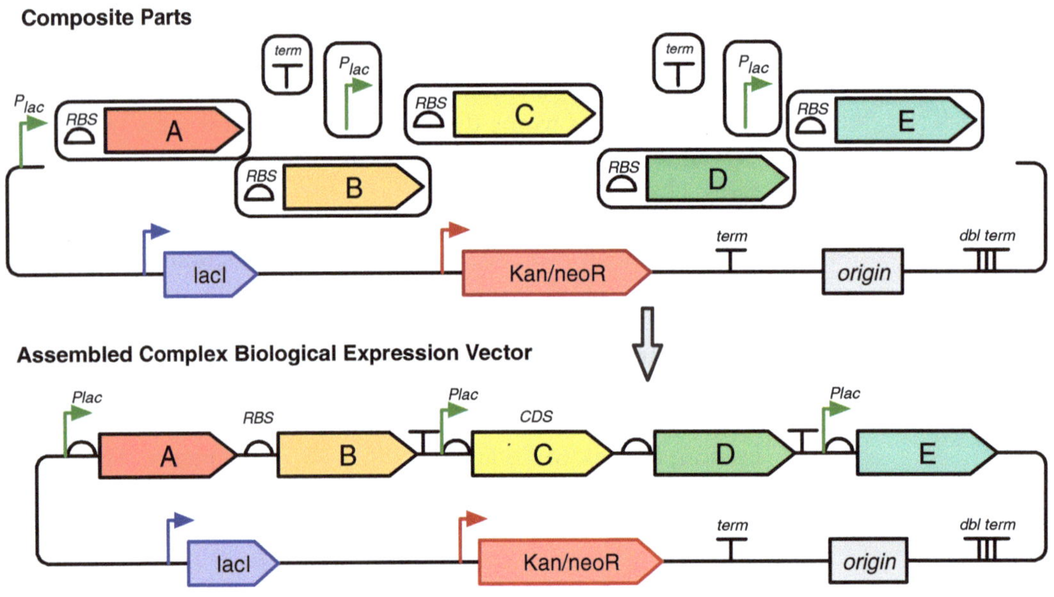

Fig. 1 Multipart DNA assembly

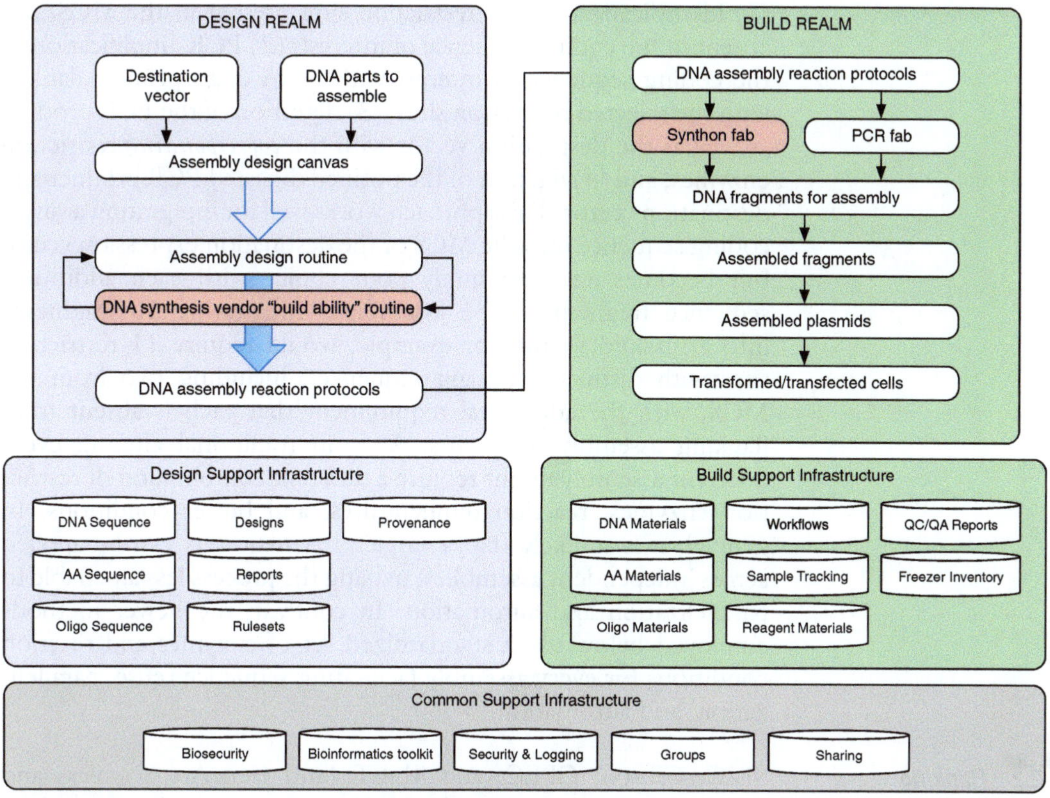

Fig. 2 Design/Build information technology elements

to engineering biology via intelligent, iterative, high content screens will outpace raw simulation.

These considerations imply that a modular but tight coupling between DNA design, cell construction, testing, and learning will allow researchers to converge quickly on a bioengineered product of interest. Modern synthetic biology software can help organize and inform this process. A schematic including common elements of a platform that can organize the bioCAD/CAM or Design/Build process is shown in Fig. 2.

2 DNA Design and Build Methods

2.1 Traditional Multiple Cloning Site Approach to DNA Assembly

In order to build libraries in ways that facilitate larger scales and automation, we avoid designing and building DNA constructs using the traditional MCS (Multiple Cloning Site) approach. To understand why, we briefly review the traditional approach. In a typical cloning vector, the MCS follows a promoter (e.g., a T7 promoter) and is in turn followed by a terminator. Integrating a protein coding sequence of interest into a plasmid vector involves:

(1) identification of two restriction sites present in the MCS, but absent in the coding sequence of interest, (2) PCR amplification of the coding sequence of interest with DNA oligo primers flanked with the selected restriction sites, (3) digestion of the PCR product as well as the destination vector with the *corresponding* restriction enzymes, and (4) ligation of the purified digested PCR product and destination vector. This approach works well for integrating a single coding sequence into the MCS of the destination expression vector but becomes an increasingly poor choice with each additional sequence fragment to be assembled. Integrating 10 fragments into a plasmid vector, for example, would require 11 restriction sites with distinct overhang sequences, including two from the MCS, with the additional requirement that each is absent from flanking assembly fragments. Also, in traditional cloning every different assembly might require a different combination of restriction enzymes, reaction temperatures, and buffer conditions. In general, it is unlikely that a single enzymatic mix can be applied across independent assemblies, making the process less amenable to parallelization and automation. In contrast, the newer methods employed below use a standardized set of enzymes and reaction conditions for every assembly facilitating sequence reuse, parallelization, and automation.

2.2 Flanking Homology Methods (SLIC, Gibson, CPEC, SLiCE; GeneArt, In-Fusion)

SLIC, Gibson, CPEC, and SLiCE (and GeneArt Seamless and In-Fusion) are related methods that offer standardized, scarless, sequence independent, multipart DNA assembly.

2.2.1 SLIC

SLIC, or sequence and ligase independent cloning [4], as its name implies, does not utilize restriction enzymes or ligase. A DNA sequence fragment to be cloned into a destination vector is PCR amplified with oligos whose 5′ termini contain about 25 bp of sequence homology to the ends of the destination vector, linearized either by restriction digest or PCR amplification as shown in Fig. 3. The linearized destination vector and the PCR product containing part A are separately treated with T4 DNA polymerase in the absence of dNTPs. In the absence of dNTPs, T4 DNA polymerase has 3′ exonuclease activity, which begins to chew back the linearized destination vector and the PCR product from 3′ to 5′. Once the termini of the linearized destination vector and the PCR product have sufficient complementary single-stranded 5′ overhangs exposed, dCTP is added to arrest the chew-back reaction. With the addition of dCTP, the T4 DNA polymerase changes activity from 3′ exonuclease to polymerase but stalls because not all dNTPs are present, retaining most, if not the entirety, of each chewed-back overhang. Alternatives to the 3′ chew-back with T4 DNA

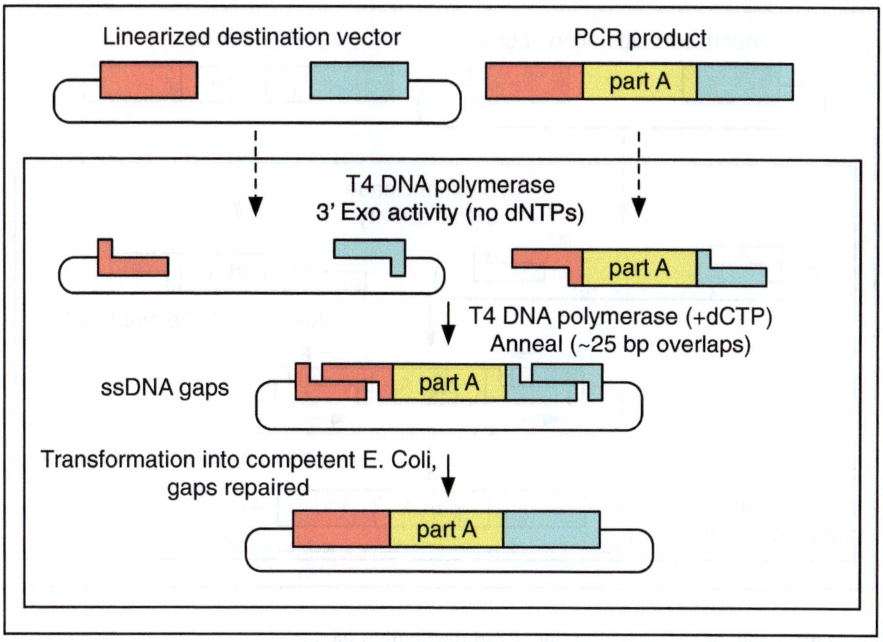

Fig. 3 SLIC assembly of part A with a linearized destination vector

polymerase in the absence of dNTPs include the use of mixed or incomplete PCR products (so this does not apply to the linearized vector backbone if it is derived from a restriction enzyme digest), which can also result in the desired 5′ overhangs, as described in the original SLIC publication [4]. The chewed-back linearized destination vector and PCR product are mixed together and annealed to each other. Since there is no ligase in the reaction, this results in a plasmid with four single-stranded gaps or nicks. Once transformed into competent *E. coli*, the gaps are repaired. Note that SLIC assembly is standardized, in that it always uses the same reaction components and conditions, scarless, since there is no sequence in the resulting assembly that is not user-designed, and sequence-independent, as the method is not (at least to a large extent, but see below) sensitive to the sequences of either the destination vector or the part to be incorporated.

2.2.2 Gibson

Gibson DNA assembly, named after the developer of the method [5] is analogous to SLIC, except that it uses a dedicated exonuclease (no dNTP addition step), and uses a ligase to seal the single-stranded nicks as shown in Fig. 4. The linearized destination vector and the PCR product containing part A are mixed together with T5 exonuclease, which chews back the linearized destination vector and the PCR product from 5′ to 3′. Phusion polymerase, which (with the annealed linearized destination vector and PCR product effectively priming each other) fills in the gaps, and ligase seals the

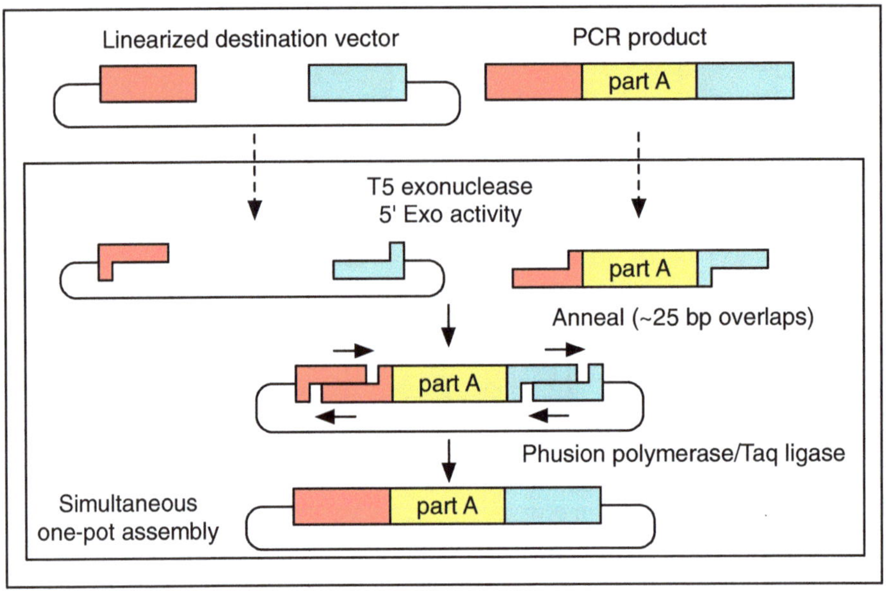

Fig. 4 Gibson assembly of part A with a linearized destination vector

four single-stranded nicks. The polymerase chases the exonuclease around the plasmid, with the polymerase eventually overtaking, as the exonuclease is gradually heat-inactivated (and Phusion is extremely fast). Like SLIC, Gibson assembly is standardized, scarless, and largely sequence-independent. Gibson is advantageous over SLIC in that it is a simultaneous one pot reaction (the two-step addition of dCTP is not required), the presence of ligase may boost assembly efficiency, and since the assembly reaction occurs at an elevated temperature relative to SLIC, there may be fewer problems when somewhat stable secondary structures occur at the ends of assembly pieces. The disadvantage of the Gibson method is that the T5 exonuclease, Phusion polymerase, and Taq ligase cocktail is more expensive than that required for SLIC. An anecdotal/empirical limitation of the Gibson method is that it works best to assemble DNA fragments that are at least 250 bp in length or longer; this is perhaps due to the likelihood that the T5 exonuclease would entirely chew through a short DNA fragment before it has a chance to anneal and prime the Phusion polymerase for extension. While the same could be said for SLIC, the timing of dCTP addition provides some control in switching from the exonuclease to the polymerase activity of T4 DNA polymerase (the use of mixed or incomplete PCR products can prevent this problem all together), although caution should be applied when using SLIC to assemble small DNA fragments. Prior to Gibson (or SLIC) assembly, it is recommended to SOE (splice by overlap extension) together neighboring assembly fragments until their cumulative size is larger than 250 bp. Fortunately, the very same PCR products

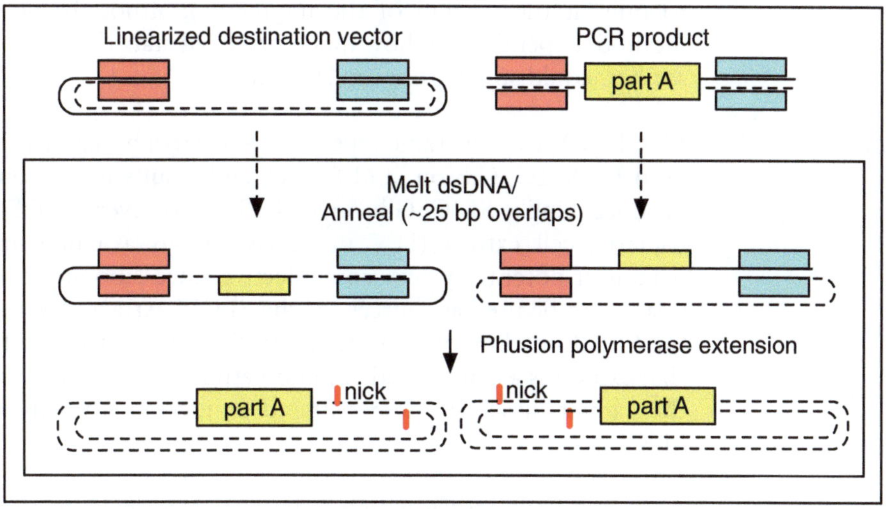

Fig. 5 CPEC assembly of part A with a linearized destination vector

designed for Gibson (and SLIC) assembly, already contain the flanking homology sequences required for SOEing. Gibson has been shown to be a good method for assembling large numbers of dsDNA fragments at once. With the help of whole-genome thermodynamic analysis software PICKY [6] up to 45 fragments have been assembled at once [7]. Modifications to the Gibson protocol have also been useful for assembling large DNA fragments with high GC content [8].

2.2.3 CPEC

CPEC, or circular polymerase extension cloning, shown in Fig. 5, is analogous to SOEing together the fragments to be assembled, except that no oligos are utilized (the linearized destination vector and PCR product prime each other, as in SLIC/Gibson assembly) and there are typically only a few thermocycles required [9]. Since there are no (or very few) reamplifications of a given template sequence, PCR-derived mutations are not propagated as much as one would anticipate for standard SOEing reactions. Like SLIC and Gibson assembly, CPEC is standardized, scarless, and largely sequence-independent. CPEC is advantageous in that, since there is no exonuclease chew-back, small sequence fragments can be assembled directly without a preliminary SOEing step, there is no dNTP addition step (unlike SLIC), there is only a single enzyme (polymerase) required (unlike Gibson), and since the CPEC assembly reaction occurs at higher temperatures than either SLIC or Gibson, stable secondary structures at the ends of assembly pieces are relatively less of a concern. The main disadvantages of CPEC is that it is more likely to result in polymerase-derived mutations than SLIC or Gibson, and mispriming events are now possible anywhere along the sequences of the fragments to be assembled (as opposed to only at the termini of the fragments), although the Gibson method, depending on how much of a head start the T5 exonuclease has, could suffer from similar drawbacks.

2.2.4 SLiCE

SLiCE [10] uses the same types of DNA starting materials as those used for SLIC, Gibson, and CPEC, and results in the same final product. Unlike SLIC, Gibson, or CPEC, however, SLiCE utilizes bacterial cell extract (i.e., an ex vivo DNA assembly method) making it potentially very cost effective since laboratory bacterial strains can be used as sources for the SLiCE extract. A variation of the SLiCE method, which utilizes PPY, a strain of *E. coli* DH10B that expresses a lambda-red recombination system, as the source of the extract, has been demonstrated to increase the efficiency of SLiCE.

2.2.5 GeneArt® Seamless Cloning

GeneArt® Seamless Cloning is a proprietary assembly methodology developed by Thermo Fisher. This assembly method uses the same types of DNA starting materials as those used for SLIC/Gibson/CPEC/SLiCE and results in the same final product. One key difference is that the recommended overlap length is only 15 bps (enabled by a room temperature assembly reaction), which may prove advantageous over SLIC/Gibson/CPEC/SLiCE from the standpoint of requiring shorter/cheaper DNA oligos and enabling combinatorial assembly designs with sequence diversity close to the ends of the sequence fragments to be assembled. On the other hand, a shorter overlap length may reduce assembly specificity, and depending on the assembly mechanism (proprietary), high self-complementarity or strong single-stranded DNA secondary structure in the overlap region may prove more problematic than for SLIC/Gibson/CPEC/SLiCE. Since the overlap length is shorter (~15 bps) than that generally recommended for SLIC/Gibson/CPEC/SLiCE (~25 bps), applying the SLIC/Gibson/CPEC/SLiCE methods to DNA fragments optimized for GeneArt® Seamless Cloning may not be successful. It is possible to use TeselaGen *j5* design parameters optimized for GeneArt® Seamless Cloning. For more information, see the GeneArt® Seamless Cloning documentation on the Thermo Fisher website.

2.2.6 In-Fusion® Cloning

In-Fusion® Cloning is a proprietary assembly methodology developed by Takara-Clontech. This assembly method uses the same types of DNA starting materials as those used for SLIC/Gibson/CPEC/SLiCE and results in the same final product. One key difference is that the recommended overlap length is only 15 bps (like GeneArt® Seamless Cloning, described above, except it operates at 50 °C like Gibson), which may prove advantageous over SLIC/Gibson/CPEC/SLiCE from the standpoint of requiring shorter/cheaper DNA oligos and enabling combinatorial assembly designs with sequence diversity close to the ends of the sequence fragments to be assembled. On the other hand, a shorter overlap length may reduce assembly specificity, and depending on the assembly mechanism (proprietary), high self-complementarity or strong single-stranded DNA secondary structure in the overlap region may prove more problematic than for SLIC/Gibson/CPEC/SLiCE. Since the overlap length is shorter (~15 bps) than that generally recommended for SLIC/Gibson/CPEC/SLiCE (~25 bps), applying the SLIC/Gibson/CPEC/SLiCE methods to DNA fragments optimized for In-Fusion® Cloning may not be successful. It is possible to use TeselaGen *j5* design parameters optimized for In-Fusion® Cloning. For more information, see the In-Fusion® Cloning User Manual and design tool, available from the Takara-Clontech website.

2.2.7 Flanking Homology Method Similarities

Despite differences in implementation, flanking homology methods all start with the same starting materials and result in the same final products. Thus, an assembly designed for CPEC will be equally applicable to SLIC or Gibson assembly. In certain situations, combinatorial SLIC/Gibson/CPEC assembly can be a very reasonable and effective choice (see, e.g., [11] for Gibson and [12] for CPEC), if the sequence identity throughout all combinations and assembly junctions is extensive enough not to be a limitation.

2.2.8 Flanking Homology Method Limitations

A major limitation to flanking homology methods is that the termini of the DNA sequence fragments to be assembled should not have stable single-stranded DNA secondary structure, such as a hairpin or a stem loop (as might be anticipated to occur within a terminator sequence), as this would directly compete with the required single-stranded annealing/priming of neighboring assembly fragments. It may be possible to mitigate this by padding these problematic termini with sequence from their neighboring assembly fragments. Repeated sequences (such as the repeated terminators and promoters in the example above) are often obstacles to SLIC/Gibson/CPEC/SLiCE assembly, since assembly is directed by sequence homology, and if two distinct assembly fragments are identical at one terminus this can lead to assemblies that do not contain all of the desired parts or may contain parts arranged in the wrong order (*see* **Note 1**). To circumvent these obstacles, which TeselaGen *j5* refers to as "assembly fragment incompatibilities," it is often necessary to perform a sequential hierarchical assembly so as not to place assembly fragments with identical termini in the same assembly reaction at the same time. When possible, it is better to substitute repeated sequences with sequence pairs that are not identical yet encode comparable biological function; this provides a benefit not only to the DNA assembly process but will also enhance the DNA stability of the resulting construct. Finally, flanking homology methods might not be the optimal choice for combinatorial assembly if sequence diversity occurs at the very ends of the sequence fragments to be assembled (within about 15 bps of the termini), since this will preclude the reuse of the same homology sequences throughout all of the combinations (*see* **Note 2**). However, in certain situations, combinatorial SLIC/Gibson/CPEC/SLiCE assembly can be a very reasonable and effective choice (*see* [11] for Gibson and [12] for CPEC), if the sequence identity throughout all combinations and assembly junctions is extensive enough not to be a limitation. These limitations, which imply that the SLIC/Gibson/CPEC/SLiCE assembly methods are not completely sequence-independent, are largely addressed by the Golden Gate assembly method.

2.3 Type IIs Endonuclease Methods (Golden Gate, MoClo, GoldenBraid)

2.3.1 Golden Gate

The Golden Gate method [13–15] offers standardized, scarless, multipart DNA assembly and is an good choice for combinatorial library construction. The Golden Gate method relies upon the use of a single type IIs endonuclease, whose recognition sites are distal from their cut sites. The example shown in Fig. 6 uses BsaI, where the recognition sequence "GGTCTC" is separated from its 4-bp overhang by a single bp, and its activity is independent of the sequences of the single bp spacer and the 4-bp overhang. The recognition site for BsaI is not palindromic and is therefore directional. In the notation used here, the recognition site is abstractly represented by a clear rectangle below the dsDNA line and the 4-bp overhang sequence is represented by a shaded box (with different

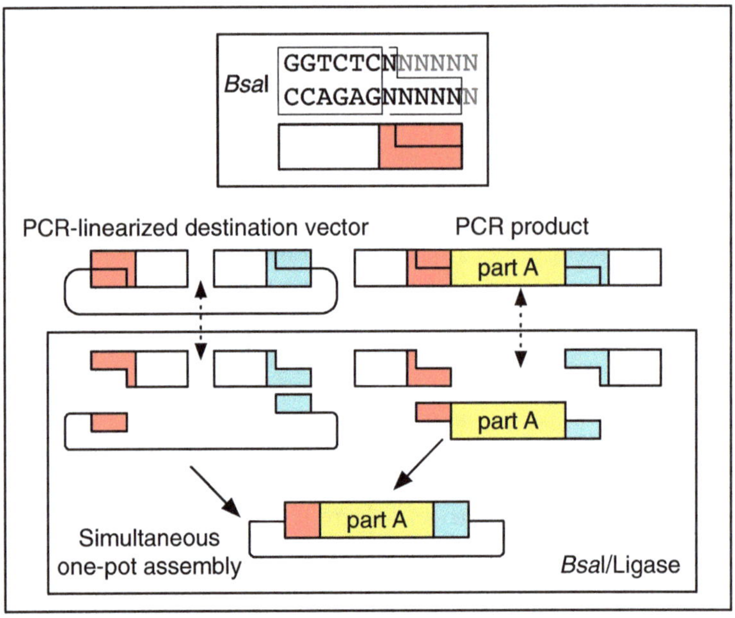

Fig. 6 Golden Gate assembly of part A with a linearized destination vector

shadings indicating different 4-bp sequences). Using this notation, the PCR product containing part A in the example is flanked by two BsaI recognition sites, both pointing inward toward part A. The linearized destination vector is similarly depicted. If the PCR product shown above is mixed with BsaI and ligase, the PCR product is (reversibly) digested, resulting in three DNA fragments, and ligated back together again. The same is true of the linearized destination vector. However, if the PCR product and the linearized destination vector (each of which contains two different 4-bp overhangs) are both mixed together with BsaI and ligase, the cut linearized destination vector will irreversibly ligate (dead-end reaction product) with the cut PCR product containing part A. This particular ligation is irreversible, because the ligation product no longer contains any BsaI recognition sequences. Thus, over time, all reactions will tend toward the desired assembly product. It should be pointed out that the sequences of the 4-bp overhangs are entirely user-specifiable. In this regard, Golden Gate assembly is scarless, since we have complete control over the sequence of the resulting assembly product.

As is true of the flanking homology methods, we can put together many parts at the same time in the same pot (multipart assembly reaction). Golden Gate assembly provides immediate access to every part to be assembled, and with only one transformation step, combinatorial diversity is achieved. Golden Gate assembly is a particularly good choice for constructing combinatorial libraries. Every part in each combinatorial bin is flanked by the same two 4-bp overhang sequences. Any two parts in a bin are completely interchangeable with respect to Golden Gate assembly, and only a single pair of oligos is required for each part across the entire assembly.

Returning to the previous DNA assembly challenge shown in Fig. 1 we can now see that many fragments can be assembled together using Golden Gate, as shown in Fig. 7. In this example, each 4-bp overhang is color-coded, (the BsaI recognition sites, while present and inwardly facing in all of the sequence fragments to be assembled, are not depicted here). We must design the 4-bp overhang sequences for each assembly junction and incorporate them into the 5′ flanking sequence of each oligo, a process that was laborious and error-prone before the advent of assembly software. Note that Fig. 7 is schematic. The assembly junctions must be between DNA fragments that are assembled in the assembly reaction but do not necessarily have to be between the schematic parts being displayed diagrammatically.

Golden Gate assembly is a particularly good choice for constructing combinatorial libraries. TeselaGen DESIGN with *j5* can calculate a protocol that will always result in a scarless assembly, even for complex combinatorial libraries. As shown in Fig. 8, every part in each combinatorial bin (the linearized destination vector is

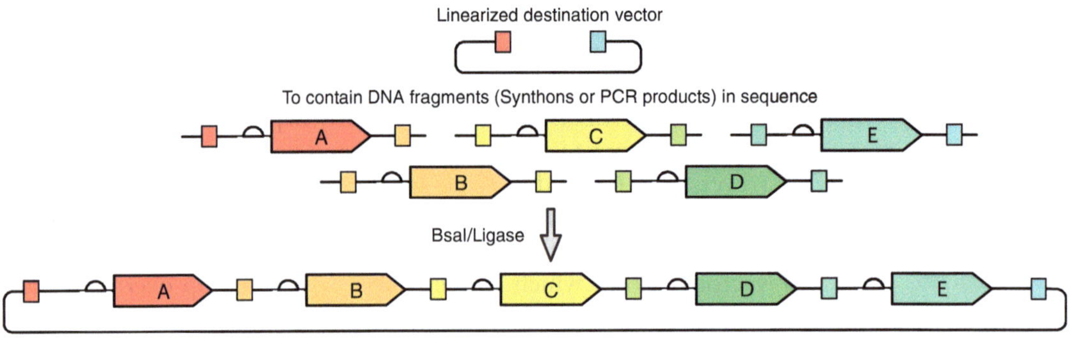

Fig. 7 Golden Gate multipart assembly

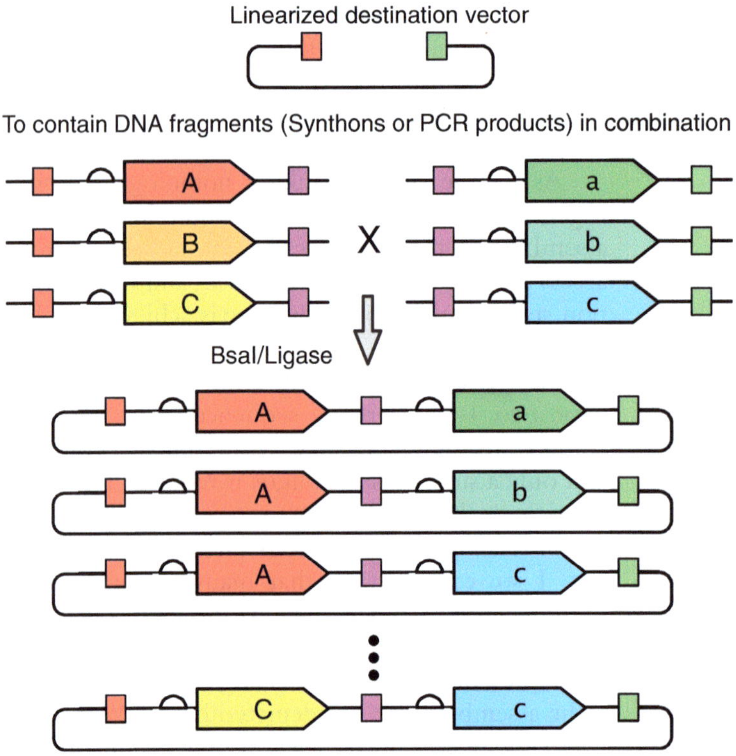

Fig. 8 Combinatorial assembly of a $1 \times 3 \times 3 = 9$ element library

the first bin, the parts labeled A, B, C in the second, and the parts labeled a, b, c in the third) are nominally flanked by the same two 4-bp overhang sequences. Any two parts in a bin are completely interchangeable with respect to Golden Gate assembly, and only a single pair of oligos is required for each part across the entire assembly. If the algorithm cannot find an identical junction across all combinations it relaxes the full reuse constraint (potential for different overhangs for each combination) but never introduces scars (*see* **Note 3**).

2.4 Flanking Homology and Type IIs Endonuclease Method Treatment by TeselaGen DESIGN Module with j5

1. After the user has selected a protocol that maps to the Type IIs assembly method we heuristically determine the most cost-effective strategy to incorporate each part into an assembly fragment prior to executing the full assembly design process. The algorithm calculates the marginal PCR cost and the marginal synthesis cost to make its determination.
   (a) Embed part in primer check. If the part length is less than the minimum allowed part length for PCR, then the part will be embedded in a reverse primer or marked for synthesis as appropriate.
   (b) PCR vs synthesis of a part. Even when synthesis is not chosen as the default strategy, if it is cheaper to synthesize a part, the algorithm will recommend synthesis.
   (c) PCR vs synthesis of next part. If the current part strategy is synthesis, even when synthesis is not chosen as the default strategy for the next part, if it is cheaper to synthesize the next part together with the current part, the algorithm will recommend synthesis.
   (d) PCR vs synthesis of previous part. If the current part strategy is synthesis, even when synthesis is not chosen as the default strategy for the previous part, if it is cheaper to synthesize the previous part together with the current part, the algorithm will recommend synthesis.
2. Progressively relieve violated constraints during primer (or flanking sequence) design. Existing programs such as Primer3 [16] can optimize the design of primers or flanking homology sequences (effectively primers for adjacent assembly pieces during Gibson and CPEC assembly). One drawback is that they provide primer pair designs only if a given set of design criteria is met. This algorithm first attempts to design optimal primers that meet all design constraints; if unable to do so, constraints are progressively relieved until an acceptable primer pair has been found. In addition to the primers (or flanking homology sequences) designed, warning messages are issued if any design constraints were violated/relieved during the design process and/or if any putative template mispriming events with above threshold melting temperatures are identified via BLAST [17].
3. Fragment matching.
   (a) For Flanking Homology: Identify flanking homology assembly piece incompatibilities; if found, design a hierarchical assembly strategy.
      - The algorithm optimizes the flanking homology overlap sequences against typical assembly design parameters.

- The algorithm checks for off target homology annealing sites and designs the assembly process as a 2-level hierarchy if such sites are found. In this way, regions of incompatibly can be buried inside the post first-level contigs input into the second-level assembly process.

(b) For Type IIs: Search for the optimal set of (Golden Gate) assembly piece overhangs.

- For all assembly junctions, the algorithm sets the nominal control position offset to zero. The maximum allowed shift in overhang position is ultimately based on oligo size. As the algorithm attempts for find optimized assembly piece overhangs, shift from neutral will occur, and oligos will be made longer to cover those shifts. As the algorithm proceeds it starts with all neutral offsets, and if any off-target pair of overhangs is incompatible (have too many aligned bps), then shifts the first overhang by +1, and then −1, checking for compatibility across all pairs of overhangs. If not compatible, then that first offset is returned to zero and the next overhang is tested to see if minimalistic offsets can bring the overhangs into a compatible state. Eventually, all combinations of minimal offsets to all overhangs are checked until we find a set that are 100% compatible, or the algorithm determines that no compatible set of overhangs exists given the maximum oligo size.

4. When PCR is called for, closely approximate the optimal distribution of PCR reactions in multiwell plates across thermocycler block annealing temperature zone gradient(s).
5. Checks for DNA fragment buildability and cost are done by submitting DNA fragments to vendors via their APIs. To date, Twist, IDT, and GenScript are developing TeselaGen compatible APIs that provide detailed information to support the design process.

2.5 Other Methods

The list of protocols exploiting modern scarless single-reaction cloning methods is long; however, the number fundamentally different DNA assembly topologies are relatively few. In Table 1 we outline some of the more popular protocols and how they map into a set of topologies that can be used to guide the design of similar protocols using modern DNA assembly software.

Table 1
Examples of industrial cloning methods suitable for designing large-scale DNA libraries

Class	Protocol	Reference	Note
Flanking homology	SLIC	[4]	See text
	Gibson	[5, 11]	See text
	CPEC	[9, 12]	See text
	GeneArt® Seamless	See ThermoFisher website	See text
	In-Fusion®	See Takara website	See text
	Yeast Assembly	[18]	The yeast *Saccharomyces cerevisiae* can take up and assemble at least 38 overlapping single-stranded oligonucleotides and a linear double-stranded vector in one transformation event.
	SLiCE	[10]	See text
	LIC	[19]	Predates SLIC. Inserts are usually PCR amplified and vectors are made linear either by restriction enzyme digestion or by PCR. Technique uses the $3' \rightarrow 5'$ exo activity of T4 DNA polymerase to create overhangs with complementarity between the vector and insert
Type IIs	Golden Gate	[15]	See text
	MoClo	[20, 21]	See text
	GoldenBraid	[22]	See text
Polymerase/ ligase	DATEL	[23]	DNA assembly method using thermal exonucleases (Taq and Pfu DNA polymerases) and Taq DNA ligase (DATEL)
Blunt end	LCR	[24, 25]	LCR assembly uses single-stranded bridging oligos complementary to the ends of neighboring DNA parts, a thermostable ligase to join DNA backbones, and multiple denaturation–annealing–ligation temperature cycles to assemble complex DNA constructs
Uracil excision	USER	[26, 27]	By varying the design of the PCR primers, the protocol can to perform one or more simultaneous DNA manipulations such as directional cloning, site-specific mutagenesis, sequence insertion or deletion and sequence assembly
Linker mediated	BASIC	[28]	Based on linker-mediated DNA assembly and provides highly accurate DNA assembly
PCR	AFEAP	[29]	The AFEAP method requires two rounds of PCRs followed by ligation of the sticky ends of DNA fragments

3 Example DNA Assemblies Using TeselaGen's DESIGN Module

TeselaGen's DESIGN software provides a unified interface and compute infrastructure for the design of DNA libraries, and the generation of instructions for how to build those DNA libraries (*see* **Note 4**). Those instructions can be generated in a way that can be optimized for both humans and automation. The platform provides a standardized system for tracking the relationships between design elements (parts and annotations), genetic designs (simple, hierarchical, or combinatorial) and DNA assembly protocols. A number of assembly reaction types are supported including Type IIs Endonuclease (Golden Gate, MoClo, etc.), Flanking Homology (Gibson, InFusion, etc.). Among features supporting the design process is the ability to create Design Templates that can be reused across designs. The system optimizes assembly reactions to take advantage of a variety of DNA sourcing options including DNA synthesis vendors such as Twist, IDT, and GenScript.

Important features of the DESIGN module:

- Capture of DNA designs that can be simultaneously combinatorial and hierarchical.
- Scarless design of large-scale DNA libraries.
- Cost optimization including part reuse where warranted.
- Sourcing material from best available options including direct links to DNA vendors.
- Design Templating to aid the design process.
- DNA, Amino Acid, Oligo sequence libraries.
- DNA Combinatorial and Hierarchal Design libraries.
- User lab groups, secure sharing, alerts, and messaging.
- Automated protocol generation for use with automation.

3.1 Design Capture

The fundamental role of the DESIGN module is the accurate capture of the designer's intent. Target genetic designs are constructed 5′ to 3′ (or N-terminus to C-terminus), left to right, by selecting parts from a parts library to columns in a whiteboard style user interface. Alternatives for any given part are listed as cell entries within a column. The user can also specify the naming scheme and the preferred DNA assembly chemical reactions.

3.2 Complex Designs

TeselaGen's DESIGN module has added support for combinatorial DNA libraries as well as hierarchical designs as shown in Fig. 9. The hierarchical capability is particularly useful for the following users:

1. Users who choose to adopt inherently hierarchical assembly methods. They are able to rapidly design complex hierarchical builds using while maximizing part reuse and minimizing cost.

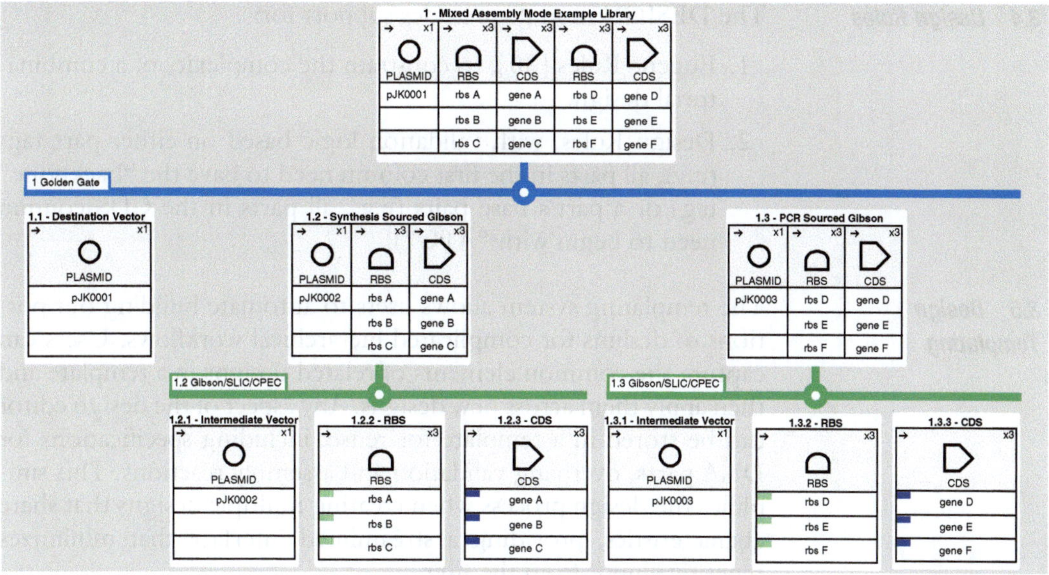

Fig. 9 Hierarchical Design Editor. A two-level combinatorial design specifies four Gibson assemblies that generate intermediates that will be reactants for a final Golden Gate assembly

2. Users who build very long pieces of scarless DNA or construct gene stacks will find the streamlined interface a convenient and reproducible way to break down very long target designs into buildable submodules.

3.3 Part Reuse and Sourcing

Part reuse is important when the size of the built library is large, and cost is a constraint. Without part reuse the cost of a library can grow linearly with the number of parts, while with part reuse it can grow logarithmically. The DESIGN module utilizes the *j5* algorithm when designing a set of combinatorial assembly reactions which automatically maximizes part reuse within a combinatorial design. Hierarchical designs can also be optimized for part reuse across designs with utilities that check if intermediate stretches of DNA already exist in inventory, modifying the build instructions accordingly so that previously built constructs are not assembled again. Users can also use this availability information to automatically break down target constructs into divisions based on available subsections instead of manual divisions. Material availability is can also be extended to query external vendors for what they can provide. This is done through a series of API integrations with vendor utilities that check DNA segments for their manufacturability, cost, and delivery times. The Design module also provides the user with greater control over the sourcing of the DNA parts used in their designs, especially with Type IIs restriction enzyme digest/ligation assemblies. When performing such an assembly, the interface automatically adds validation to the input parts to ensure that they are sourced with the appropriate flanking digest sites.

3.4 Design Rules

The DESIGN module provides support for:

1. Eugene Rules [30], to constrain the complexity of a combinatorial design.
2. Design Rules, with validation logic based on either part tags (e.g., all parts in the first column need to have the "backbone" tag) or a part's base pairs (e.g., all parts in the CDS column need to begin with "ATG").

3.5 Design Templating

The templating system allows users to automate building out portions of designs for complicated hierarchical workflows. Users can capture the common elements of related designs in a template and then apply them across new designs. Any aspect of the design editor can be stored in a template for reuse, including specifications for DNA parts, overhang validation and assembly reactions. This simplifies the design process when creating multiple designs that share characteristics, providing a streamlined interface that minimizes repetitive input from the user.

3.6 Example 1: Combinatorial DNA Library Design

We will use the combinatorial design view of the Design Editor whiteboard to make the design simple and compact. When we are finished with our design it will look something like the design shown in Fig. 10.

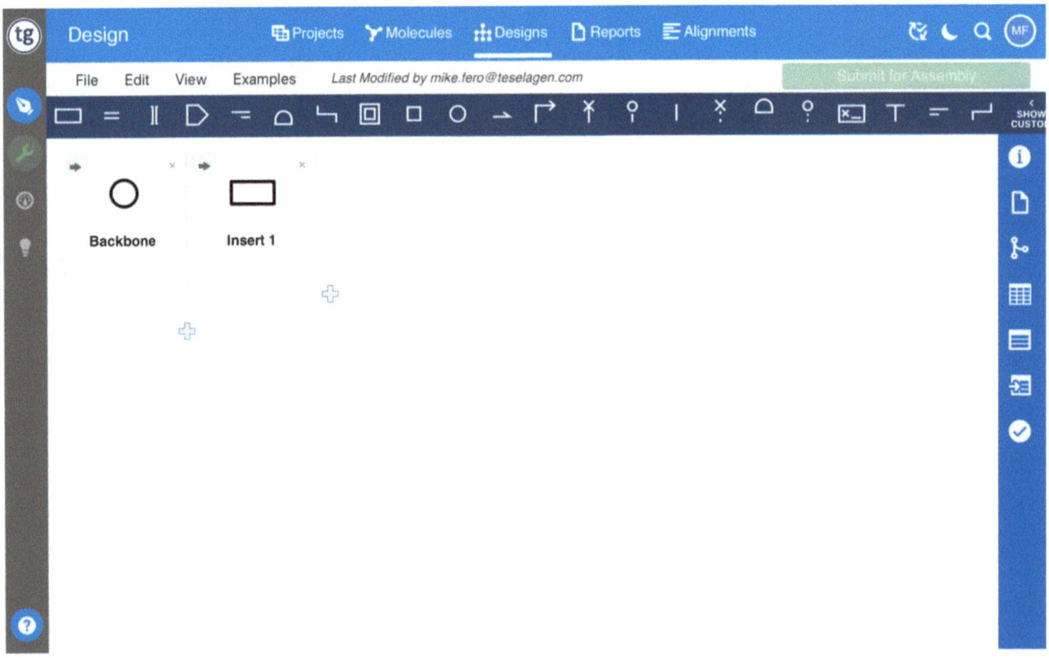

Fig. 10 A combinatorial DNA library design

3.6.1 Sketch Out the DNA Design

We want to divide each construct into the following components:

1. Vector backbone.
2. N-terminal Signal Peptide (2 variants).
3. Gly-Ser Linker (2 variants).
4. GFP (1 variant).
5. ssrA 5-prime degradation tag (2 variants),
6. ssrA 3-prime degradation tag (1 variant).

We will start by creating a new design called Small DNA Library. From the main page, click Designs → Designs → New Design. To add a new component column or variant row, click on the corresponding plus sign. Details of each bin (or component) can be viewed or modified by clicking the bin icon on the inspector panel to the right. Since each construct has six components, let us add four more bins to the current design. In addition, the second, third, and fifth parts each have two variants; therefore, we want to add one more row to the current design.

TeselaGen design editor has a variety of symbols representing different genetic elements that accommodate the needs of visualizing complex DNA constructs. In this design, the second, third, and fourth components are coding sequences; hence, it makes sense to use a "CDS" (coding sequence) icon for these components. For the last two components, we can use a "Protein Stability Element" icon. To change a bin's symbol, click on that bin and choose a symbol from the list of SBOL Visual glyphs [31]. In addition, we can change the name of each component to a more descriptive name by changing its name in the bin's detail tab.

3.6.2 Assign DNA Parts

Now that the design has been sketched, we assign the actual DNA parts. From our DNA Parts Library, we assign the DNA to cells under the proper column headings. Do this by double-clicking a cell you want to assign a DNA part to, the GFP cell for example. A dialog containing the DNA parts library will open up, which allows us to search for "GFP," find the DNA part we are looking for and assign it. For the components with variants (e.g., second, third, and fifth parts), assign the variants to the cells of the same column but on different rows. For example, for the N-terminal signal peptide, the DNA part could be either a BMC_nterm_sig_pep or a ccmN_nterm_sig_pep. Once all the cells have been assigned to actual DNA Parts, the "Submit for Assembly" button turns green and becomes enabled.

3.6.3 Design Rules and Parameters

Here is an important design consideration. The 5-prime and 3-prime degradation tags are rather short. We have chosen to do some PCR to pull some of the desired DNA out of their host vectors, so why not embed these tags into the forward and reverse primers? We can tell the assembler to do this by assigning a "Forced Assembly Strategy." You can select the forced assembly strategy options for any part by selecting the part of interest, then selecting the part icon on the inspector panel on the right to view the part's details. For our example, let us click select the "ssrA_tag_3prime" cell, then pick "Embed in primer forward" from the "Forced Assembly Strategy" dropdown menu. For the two DNA part options in the ssrA_5primeTag column, pick "Embed in primer reverse." The design editor uses colored markers to indicate different assembly strategies. In our example, the "Embed in primer forward" has the green color bar, while "Embed in primer reverse" has the purple color bar as seen in Fig. 10.

We can add one more feature to this design. It turns out that some of these parts are contiguous in their hosts. The TeselaGen assembler is smart enough not to break everything apart, just to put it all back together again. However, the assembler will do a cost tradeoff between direct DNA synthesis and PCR, based on the cost of DNA synthesis. To tell the assembler not to consider the cost tradeoff and just go with DNA synthesis, you can impose a "Direct Synthesis Firewall" to a column. To do this, select a bin and click on the bin icon on the right panel to view its details. Check the option for "Direct Synthesis Firewall." You should see a red line appear on the right of the selected bin indicating the direct synthesis firewall (DSF) is in effect. In our example we do this for columns 2 and 5. Hence, the red bars in the design shown in Fig. 10.

Finally, you can modify the assembly method by selecting the method from the drop-down menu in the "Assembly Reaction Details." In this design we have chosen Golden Gate as our assembly method.

3.6.4 Submit for Assembly

Navigate to the green "Submit for Assembly" button at the upper right corner of the design editor to run the assembler and build your library.

The DESIGN module does not require you to figure out how to build a library in the traditional artisanal fashion. Once you capture your design in Design Editor, the DNA Assembler takes over and generates all the information you need to build your library. While the DNA Assembler is working a task monitor will appear at the top right of the interface.

3.6.5 Interpret Output

After the DNA Assembler finishes the job, you can view the results by clicking on the Assembly Reports icon on the right panel and navigating to the report of interest. The report starts with information about your design, the assembly method, the time it was run, the export format options, and whether any warning or errors were encountered during the assembly. Following this descriptive metadata, the report is sectioned as follows: prebuilt constructs, assembled constructs, input sequences, input parts, oligo synthesis, DNA synthesis, PCR reactions, DNA pieces to be assembled, combination of assembled pieces. Let us look at each section in more detail.

1. Prebuilt Constructs: These are constructs that have been built and are available in the library. These are particularly useful for hierarchical designs as they allow us to build complex DNA constructs from simpler constructs that have been built before. In our example, we built our constructs from scratch so there are no prebuilt constructs to display.
2. Assembled Constructs: Recall that we used the combinatorial design editor to create eight constructs. Those are the assembled constructs or output constructs. You can view an assembled construct in the Vector Editor by double-clicking that construct. You can also save the assembled constructs to the DNA sequence library.
3. Input Sequences: As the name suggests, input sequences are the source sequences of the parts in your assembled constructs. They are present in the "DNA Sequences" library.
4. Input Parts: The input parts are the segments of input sequences used in the assembly. They are not necessarily the fragments to be assembled together. This is because the DNA Assembler is smart enough to know not to break up contiguous parts, but to leave them intact to minimize the number of assembly fragments and maximize reuse.
5. Oligo Synthesis: The section tells you what you need to go out and order from your favorite synthesis provider. You have the options to either save it to the oligo library or export as a CSV file. TeselaGen provides direct links to these providers to simplify ordering.
6. DNA Synthesis: This section lists the DNA pieces that need to be directly synthesized with similar properties as the oligos. In our example, there is no direct synthesis.
7. PCR Reactions: This section lists the PCR reactions that need to be done to generate the assembly pieces.
8. DNA Pieces to be Assembled: This section lists the fragments to be put together in the final assembly reactions to give the desired constructs. The parts of each fragment can be viewed from the last section—Combination of Assembly Pieces.
9. Combination of Assembly Pieces: This section lists what goes where to make up the final DNA library. In an automated laboratory setting this list gets translated into a worklist for the robots by the TeselaGen BUILD module, see Fig. 2.

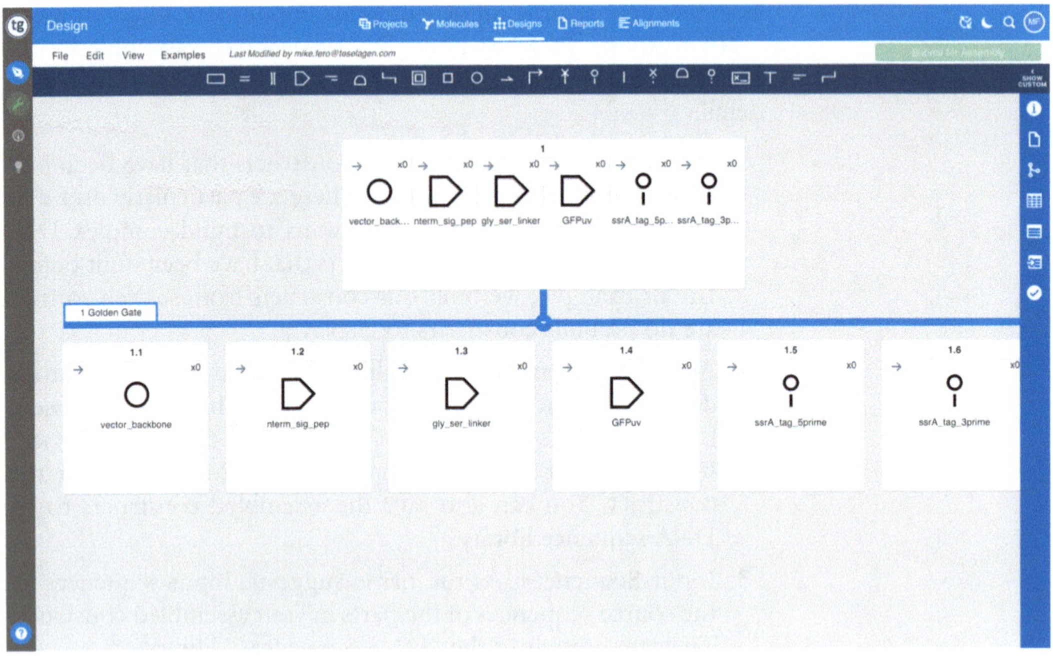

Fig. 11 A hierarchical DNA library design

3.7 Example 2: Hierarchical DNA Library Design

Let us take a look at how to use the TeselaGen DESIGN module to create a DNA library based on a hierarchical approach. In this example we will be use the hierarchical design editor to assemble a library that consists of a backbone, a promoter, 4 ribosome binding sites, and 4 gene using Golden Gate assembly method. Additionally, we want to also specify in the design how the assembly pieces for the final construct was sourced from intermediate constructs via Golden Gate and Gibson assembly. This is a simple design with two levels of hierarchy, but the ideas are extensible to much larger designs with an arbitrary number of levels in the hierarchy. When we have completed the design, it will look something like that shown in Fig. 11.

3.7.1 Sketch Out the DNA Design

We will start by creating a new design called Simple Hierarchical Design. From the main page, click Designs → Designs → New Design. Save the new design as Simple Hierarchical Design. Make sure that you are in an appropriate view mode. From the design editor page, click View → View Mode → Vertical. This view mode separates input and output DNA into their own "cards" with connecting colored lines that symbolize the assembly reaction. The topmost card represents the final DNA assembly product with the assembly reaction beneath it. The tab on the assembly reaction colored lines identifies the reaction. Input cards representing DNA assembly reactants are arrayed below. This process can continue until DNA fragments that represent the building blocks for the entire hierarchical assembly are defined.

Note that the design paradigm is "top down." Users specify the final assembled construct they would like to build and break that construct into intermediate steps until they reach the lowest level of the hierarchy, the building block parts. Once a design is specified in this way, the software builds from bottom up, generating all of the instructions for the reactions and steps that will result in the desired construct. Note also that there are different ways to approach a hierarchal design. One is a biology centric approach, defining fragments of DNA as "parts" and helping the user think of the design process in terms of functional DNA parts, many of which may already be stored in plasmids and are available to source using simple PCR reactions. Another is a factory approach suitable for service centers or vendors, where the approach might be to just do binary splits of the DNA from the top down until we get to fragment sizes that can be synthesized, then reversing the process through standardized assembly methods. This introductory example favors the biology centric approach in order to stay relevant to most bench scientists.

3.7.2 Layout the Target Construct

When working in the hierarchical view we start with the final goal in mind and then break it down into its constituent parts. For this example, we will want the following bins in our final assembly product card:

1. Backbone (Origin of Replication icon).
2. Promoter (Promoter icon).
3. Ribosome binding site (RBS)(RBS icon).
4. Gene (CDS icon).
5. Ribosome binding site (RBS icon).
6. Gene (CDS icon).
7. Ribosome binding site (RBS icon).
8. Gene (CDS icon).
9. Ribosome binding site (RBS icon).
10. Gene (CDS icon).

Add bins to the topmost card by right-clicking the card and choosing Insert > Insert Bin Right until there are a total of four bins. Next, click on a bin and give it an appropriate icon by clicking the corresponding icon in SBOL glyph ribbon above the design editor canvas. Give the bins a name by clicking on the bin and opening the Bin Details inspector panel from the design editor toolbar along the right side of the screen.

3.7.3 Add First-Level Assembly Reaction

Next, change the assembly reaction to be a Golden Gate reaction instead of the default Mock Assembly (mock assembly does a simple concatenation of sequences in order to make a quick check of the output library). Right click on the reaction tab on the line that connects different elements of the design tree and choose Change Assembly Reaction. From the Assembly Reaction Parameters window, select Golden Gate as the Assembly Method, and Default as the Parameter Preset. Give the reaction a name (e.g., "Golden Gate") and then click the "Next" button. You can also set custom Assembly Report Naming Templates which will affect how the app names various items in the assembly report. Also, if you have a version of this reaction form that you would like to reuse in later designs, you can save the form as an Assembly Reaction Preset. The preset will appear in the reaction preset library and can be used to quickly fill out the reaction form with a single click.

We will be defining the reactants by splitting up the bins of the product card. In our case we want to split the bins in the following way: the promoter, the fourth ribosome binding site (RBS) and the fourth gene are sourced from the vector backbone, each of the other pairs of RBS and gene are sourced together from a different plasmid. To do this splitting, click on the vertical dash lines as shown in Fig. 12 and check the box for "Make all reactants circular (sourced on a backbone)."

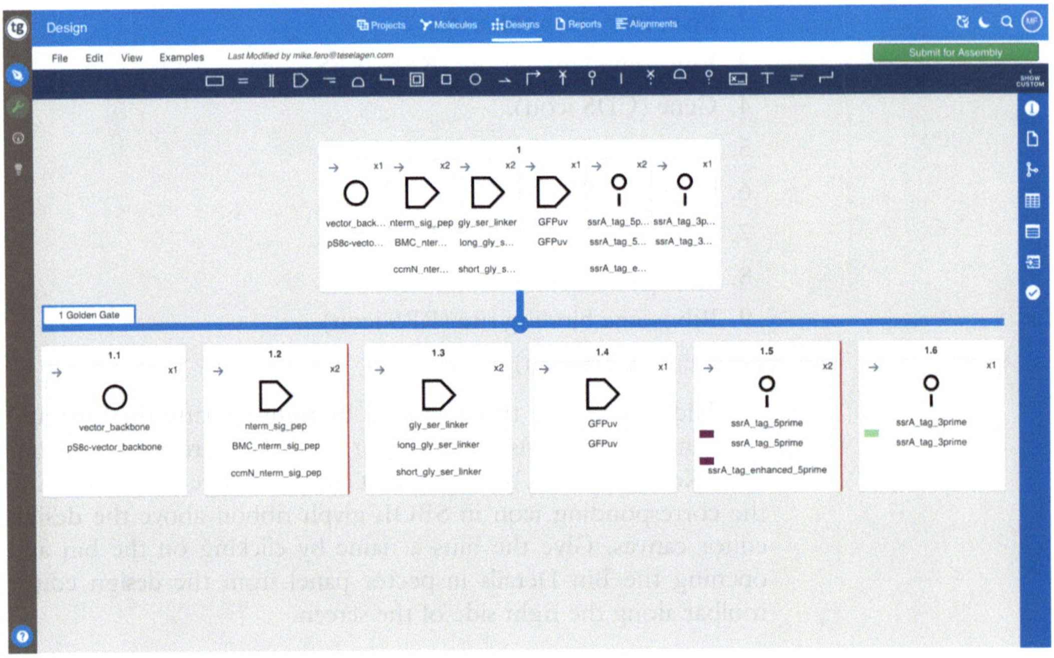

Fig. 12 Assembly reaction definition

3.7.4 Add Second-Level Assembly Reactions

Next let us add our final layer of assembly reactions. This layer describes how each of the intermediate cards (1.1, ..., 1.4) are assembled. For example, construct in card 1.1 is assembled via Gibson assembly, while the rests are assembled via Golden Gate. Either right-click the card and choose Add Assembly Reaction or click the [+] button underneath the card. From there, give the reaction a name and assembly method of your choice with the Default Parameter Preset and default Output Naming Templates.

When you are defining the reactant groups, let us choose Split All Bins. Do this for all four cards. You can hide assembly trees by clicking the +/– circle in the middle of a reaction's colored line. This is helpful if you want to focus only on a specific region of a design.

3.7.5 Add Parts to Top Card and Intermediate Cards

Now that the design is specified, we are ready to add DNA parts. In the top Target Construct card, either double click a cell or right-click a cell and choose Insert → Insert Part to add your DNA Parts to the design. Add your own backbone, promoter, genes, and RBS parts in this manner. Once the parts are assigned to the cells in your top card, it is automatically filled in the cells of the cards below. For intermediate cards that contain bins that are not propagated to the top card (e.g., card 1.2, 1.3, and 1.4 have a backbone that is not part of the target card), you need to assign parts for these bins as well. If you had previously imported this example design into your library and are rebuilding it from scratch, then the parts should be in your library. If not, then you may need to import this example design first or use your own data. While a normal Part will usually suit DNA designer's needs, there are several other ways of adding DNA to a design:

- *Part*—An annotation on a source sequence with a start and stop index.
- *Unmapped Part*—A part name that isn't associated with any DNA yet.
- *Base Pairs*—A part unassociated with a sequence consisting of user defined base pairs.
- *Assembly Piece*—A part that already has flanking homology regions on it, when used in a Gibson/SLiC/CPEC reaction our assembly software will design the overlapping ends to conform to the assembly piece part.
- *Sequence*—A convenience method of creating and inserting a part that spans an entire sequence.
- *Part Set*—A group of multiple parts tied together in one UI element, useful to reduce clutter in the design if inserting 100+ parts.

3.7.6 Submit for Assembly

Once all the parts have been inserted, we are ready to Submit for Assembly. The design should look like that shown in Fig. 11. Clicking the green button at top right will send the design to the server to generate the assembly report. Depending on how many combinations are in your design, this process may take a few minutes. If the design is not ready to submit, the green "Submit for Assembly" button will be disabled. You will need at least one part in every bin and every part passing automated validation checks in order to submit to the assembler.

In this design we have five individual assemblies, 4 Golden Gate and 1 Gibson reactions. We will get a separate assembly report for each of these reactions, all linked together in a folder in the Assembly Reports section of the Inspector Panel along the right. Each report can be interpreted individually as explained in Example 1.

3.8 Conclusions

At TeselaGen we are building a four-part AI-guided enterprise platform for bio molecule development that mirrors the design–build–test–evolve ethos of synthetic biology. In this chapter we have described important features of the first of these four modules, the TeselaGen DESIGN module. We have shown that we have been able to direct large scale DNA library construction with the DESIGN module at numerous customer sites, allowing our users to (1) Capture combined and individual combinatorial + hierarchical designs, (2) build large-scale scarless DNA libraries, (3) cost optimize their design/build, (4) source material from best available options including direct links to vendors, (5) allow design templating to aid the "design once, build many" workflows that save time and effort.

Beyond the scope of this chapter, but, nevertheless, interesting to the future of automated and optimized synthetic biology, is the BUILD module that guides communication of protocols to lab workers and automation, the TEST module which gathers high value phenotypic data from analytic instruments, and the EVOLVE module that applies machine learning to optimize workflows, cell line and microbial strain development, and scale-up to commercial production (*see* **Note 5**).

4 Notes

1. Sequence repeats, or highly homologous sequences at the termini of assembly pieces can be problematic for assembly. It may be desirable to include a given part more than once in the same assembly (e.g., a repeated terminator or promoter). However (aside from decreasing the physical stability of the resulting construct (via in vivo recombination processes)), these sequence repeats can be debilitating for the SLIC/Gibson/CPEC/SLiCE assembly methods and should be avoided

where possible. Work-arounds include identify parts with similar function but different DNA sequence (e.g., two sequences encoding the same protein with different codons). If the repeated sequences are not located at the termini of the assembly pieces, they might not significantly affect SLIC or Gibson assembly, but they will be problematic for CPEC or SLiCE assembly. In certain situations, it may be better to perform the assembly with the Golden Gate method, which is not as affected by sequence repeats. *j5* detects highly homologous sequences, automatically alerting the user to these potential problems when designing DNA assemblies.

2. Assembly piece termini with stable secondary structure can be problematic for assembly. If the terminus of an assembly piece has very stable secondary structure (which can be accessed via the DINAMelt Quikfold server, or other related software), as would be anticipated for a terminator, it will not be able to base-pair/anneal with the neighboring assembly piece, and thus inhibit assembly. A work-around is to add sufficient flanking sequence so that the problematic section with secondary structure is no longer at the terminus. In certain situations, it may be better to perform the assembly with the Golden Gate method, which is not as affected by termini with stable secondary structure.
3. For Type IIs Methods (Golden Gate, etc.) there is one situation that will potentially be problematic for assembly: BsaI (or other selected type IIs endonuclease) recognition sites are present within the DNA fragments to assemble (not only at the assembly piece termini). In this case, it is possible to generate (silent) point mutations to disrupt these sites. Even with the undesired BsaI sites present, assembly may still occur (since the digest/ligation is a reversible-process), but the efficiency will be decreased.
4. A variety of detailed protocols compatible with *j5* can be found on the TeselaGen and JBEI websites and Protocols.io.
5. Functional testing is a key bottleneck in high-throughput approaches to screening large libraries for constructs that maximize activity of an enzyme or desired metabolic product. An exhaustive test of a pathway with ten variants across eight parts yields 10^8 possible constructs to be assayed. Obviously, there are two immediate tasks at hand: (1) reduce the complexity of the library and (2) enable building, testing and screening the remaining irreducible set of constructs in the most time and cost-effective way possible. Modern synthetic biology efforts are addressing both these issues, using rules to reduce complexity and automation and intelligent, iterative functional testing to converge using an optimal search routine. It has

proven to be very valuable to integrate design of experiments and statistical learning approaches with construct design/fabrication to minimize the number of constructs screened per round while maximizing information learned.

Acknowledgments

Michael Matena and Adam Thomas for development of TeselaGen's Hierarchical Design Editor; Thomas Rich and Tiffany Dai for development of TeselaGen's Open Vector Editor; Rodrigo Pavez, Tim Thimmaiah, and Nick Elsbree for TeselaGen software platform development to support DNA design; and Katy Basile and Matthew Gibson for intellectual property consultation and support. Portions of this work were funded by NSF 1430986.

References

1. Endy D (2005) Foundations for engineering biology. Nature 438(7067):449–453
2. Arkin A (2008) Setting the standard in synthetic biology. Nat Biotechnol 26(7):771–774
3. Hillson NJ, Rosengarten RD, Keasling JD (2012) j5 DNA assembly design automation software. ACS Synth Biol 1:14–21
4. Li MZ, Elledge SJ (2007) Harnessing homologous recombination in vitro to generate recombinant DNA via SLIC. Nat Methods 4 (3):251–256
5. Gibson DG (2009) Enzymatic assembly of DNA molecules up to several hundred kilobases. Nat Methods 6(5):343–345
6. Chou HH, Hsia AP, Mooney DL, Schnable PS (2004) Picky: oligo microarray design for large genomes. Bioinformatics 20:2893–2902
7. Birla BS, Chou HH (2015) Rational design of high-number dsDNA fragments based on thermodynamics for the construction of full-length genes in a single reaction. PLoS One 10(12): e0145682
8. Li L, Jiang W, Lu Y (2018) A modified Gibson assembly method for cloning large DNA fragments with high GC content. Methods Mol Biol 1671:203–209
9. Quan J, Tian J (2009) Circular polymerase extension cloning of complex gene libraries and pathways. PloS one 4(7):6441
10. Zhang Y, Werling U, Edelmann W (2012) SLiCE: a novel bacterial cell extract-based DNA cloning method. Nucleic Acids Res 40 (8):e55
11. Ramon A, Smith HO (2011) Single-step linker-based combinatorial assembly of promoter and gene cassettes for pathway engineering. Biotechnol Lett 33(3):549–555
12. Quan J, Tian J (2011) Circular polymerase extension cloning for high-throughput cloning of complex and combinatorial DNA libraries. Nat Protoc 6(2):242–251
13. Engler C, Gruetzner R, Kandzia R, Marillonnet S (2009) Golden Gate shuffling: a one-pot DNA shuffling method based on type IIs restriction enzymes. PLoS One 4(5):e5553. https://doi.org/10.1371/journal.pone.0005553
14. Engler C, Marillonnet S (2011) Generation of families of construct variants using golden gate shuffling. Methods Mol Biol 729:167–181
15. Engler C, Kandzia R, Marillonnet S (2008) A one pot, one step, precision cloning method with high throughput capability. PLoS One 3 (11):3647
16. Rozen S, Skaletsky H (2000) Primer3 on the WWW for general users and for biologist programmers. Methods Mol Biol 132:365–386
17. Zhang Z, Schwartz S, Wagner L, Miller W (2000) A greedy algorithm for aligning DNA sequences. J Comput Biol 1(1-2):203–214
18. Gibson DG (2009) Synthesis of DNA fragments in yeast by one-step assembly of overlapping oligonucleotides. Nucleic Acids Res 37 (20):6984–6990
19. Aslanidis C, de Jong PJ (1990) Ligation-independent cloning of PCR products (LIC-PCR). Nucleic Acids Res 18 (20):6069–6074
20. Weber E, Engler C, Gruetzer R, Werner S, Marillonnet S (2011) A modular cloning

system for standardized assembly of multigene constructs. PLoS One 10(1371):0016765

21. Werner S, Engler C, Gruetzner R, Marillonnet S (2012) Fast track assembly of multigene constructs using Golden Gate cloning and the MoClo system. Bioeng Bugs 3(1):38–43
22. Sarrion-Perdigones A, Falconi E, Zandalinas S, Juárez P, Fernández-del-Carment A, Granell A, Orzaez D (2011) GoldenBraid: an iterative cloning system for standardized assembly of reusable genetic modules. PLoS One 6(7): e21622
23. Jin P, Ding W, Du G, Chen J, Kang Z (2016) DATEL: a scarless and sequence-independent DNA assembly method using thermostable exonucleases and ligase. ACS Synth Biol 5 (9):1028–1032
24. de Kok S, Stanton LH, Slaby T, Durot M, Holmes VF, Patel KG, Platt D, Shapland EB, Serber Z, Dean J, Newman J, Chandran SS (2014) Rapid and reliable DNA assembly via ligase cycling reaction. ACS Synth Biol 3 (2):97–106
25. Chandran S (2017) Rapid assembly of DNA via ligase cycling reaction (LCR). Methods Mol Biol 1472:105–110
26. Bitinaite J, Rubino M, Varma KH, Schildkraut I, Vaisvila R (2007) USER™ friendly DNA engineering and cloning method by uracil excision. Nucleic Acids Res 35 (6):1992–2002
27. Bitinaite J, Nichols NM (2009) DNA cloning and engineering by uracil excision. Curr Protoc Mol Biol Chapter 3:Unit 3.21
28. Storch M, Casini A, Mackrow B, Ellis T, Baldwin GS (2017) BASIC: a simple and accurate modular DNA assembly method. Methods Mol Biol 1472:79–91
29. Zeng F, Zang J, Zhang S, Hao Z, Dong J, Lin Y (2017) AFEAP cloning: a precise and efficient method for large DNA sequence assembly. BMC Biotechnol 17(1):81
30. Bilitchenko L, Liu A, Cheung S, Weeding E, Xia B, Leguia M, Anderson JC, Densmore D (2011) Eugene – a domain specific language for specifying and constraining synthetic biological parts, devices, and systems. PLoS One 6(4):e18882
31. Quinn J, Cox RS, Adler A, Beal J, Bhatia S, Cai Y, Chen J, Galdzicki M, Clancy K, Hillson N, Le Novère N, Maheshwari A, McLaughlin JA, Myers C, Umesh P, Pocock M, Rodriguez C, Soldatova L, Stan G-BV, Swainston N, Wipat A, Sauro HM (2015) SBOL visual: a graphical language for genetic designs. PLoS Biol 13(12):e1002310

Chapter 3

The AssemblX Toolkit for Reliable and User-Friendly Multigene Assemblies

Fabian Machens and Lena Hochrein

Abstract

The implementation of complex cloning projects covering the assembly of entire biological pathways or large genetic circuits poses a major challenge in the field of biotechnology and synthetic biology, as such projects can be costly and time-consuming. To overcome these difficulties, we developed the software-assisted AssemblX toolkit, which allows even unexperienced users to design, build, and subsequently test large DNA constructs. Currently, AssemblX allows the assembly of up to 25 functional units (e.g., genes), from 75 or more subunits (e.g., promoters, coding sequences, terminators). At the first assembly level, AssemblX uses overlap-based, scar-free, and sequence-independent cloning methods. This allows the unrestricted design at the gene level without the need for laborious parts domestication. The standardized, polymerase chain reaction–free, and virtually sequence-independent assembly into multigene modules relies on rare cutting homing endonucleases and computationally optimized overlap sequences. Selection and marker switching strategies ensure an effective process, and the assembly product can be transferred to any desired expression host.

Key words DNA assembly, Cloning, Multipart, Pathway assembly, Sequence independent, Software-assisted

1 Introduction

Innovative methods for multipart DNA assemblies are essential for carrying out biotechnological pathway engineering projects, the construction of synthetic genomes, or the design of regulatory circuits within cells [1–5]. Although cloning single DNA parts into target vectors based on restriction enzyme digestion and ligation became a routine technique in laboratories long ago, the assembly of multigene constructs or even whole genomes by de novo assembly of DNA fragments from different sources is still challenging [6]. The use of traditional cloning techniques, based on iterative restriction and ligation steps, often leaves unwanted

Fabian Machens and Lena Hochrein contributed equally to this work.

Sunil Chandran and Kevin W. George (eds.), *DNA Cloning and Assembly: Methods and Protocols*, Methods in Molecular Biology, vol. 2205, https://doi.org/10.1007/978-1-0716-0908-8_3, © Springer Science+Business Media, LLC, part of Springer Nature 2020

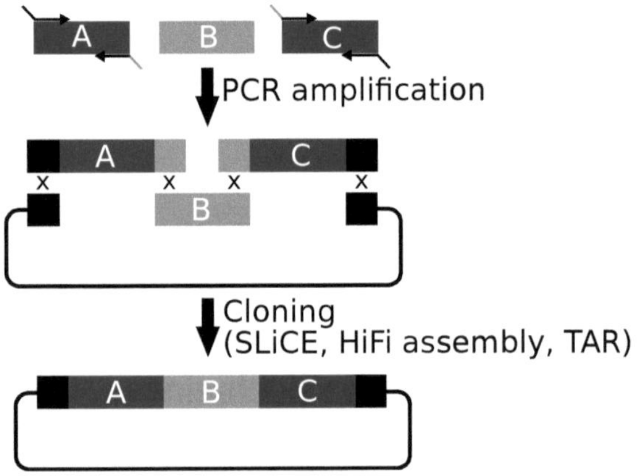

Fig. 1 Overlap-based cloning methods. The overlaps between the individual subunits can be added via PCR, using primers giving overlaps to the neighboring fragments in their 5′ regions. After amplification of part A and part C with appropriate primers, the three parts can be assembled into a vector backbone. (Reproduced with modifications from Hochrein et al., AssemblX: A user-friendly toolkit for rapid and reliable multi-gene assemblies. Nucleic Acids Research (2017), 45(10) by permission of Oxford University Press)

scar sites at the junction between two DNA parts and frequently makes it necessary to 'domesticate' assembly parts by eliminating undesired cutting sites within the DNA fragments to be assembled. This can be time consuming and laborious, especially for large DNA assemblies. To overcome the drawbacks of using restriction enzymes (REs), sequence-independent cloning methods like in vitro Gibson assembly [7], bacterial cell extract-based SLiCE [8] and in vivo transformation-associated recombination (TAR) in *Saccharomyces cerevisiae* [9, 10] have been developed. These cloning methods require short overlapping sequences shared by neighboring DNA parts to allow for the homology-directed assembly of multiple DNA fragments. The individually designed overlaps are commonly added to the assembly fragments via PCR (Fig. 1).

The recently developed AssemblX strategy combines the advantages of traditional and new cloning methods in a streamlined process and enables the user to assemble up to 25 functional units (e.g., transcriptional units), consisting of many more subunits [11]. The assembled genes can either be directly expressed in *S. cerevisiae*, or the final multigene module can be subcloned into destination vectors allowing expression in virtually any organism. The AssemblX workflow enables users to perform scarless, sequence-independent de novo assemblies starting at the subgene level, as well as standardized, modular and PCR-free multigene assemblies into complex genetic constructs. Parallel assemblies speed up the cloning process, which makes use of cost-efficient cloning methods. An AssemblX web tool supports the user in all

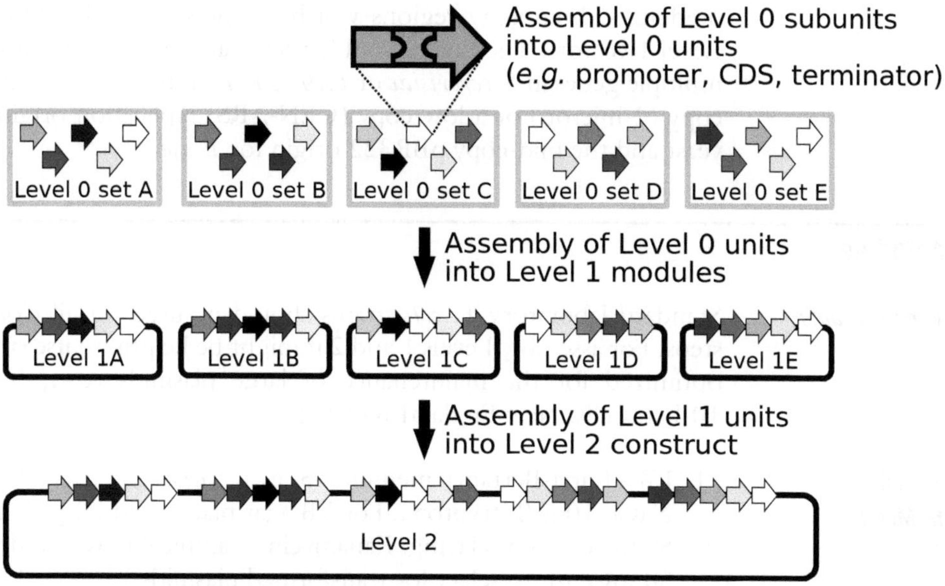

Fig. 2 The AssemblX workflow for the assembly of 25 units. Level 0 units, here with three DNA subunits each, are assembled into five different sets of Level 0 vectors, resulting in 25 Level 0 vectors. In the next step, Level 0 units from one Level 0 set are assembled in one corresponding Level 1 module. The five Level 1 modules, each containing five Level 0 units, are then assembled into one single Level 2 vector. (Reproduced with modifications from Hochrein et al., AssemblX: A user-friendly toolkit for rapid and reliable multi-gene assemblies. Nucleic Acids Research (2017), 45(10) by permission of Oxford University Press)

steps and generates a protocol with detailed instructions for all assembly steps.

The assembly of up to 25 transcriptional units into a single destination vector is organized in three cloning hierarchies, for which we developed level-specific assembly plasmids: Entry Level 0, and Assembly Levels 1 and 2 (Fig. 2).

To start the assembly, each single Level 0 unit, which is usually a transcriptional unit composed of a promoter, a coding sequence (CDS) and a terminator, is assembled in a different positioning vector (Level 0 vector). However, it is possible to assemble DNA fragments of every type, regardless of their biological functions, into Level 0 units.

Each individual Level 0 backbone provides a predefined combination of two ′homology regions′, which flank the assembled Level 0 unit. These homology regions will later direct the assembly of multiple Level 0 units in a Level 1 vector holding up to five Level 0 units (Fig. 2). Level 0 vectors are therefore organized into five sets (A to E) and the vectors in one set allow the directed assembly of up to five Level 0 units into a corresponding Level 1 backbone (Level 1A to Level 1E). If necessary, up to five Level 1 modules form different sets (A to E) can be combined into a single Level 2 vector. Again, this is achieved via overlap-based cloning using predefined homology regions which are present in the different Level 1 vectors. The final assembly vector allows the expression of multiple genes in *S. cerevisiae* or *Escherichia coli* using either a high-copy (2-micron) or a low-copy (CEN-ARS) replication origin for yeast and the mid-copy pBR322 origin for *E. coli*.

2 Materials

2.1 E. coli *Strains*

Standard laboratory *E. coli* strains should be used for all cloning steps. For assembly Levels 1 and 2 it might be helpful to use strains optimized for the maintenance of large plasmids (e.g., NEB® 10-beta cells; New England Biolabs).

2.2 E. coli *Growth Media*

1. LB (Luria-Bertani) medium: 5 g/L yeast extract, 10 g/L NaCl, 10 g/L tryptone. For LB agar plates, add 20 g/L agar. Sterilize by autoclaving. Kanamycin or ampicillin were added at 50 mg/mL to select for transformed plasmids.
2. 2× YT medium: 16 g/L tryptone, 10 g/L yeast extract, 5 g/L NaCl, pH 7.0.

2.3 *Yeast Strains*

We recommend using *S. cerevisiae* YPH500 (ATCC® 76626™) for the AssemblX procedure, as it has all auxotrophies required for a full Level 2 assembly with five Level 1 modules. However, other yeast strains may also be used, depending on their auxotrophies.

2.4 *Yeast Growth Media*

All yeast media can be sterilized by autoclaving. Glucose should be autoclaved separately and added to the medium after sterilization. To this end, dissolve all media components except for glucose, raise the volume to 950 mL (900 mL for 2× YPDA), and autoclave. After autoclaving, add 50 mL (100 mL for 2× YPDA) of sterile 40% (w/v) glucose.

1. YPDA medium: 10 g/L yeast extract, 20 g/L peptone, 80 mg/L adenine hemisulfate, 20 g/L glucose. For YPDA agar plates, add 20 g/L agar.
2. 2× YPDA: 20 g/L yeast extract, 40 g/L peptone, 80 mg/L adenine hemisulfate, 40 g/L glucose. For 2× YPDA agar plates, 20 g/L agar
3. SD media: Based on your supplier's recommendation, mix the indicated amount of yeast nitrogen base without amino acids with the required dropout formulation containing all necessary amino acids except for those required for selection: For example, if transforming a Level 0 plasmid carrying the *URA3* gene, use a -Ura dropout mixture. Adjust pH to 5.6–5.8 using NaOH. Add glucose to a final concentration of 2% (w/v) after autoclaving.

2.5 Solutions for TAR Cloning

1. 50% (w/v) PEG 3350: Dissolve 50 g of PEG 3350 in 30 mL ddH_2O, use a heating plate if necessary. Adjust the final volume to 100 mL in a measuring cylinder. Sterilize by autoclaving and store at room temperature. Seal the container with parafilm to prevent evaporation.
2. 1 M LiAc: Dissolve 10.2 g of lithium acetate dihydrate in 100 mL ddH_2O. Sterilize by autoclaving and store at room temperature.
3. Single-stranded carrier DNA: Dissolve 200 mg of salmon sperm DNA in 100 mL TE (10 mM Tris–HCl, 1 mM Na_2EDTA, pH 8.0) by stirring at 4 °C for several hours. Dispense aliquots of 1.0 mL into 1.5-mL tubes and store at −20 °C. Before the first use, denature an aliquot of the carrier DNA in a boiling water bath or at 95 °C on a thermoblock for 5 min and place on ice immediately afterwards. Vortex well before use. Denatured carrier DNA can be stored at −20 °C, repeated boiling is not necessary if the DNA is thawed on ice prior to use.

2.6 Cloning Vectors

All plasmids required for the AssemblX procedures are freely available through the authors. All sequences are deposited at the NCBI GenBank (Accession numbers KY131987–KY132089) and can also be accessed at www.AssemblX.org. *See* Tables 1 and 2 for a list of all available plasmids.

2.7 Primers

The following primers are recommended for colony PCR screening of *E. coli* and yeast clones during Level 0 assemblies:

(a) Lev0_fwd 5′ AACAAATAGGGGTTCCGC 3′.

(b) Lev0_rev 5′ GCGGTGGTTTTTTTGTTTG 3′.

2.8 Special Reagents (Optional)

1. SLiCE buffer [8]: 500 mM Tris–HCl (pH 7.5 at 25 °C), 100 mM $MgCl_2$, 10 mM ATP, 10 mM DTT.
2. SLiCE lysate: Lysate was prepared as described elsewhere [8], with some modifications [12]: *E. coli* DH5α cells were grown overnight in a 100-mL baffled flask in 20 mL 2× YT medium at 37 °C and 200 rpm. Hundred mL 2× YT medium in a 500-mL baffled flask were inoculated with 2 mL of the preculture and incubated at 37 °C and 230 rpm until an $OD_{600nm} = 4.5$ was reached. Cells were harvested by centrifugation at 5,000 × *g* for 20 min at 4 °C, washed with 500 mL ice-cold ddH_2O, and pelleted again by centrifugation at 5,000 × *g* for 20 min at 4 °C. Afterwards, the cell pellet (about 0.25 g wet weight) was resuspended in 300 mL CelLytic B Cell Lysis Reagent (Sigma-Aldrich, St. Louis, USA) and incubated for 10 min at room temperature to allow cell lysis. To pellet insoluble cell debris, lysates were centrifuged for 2 min at 20,000 × *g* at room

Table 1
Complete list of available Level 0 and Level 1 vectors

Level 0 vector set	Vector name	HR1	HR2	Corresponding Level 1 vectors (HR1/HR2)
A	pL0A_0-1	A0	A1	pL1A-hc / pL1A-lc (A0/AR)
	pL0A_1-2	A1	A2	
	pL0A_2-3	A2	A3	
	pL0A_3-4	A3	A4	
	pL0A_0-R	A0	AR	
	pL0A_1-R	A1	AR	
	pL0A_2-R	A2	AR	
	pL0A_3-R	A3	AR	
	pL0A_4-R	A4	AR	
B	pL0B_0-1	B0	B1	pL1B-hc / pL1B-lc (B0/BR)
	pL0B_1-2	B1	B2	
	pL0B_2-3	B2	B3	
	pL0B_3-4	B3	B4	
	pL0B_0-R	B0	BR	
	pL0B_1-R	B1	BR	
	pL0B_2-R	B2	BR	
	pL0B_3-R	B3	BR	
	pL0B_4-R	B4	BR	
C	pL0C_0-1	C0	C1	pL1C-hc / pL1C-lc (C0/CR)
	pL0C_1-2	C1	C2	
	pL0C_2-3	C2	C3	
	pL0C_3-4	C3	C4	
	pL0C_0-R	C0	CR	
	pL0C_1-R	C1	CR	
	pL0C_2-R	C2	CR	
	pL0C_3-R	C3	CR	
	pL0C_4-R	C4	CR	
D	pL0D_0-1	D0	D1	pL1D-hc / pL1D-lc (D0/DR)
	pL0D_1-2	D1	D2	
	pL0D_2-3	D2	D3	
	pL0D_3-4	D3	D4	
	pL0D_0-R	D0	DR	
	pL0D_1-R	D1	DR	
	pL0D_2-R	D2	DR	
	pL0D_3-R	D3	DR	
	pL0D_4-R	D4	DR	
E	pL0E_0-1	E0	E1	pL1E-hc / pL1E-lc (E0/ER)
	pL0E_1-2	E1	E2	
	pL0E_2-3	E2	E3	
	pL0E_3-4	E3	E4	
	pL0E_0-R	E0	ER	
	pL0E_1-R	E1	ER	
	pL0E_2-R	E2	ER	
	pL0E_3-R	E3	ER	
	pL0E_4-R	E4	ER	

The Level 0 vectors are grouped into five sets. Each set is compatible with a distinct Level 1 vector, which is available in either high- or low-copy format (hc/lc). HR: homology region. Reproduced with modifications from Hochrein et al., AssemblX: A user-friendly toolkit for rapid and reliable multi-gene assemblies. Nucleic Acids Research (2017), 45 (10) by permission of Oxford University Press

Table 2
Available Level 2 vectors

Level 1 modules used for assembly	Auxotrophic markers used in level 1 modules	Corresponding level 2 vectors (HR1/HR2)
A	URA3	n. a.
A,B	URA3, LEU2	pL2_AB_hc/pl2_AB_lc (A0_C0)
A,B,C	URA3, LEU2, HIS3	pL2_AC_hc/pL2_AC_lc (A0_D0)
A,B,C,D	URA3, LEU2, HIS3, LYS2	pL2_AD_hc/pL2_AD_lc (A0_E0)
A,B,C,D,E	URA3, LEU2, HIS3, LYS2, TRP1	pL2_AE_hc/pl2_AE_lc (A0_F0)

Depending on the number of Level 1 modules to assemble, different Level 2 vectors must be used. Each vector is available in high- and low-copy format (hc/lc). Depending on the assembled Level 1 modules, the resulting Level 2 construct can be selected with up to five auxotrophic markers. Reproduced with modifications from Hochrein et al., AssemblX: A user-friendly toolkit for rapid and reliable multi-gene assemblies. Nucleic Acids Research (2017), 45(10) by permission of Oxford University Press

temperature. The supernatant was carefully removed from the cell debris, mixed with an equal volume of glycerol, aliquoted into 50-μL portions in 0.2-mL thin-walled plastic tubes and immediately frozen in liquid nitrogen and stored at −80 °C.

3. NEBuilder® HiFi DNA Assembly mix (New England Biolabs) [9].

3 Methods

3.1 Planning the Multigene Construct Using the AssemblX Web Tool

1. Decide on the number of Level 0 units required for the intended construct and define their sequences. The number of Level 0 units automatically determines the number of Level 1 vectors required, as one Level 1 vector can hold up to five Level 0 units.
2. Use the AssemblX web tool at www.AssemblX.org for the detailed planning of your constructs (*see* **Note 1**). On the home screen, first activate as many Level 0 units as needed for your assembly. This must be done consecutively from Level 0 A to Level 0 E, without skipping a whole Level 0 set or a single Level 0 vector in one set. It is, however, possible to put less than five Level 0 units into one set (the web tool will automatically choose the correct Level 0 backbone which gives the required homology for the assembly into Level 1). Click on the individual Level 0 units for detailed planning at the sequence level. Follow the instructions provided on the web-page to upload the required sequence information for each part and to organize the individual parts into a single Level 0 unit

(*see* **Note 2**). The web tool will later provide the necessary primer sequences that allow the scar-free overlap-based assembly of all fragments in the order that is defined here. It is therefore not required to include any cloning adapters in the sequences that are uploaded here. Repeat the Level 0 design process for all required Level 0 assemblies. Subsequently, back on the home screen, choose the relative orientation of the Level 0 units towards each other. Here, it can also be decided if the Level 1 and Level 2 backbones are used with either the high-copy 2-micron origin of replication or the low-copy CEN/ARS region. This will only affect the plasmid copy number in yeast, but not in *E. coli*. Note, that the user only needs to plan the individual Level 0 units. All subsequent Level 1 or Level 2 assemblies are automatically designed.

3. Submit the sequence information and wait until you receive the results file, which will contain Genbank files for the final (Level 1 or Level 2) and all intermediate Level 0 (and Level 1) constructs. Detailed instructions for the cloning approach for every Level 0, Level 1 and Level 2 assembly are provided in a single Excel file. This file is divided into three sheets for the assembly levels. Check the results file for any error messages or warnings. If no problems arise, check the list of primers that are required for the assembly and order them (*see* **Note 3**).

3.2 Performing Level 0 Assemblies

1. Check the Level 0 assembly instruction provided by the web tool to identify all required Level 0 backbones. Depending on the number of Level 0 units that need to be assembled, you will need to use up to 25 different Level 0 backbones. Note that the last Level 0 unit in each Level 0 set will always be cloned into a Level 0 vector of type "XR" (e.g., pL0A3-R), giving the overlap to the respective Level 1 vector (reconstitution of a yeast selection marker).
2. For plasmid isolation, grow the required *E. coli* strains overnight in LB medium supplied with kanamycin. Use a standard plasmid mini preparation kit to isolate the required plasmids.
3. Digest 1 μg of each plasmid with HindIII to release the *ccdB* cassette from the backbone and to expose the terminal homology regions that are necessary for the overlap-based assemblies (Fig. 3). Inactivate HindIII by heating according to the supplier's instructions. *Optional*: Gel purify the vector backbones using a gel elution kit (*see* **Note 4**). The linearized backbones can be stored at −20 °C and may be used for many future assembly reactions.
4. According to the assembly instructions, PCR amplify all necessary assembly parts for each Level 0 assembly. The instructions contain information on the primer pair and the template for

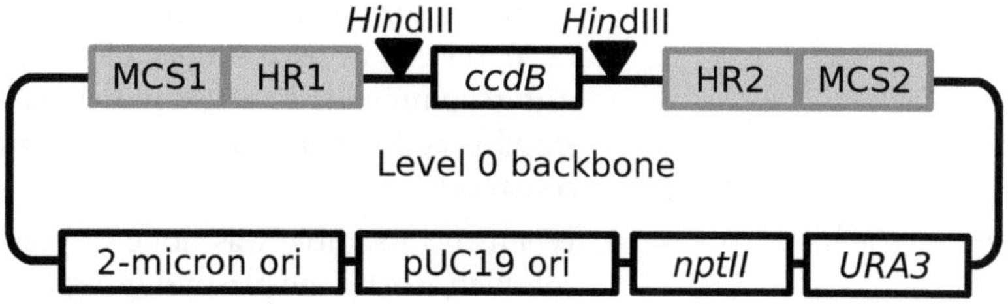

MCS1 = *AscI, SbfI, SwaI, FseI, PmeI*
MCS1 = *SwaI, AscI, PmeI, SbfI, FseI*

Fig. 3 Design of Level 0 vector backbones. The backbone is a yeast shuttle vector with a pUC19 replication origin, the kanamycin resistance gene *nptII* for selection in *E. coli*, a high-copy 2-micron origin for replication and the *URA3* marker gene for selection in *S. cerevisiae*. Each Level 0 vector possesses a different combination of homology regions (HR). HR1 gives homology to the Level 1 vector backbone or to the HR2 of the adjacent Level 0 vector. HR2 gives homology to the next Level 0 vector or to the Level 1 vector backbone. These regions are separated by a HindIII-flanked *ccdB* expression cassette. The HindIII sites are used to release the *ccdB* cassette and to allow the subsequent insertion of the individual subunits in-between the homology regions by scar-free, overlap-based cloning. The homology regions are flanked by two multiple cloning sites (MCS), containing the recognition sites of the five 8-base cutters AscI, SbfI, SwaI, FseI, and PmeI to release Level 0 constructs for further cloning into Level 1. (Reproduced with modifications from Hochrein et al., AssemblX: A user-friendly toolkit for rapid and reliable multi-gene assemblies. Nucleic Acids Research (2017), 45(10) by permission of Oxford University Press)

each part in all necessary assemblies. Use a proofreading polymerase to prevent PCR-borne errors. The resulting PCR fragments will contain the terminal homologies that allow an ordered assembly reaction.

5. Analyze a small aliquot of each PCR reaction by agarose gel electrophoresis. If multiple bands appear, gel-purify the correct amplicon. PCR products, visible as a single specific band after agarose gel electrophoresis, can be used in assembly reactions without further purification, although it is sometimes advantageous to column-purify the products.
6. At this point, assemble PCR products using a cloning technique of choice (*see* **Note 5**). For detailed instructions on the different cloning methods refer to Subheadings 3.6 and 3.7 and follow the instructions provided.
7. After *E. coli* transformation, select for assembled plasmids on LB plates supplied with kanamycin, or on SD-Ura medium when performing TAR cloning in yeast.
8. *Optional*: It is recommended to analyze each assembly with (yeast) colony PCR prior to picking clones for plasmid

isolation. For shorter assembly units, primers Lev0_fwd and Lev0_rev can be used (Subheading 2.7). For assemblies >3 kb, it is recommended to use individually designed primers that amplify at least one assembly junction within the intended construct.

9. When the assembly was done in yeast: Pick potential positive colonies, isolate the plasmid using a yeast plasmid isolation kit. Subsequently retransform the plasmid to *E. coli* and select on LB medium containing kanamycin. Proceed with plasmid isolation from *E. coli*.
10. Inoculate overnight cultures in LB + kanamycin and do plasmid mini preparations using standard techniques.
11. *Optional*: Analyze isolated plasmids by restriction digestions using suitable enzymes.
12. To verify correct assemblies, analyze the plasmids by Sanger sequencing. Care should be taken that the assembly junctions in particular are sequenced in good quality.
13. Once all necessary Level 0 clones are sequence verified, proceed to Level 1 assemblies.

3.3 Performing Level 1 Assemblies

1. Linearize the required Level 1 vector(s) by PacI digestion. Heat-inactivate the enzyme (*see* **Note 6**).
2. Digest Level 0 constructs with the enzyme suggested in the protocol produced by the AssemblX web tool. The Level 0 backbones provide sites for the 8-mer cutters AscI, SbfI, SwaI, FseI, and PmeI on both sites of the assembled Level 0 unit. Gel-purify the digestion fragment representing the assembly part. Alternatively, it is possible to PCR-amplify the assembly part, including the homology regions.
3. Refer to the assembly instructions to set up all required assembly reactions with the PacI-linearized Level 1 vector(s) and the Level 0 assembly pieces. The Level 1 vector is compatible with SLiCE cloning, NEBuilder® HiFi-DNA assembly, and TAR cloning (*see* **Notes 6** and **7**, Fig. 4). *See* Subheadings 3.6–3.8 for detailed instructions on each method. The bacterial selection marker for Level 1 assemblies is ampicillin resistance. Depending on the Level 1 vector used, the yeast selection marker varies (Table 2). In yeast, auxotrophic selection for the Level 1 vector only works for correctly assembled Level 1 modules. The empty Level 1 vector cannot be selected for in yeast. This is due to the nature of the homology between the last Level 0 vector and the Level 1 vector (Fig. 4).
4. Verify clones by (yeast) colony PCR as described before, isolate plasmids, perform analytic digestions, and sequence the assembly junctions (*see* **Note 8**).

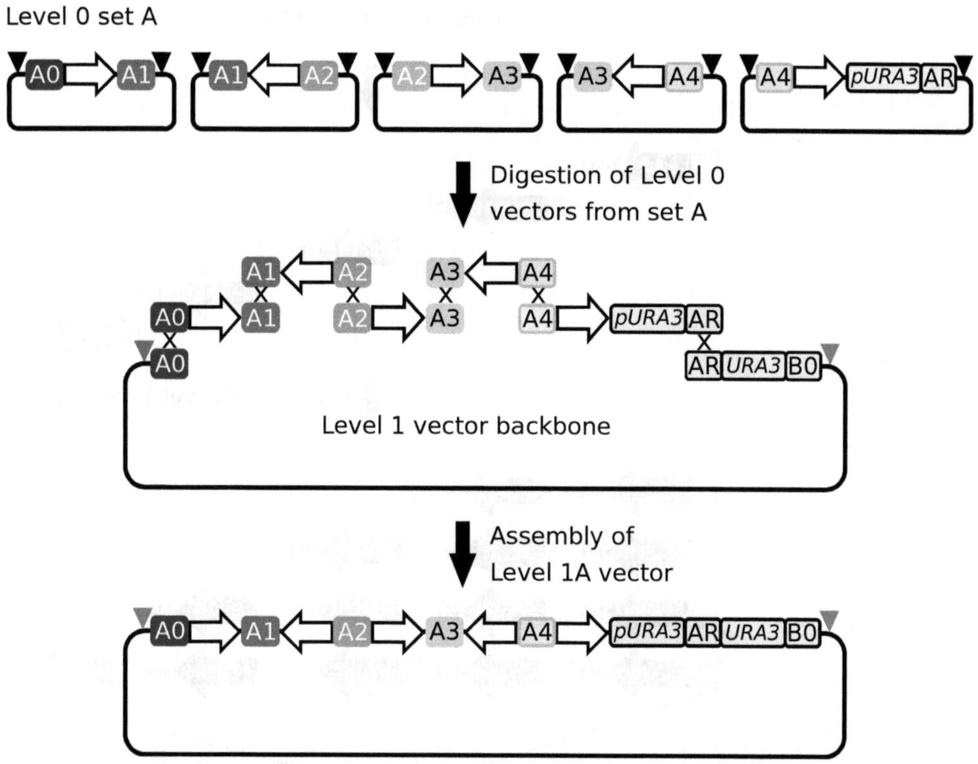

Fig. 4 Level 1 assembly scheme. Five Level 0 units of Level 0 set A are released from Level 0 backbones by digestion with one of the designated 8-base cutters of the MCS (black triangle). During in vivo or in vitro assembly, overlaps between the different Level 0 units and the Level 1 destination vector ensure a directed assembly into a single circular DNA construct. The *URA3* promoter is located in front of the homology region AR. AR itself represents the first 50 nucleotides of the *URA3* CDS. The complete *URA3* CDS and the appropriate terminator are located on the Level 1 vector backbone. In this way, a functional selection marker is created by a single successful Level 1 assembly. Complete Level 1 modules are flanked by I-SceI sites (gray triangle). (Reproduced with modifications from Hochrein et al., AssemblX: A user-friendly toolkit for rapid and reliable multi-gene assemblies. Nucleic Acids Research (2017), 45(10) by permission of Oxford University Press)

5. If the intended construct contains only a single Level 1 module, the assembly procedure is complete at this point. The Level 1 backbone can be maintained and used for expression in yeast, as well as in *E. coli*. Subcloning in vectors for alternative hosts is possible (*see* Subheading 3.5).

3.4 Performing Level 2 Assemblies

1. Digest the required Level 2 vector with PacI. Heat-inactivate the enzyme (*see* **Note 9**).
2. Digest the required Level 1 constructs with I-SceI to release the assembly units and to expose the homology regions which are necessary for the assembly reaction (Fig. 5). We recommend the following reaction setup: 2 μg of Level 1 vector, 5 μL 10× CutSmart buffer (NEB), 2 μL I-SceI (New England Biolabs).

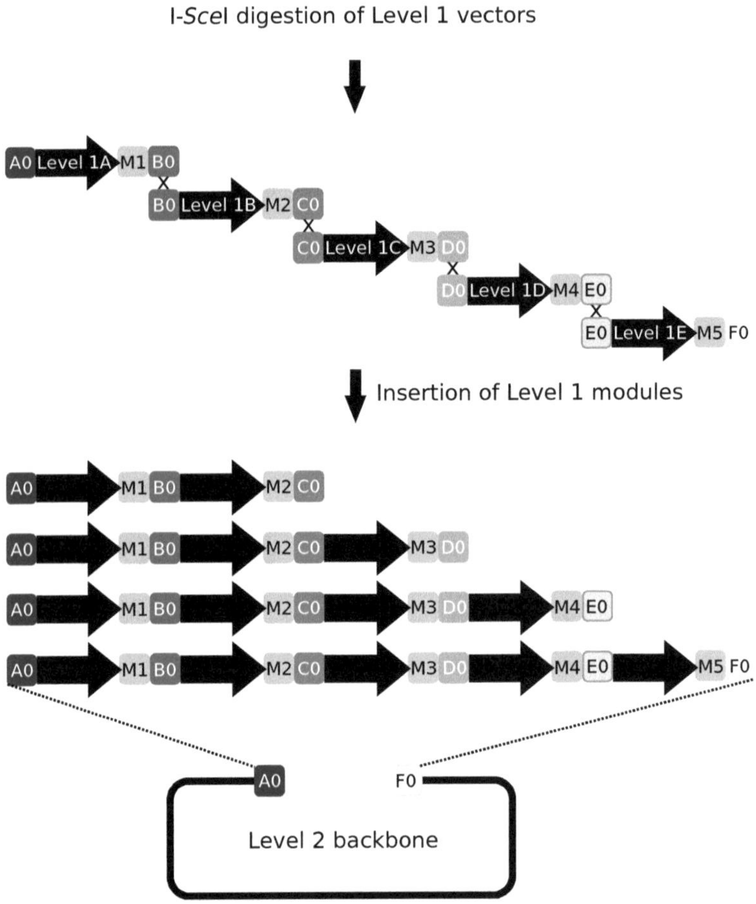

Fig. 5 Level 2 assembly scheme. Level 1 modules are released from Level 1 vectors by digestion with I-SceI. During assembly, overlaps between the different modules and the Level 2 destination vector ensure a directed assembly into a single circular DNA construct. Depending on the number of Level 1 modules to be assembled, different Level 2 vectors for the assembly of two to five Level 1 modules are used. M1: *URA3*, M2: *LEU2*, M3: *HIS3*, M4: *LYS2*, M5: *TRP1*. (Reproduced with modifications from Hochrein et al., AssemblX: A user-friendly toolkit for rapid and reliable multi-gene assemblies. Nucleic Acids Research (2017), 45 (10) by permission of Oxford University Press)

3. Digest overnight at 37 °C, inactivate at 65 °C for 20 min, and gel-purify the desired module (*see* **Note 10**).
4. Combine all necessary parts in a TAR reaction, using 100 ng of each Level 1 module and 100 ng of the destination vector (*see* Subheading 3.6). Select positive yeast clones on appropriate dropout medium. *See* Table 2 for selection of the correct dropout combination.

5. Verify clones by yeast colony PCR, isolate plasmids, perform analytic digestions, and sequence the assembly junctions (*see* **Note 11**).
6. After successful assembly, the whole construct can be transferred to different integrative yeast vectors or any compatible vector for the destination organism (*see* Subheading 3.5).

3.5 Transferring the Multi-Gene Assembly to Different Destination Vectors

If the final Level 1 or Level 2 assembly is intended for a host different from *E. coli* or *S. cerevisiae*, please follow one of the following three options. For all three options, it is necessary to linearize the expression vector for the intended host. This may be done by restriction digestion or preferably by PCR amplification. Once the vector is linearized, one of the following three options may be used.

Option 1: Convert your expression vector into an AssemblX Level 2 vector

1. Make sure that your expression vector does not contain a PacI site.
2. Design primers that amplify the yeast replication origin and the appropriate pair of homology regions from one of the available AssemblX Level 2 vectors. As a template, choose the same Level 2 vector you would have used if yeast was your final host for expression. In case you are uncertain, consult the 'Level 2' section of the AssemblX protocol that was produced by the web tool.
3. Equip primers with overlaps compatible to the linearized expression vector and use these primers to amplify the cassette described above.
4. Combine the linearized expression vector and the PCR-amplified cassette containing the appropriate homology regions in a suitable assembly reaction (e.g., NEBuilder® HiFi DNA assembly or SLiCE).
5. Transform into *E. coli* and identify positive clones by colony PCR and restriction analysis.
6. Use the modified expression vector instead of the designated AssemblX Level 2 vector for the final TAR-mediated Level 2 assembly.
7. Proceed with construct verification and isolation. Transform verified constructs into your final host.

Option 2: Subclone AssemblX Level 2 module into your expression vector via recombination.

1. Make sure that your desired expression vector does not contain a PacI site.

2. Design primers that amplify the appropriate pair of homology regions from one of the available AssemblX Level 2 vectors. As a template, choose the same Level 2 vector you would have used if yeast was your final host for expression. In case you are uncertain, take a look at the 'Level 2' section of the AssemblX protocol that was produced by the web tool.
3. Equip primers with overlaps compatible to your linearized expression vector and use these primers to amplify the cassette described above.
4. Combine the linearized expression vector and the PCR-amplified cassette containing the appropriate homology regions in an appropriate assembly reaction (e.g., NEBuilder® HiFi DNA assembly or SLiCE).
5. Transform into *E. coli* and identify positive clones by colony PCR and restriction analysis.
6. Use the modified expression vector to subclone your AssemblX Level 2 construct. To this end, digest your verified Level 2 construct with I-SceI, purify the fragment and perform in vitro recombination (e.g., NEBuilder® HiFi DNA assembly, SLiCE) with PacI-digested modified expression vector.
7. Identify positive clones by colony PCR and restriction analysis and proceed with transformation into your final host.

Option 3: Subclone AssemblX Level 1 or 2 module into your expression vector via restriction and ligation.

1. For this option, make sure that your expression vector does not contain an I-SceI site.
2. Design primers that amplify the I-SceI–*ccdB*–I-SceI cassette from plasmid pL1A_hc.
3. Equip primers with overlaps compatible to your linearized expression vector and use these primers to amplify the cassette described above.
4. Combine the linearized expression vector and the PCR-amplified I-SceI–*ccdB*–I-SceI cassette containing the appropriate homology regions in an assembly reaction (e.g., NEBuilder® HiFi DNA assembly or SLiCE).
5. Transform into *E. coli* and identify positive clones by colony PCR and restriction analysis.
6. Use the modified expression vector to subclone your AssemblX Level 1 or 2 construct. To this end, digest your verified construct with I-SceI, purify the fragment and perform a ligation with I-SceI-digested modified expression vector.
7. Identify positive clones by colony PCR and restriction analysis and proceed with transformation into your final host.

3.6 TAR Cloning

For TAR cloning, *S. cerevisiae* is transformed with linear DNA fragments sharing end homologies. In vivo recombination of these overlapping regions results in circular plasmids that will be propagated in the yeast cells. Yeast transformation is usually done according to the LiAC/PEG protocol, which is provided below [13]. However, every other yeast transformation protocol should be applicable as well. For all yeast manipulations, it is recommended to work under a laminar flow hood. All reagents should be autoclaved and opened only in a sterile environment.

1. Streak out *S. cerevisiae* strain YPH500 on YPDA agar plates (*see* **Note 12**). Incubate at 30 °C for 3–5 days.
2. Use a fresh colony to inoculate 5 mL of YPDA medium in a small flask. Incubate shaking overnight (220 rpm, 30 °C).
3. Measure the OD at 600 nm: Pipet 10 μL of cell suspension into 1.0 mL of water, mix thoroughly and measure the OD at 600 nm.
4. Use the preculture to inoculate 50 mL of prewarmed 2× YPDA medium to a final OD_{600} of 0.5.
5. Incubate with shaking for about 4–5 h (220 rpm, 30 °C).
6. While waiting for the cells to grow: Prepare a DNA mix for each individual assembly. In 34 μL final volume, mix 100 ng of each assembly fragment plus 100 ng of the linearized target vector.
7. Determine the OD_{600} of the main culture: The OD_{600} should be at least 2 before you continue.
8. Centrifuge the culture at 3,000 × *g* for 5 min at RT.
9. Discard the supernatant and resuspend the cell pellet in 25 mL sterile ddH2O. Centrifuge again, as in **step 8**.
10. Repeat **step 9**.
11. After the second washing step, resuspend the cell pellet in 1 mL sterile ddH_2O.
12. Transfer the cell suspension to a 1.5-mL tube and centrifuge in a benchtop centrifuge at 13,000 × *g* for 30 s.
13. Completely remove the supernatant using a micropipette.
14. Resuspend the cell pellet in 1 mL ddH_2O by pipetting up and down.
15. For each transformation sample, aliquot 100 μL of the cell suspension into a fresh 1.5-mL tube and centrifuge in a benchtop centrifuge at 13,000 × *g* for 30 s.
16. Completely remove the supernatant using a micropipette. Each cell pellet will be used for a single transformation reaction.
17. According to the number of transformations, prepare the following transformation mix, multiplied by the planned number

of reactions plus one: 240 μL 50% PEG, 36 μL LiAc, 50 μL single-stranded carrier DNA. Mix well.

18. Add 326 μL of transformation mix to each cell pellet.
19. Add 34 μL of DNA mix, containing all fragments for one assembly.
20. Mix very well using a vortex mixer. The cells should be completely resuspended in the transformation mix.
21. Heat-shock the cells in a 42 °C water bath or thermoblock for 40 min (*see* **Note 13**). Vortex the cells every 10–15 min.
22. Centrifuge the cells at 13,000 × *g* for 30 s and remove the supernatant using a micropipette.
23. Resuspend the cell pellets in 1 mL ddH_2O each.
24. Using sterile glass beads, spread 50–200 μL of the transformation mixture onto appropriate selection medium.
25. Air-dry the plates for 5 min.
26. Incubate the plates for 3–5 days at 30 °C.

3.7 SLiCE Cloning

1. Prepare a bacterial lysate and 10× SLiCE buffer following the published protocol [8]. We recommend storing the lysate in small aliquots at −80 °C, as repeated freeze-and-thaw cycles strongly reduce cloning efficiency.
2. Set up the required assembly reactions by mixing all necessary assembly parts equipped with end homologies plus the HindIII-linearized Level 0 vector. A standard reaction setup for an assembly reaction with three fragments is as follows: 50–100 ng of linearized vector backbone, each insert fragment in a ten-fold molar excess relative to the backbone, 1 μL 10× SLiCE buffer, 1 μL SLiCE lysate in a final volume of 10 μL. Incubate the reaction at 37 °C for 60 min.
3. Following incubation, we usually transform 2 μL of the assembled product into highly competent *E. coli* cells. Cloning efficiency is usually sufficient for the assembly of up to three fragments. The SLiCE reaction tolerates short nonhomologous termini, for example overhangs created by digestion with restriction enzymes.

3.8 Setting Up an NEBuilder® HiFi DNA Assembly Reaction

The commercially available NEBuilder® HiFi DNA assembly master mix is a highly efficient version of the Gibson assembly that tolerates short nonhomologous termini, for example overhangs resulting from restriction digestion. We suggest using this method for assembly of up to five fragments.

1. Set up the required assembly reactions by mixing all necessary assembly parts equipped with end homologies plus the linearized target vector. A standard reaction setup for an assembly

with 2–3 fragments is as follows: 50–100 ng of linearized vector backbone, all insert fragments, each in twofold molar excess, 10 μL of 2× NEBuilder® HiFi DNA assembly master mix in a final volume of 20 μL (*see* **Note 14**). Incubate for 15 min at 50 °C. For assemblies with 4–6 fragments use all inserts in equimolar amounts relative to the backbone and incubate for 60 min at 50 °C.

2. Usually, 2 μL assembly product is transformed into highly competent *E. coli* cells using standard procedures. Depending on the assembly level, transformed cells are either selected with kanamycin (Level 0) or ampicillin (Level 1 + 2).

4 Notes

1. To use the AssemblX web tool, please first create a user account for J5 at www.j5.jbei.org [14]. Once you have a J5 account, you can simply use the same user credentials to login to the AssemblX web tool.
2. A detailed video introduction to the web tool can be found here:
   https://www.youtube.com/channel/UCowwRbGlqX_CMbU_-8uv1OQ
3. Due to the terminal homology regions, some primers can be quite long which may prompt your oligo supplier to recommend additional purification steps like HPLC. In our experience, however, it is sufficient to order standard desalted primers in most cases.
4. While digestion is the standard procedure for Level 0 backbone linearization, PCR amplification is sometimes beneficial in terms of cloning efficiency. For PCR amplification, use primers binding to the respective homology regions in the Level 0 backbone. Make sure that the complete homology regions are amplified.
5. For Level 0 assemblies, in vitro NEBuilder® HiFi-DNA assembly (or equivalent) is recommended. Alternatively, SLiCE or TAR cloning are possible, with TAR being the most reliable method with the drawback of slower progress due to the slower doubling time of yeast as compared to *E. coli*.
6. While heat inactivation is sufficient for Level 1 assemblies using TAR, overnight digestion, dephosphorylation and subsequent gel purification are recommended to reduce background of uncut vector in subsequent in vitro assembly reactions.
7. For Level 1 assemblies with three or more inserts we strongly recommend using TAR.

8. If internal homologous sequences occur in the desired Level 1 module, for example if the same signal peptide is employed in all five Level 0 constructs belonging to the same Level 1 module, a careful analysis of positive clones arising from a TAR reaction for sequence integrity is necessary. In such cases, we recommend using NEBuilder® HiFi-DNA assembly instead.
9. After the assembly of Level 0 vectors it is generally sufficient to sequence only the junctions between assembly parts, as all parts are usually produced by restriction digestion. However, Level 0 units that were PCR-amplified prior to Level 1 assembly should be resequenced in the Level 1 module.
10. Overnight digestion, and subsequent gel purification are usually **not** necessary for the following in vivo assembly.
11. For large Level 1 modules (> 10 kb) it is highly recommended to use a silica bead suspension-kit, rather than a typical spin column-based procedure.
12. For very large Level 2 assemblies, plasmid isolation from yeast and retransformation into *E. coli* is not always straightforward. However, standard plasmid isolation kits for yeast can usually be used successfully if the following guidelines are followed: Transform the complete plasmid DNA isolated from yeast into highly competent *E. coli* cells suitable for the propagation of large constructs (e.g., NEB 10-beta cells). Use electroporation rather than heat shock transformation and extend recovery in SOC medium to 90 min. Select the complete volume of the transformation on several LB plates. Grow the cells at 30 °C.
13. We have successfully used different yeast strains for the AssemblX protocol. Essentially, every yeast strain harboring the required auxotrophies for plasmid selection is appropriate, although assembly efficiencies may vary. Depending on the yeast strain, it may be necessary to adjust the heat shock time for optimal transformation efficiency.
14. The final reaction volume can be scaled down to 5 μL for simple assemblies.

Acknowledgments

The research that allowed developing the AssemblX toolkit was funded by the Federal Ministry of Education and Research of Germany (BMBF; FKZ 031A172). The authors greatly acknowledge Nathan Hillson (Joint BioEnergy Institute, Emeryville, CA, USA) for support with the implementation of j5 into the AssemblX web tool.

References

1. Si T, Luo Y, Xiao H et al (2014) Utilizing an endogenous pathway for 1-butanol production in *Saccharomyces cerevisiae*. Metab Eng 22:60–68
2. Runguphan W, Keasling JD (2014) Metabolic engineering of Saccharomyces cerevisiae for production of fatty acid-derived biofuels and chemicals. Metab Eng 21:103–113
3. Annaluru N, Muller H, Mitchell LA et al (2014) Total synthesis of a functional designer eukaryotic chromosome. Science 344:55–58
4. Kiani S, Beal J, Ebrahimkhani MR et al (2014) CRISPR transcriptional repression devices and layered circuits in mammalian cells. Nat Methods 11:723–726
5. Gander MW, Vrana JD, Voje WE et al (2017) Digital logic circuits in yeast with CRISPR-dCas9 NOR gates. Nat Commun 8:15459
6. Cohen SN, Chang AC, Boyer HW et al (1973) Construction of biologically functional bacterial plasmids *in vitro*. Proc Natl Acad Sci U S A 70:3240–3244
7. Gibson DG (2011) Enzymatic assembly of overlapping DNA fragments. Methods Enzymol 498:349–361
8. Zhang Y, Werling U, Edelmann W (2012) SLiCE: a novel bacterial cell extract-based DNA cloning method. Nucleic Acids Res 40:e55
9. Gibson DG (2009) Synthesis of DNA fragments in yeast by one-step assembly of overlapping oligonucleotides. Nucleic Acids Res 37(20):6984–6990
10. Kouprina N, Larionov V (2008) Selective isolation of genomic loci from complex genomes by transformation-associated recombination cloning in the yeast Saccharomyces cerevisiae. Nat Protoc 3:371–377
11. Hochrein L, Machens F, Gremmels J et al (2017) AssemblX: a user-friendly toolkit for rapid and reliable multi-gene assemblies. Nucleic Acids Res 45:e80
12. Messerschmidt K, Hochrein L, Dehm D et al (2016) Characterizing seamless ligation cloning extract for synthetic biological applications. Anal Biochem 509:24–32
13. Gietz RD, Schiestl RH (2007) High-efficiency yeast transformation using the LiAc/SS carrier DNA/PEG method. Nat Protoc 2:31–34
14. Hillson NJ, Rosengarten RD, Keasling JD (2011) j5 DNA assembly design automation software. ACS Synth Biol 1:14–21

Part II

High Throughput Workflows

Chapter 4

High-Throughput PCR for DNA Part Generation

David Reif

Abstract

This protocol describes a high-throughput approach to PCR for the generation of over a thousand amplicons in parallel. Modern liquid handling robotics are used to accelerate reaction setup, miniaturize reaction volumes, and dramatically reduce reagent and consumable cost. Although the focus is on generating DNA parts for use in DNA assembly techniques, this methodology can be applied to any workflow where the parallel production of hundreds or thousands of PCR amplicons is required.

Key words Synthetic biology, PCR, High-throughput, Automation, Capillary electrophoresis, Acoustic dispensing

1 Introduction

Advances in automation and acoustic liquid handling have miniaturized common molecular biology protocols while simultaneously expanding sample throughput. Polymerase chain reaction (PCR) is an essential tool of the molecular biologist and one such method that has been amenable to dramatic downscaling [1]. This protocol describes how to quickly perform thousands of reactions in parallel with a high percentage of success and effectively validate them for use with downstream DNA assembly methods. While developed with assembly by yeast homologous recombination in mind [2], this protocol should be applicable to other assembly methods as well, or as the basis for any other process requiring a large number of validated PCR amplicons.

To facilitate management of the large number of reactions, this protocol emphasizes a plate-based approach, using 384-well plates where possible. Full plate transfers and liquid transfers greater than 2 μL are performed by a traditional robotic liquid handler (e.g., Biomek FX), while all other liquid transfers are done via acoustic liquid handlers (e.g., LabCyte Echo) to accelerate reaction setup and minimize reagent usage.

Sunil Chandran and Kevin W. George (eds.), *DNA Cloning and Assembly: Methods and Protocols*, Methods in Molecular Biology, vol. 2205, https://doi.org/10.1007/978-1-0716-0908-8_4,

Specific steps of the protocol take into consideration the intended application (i.e., generation of parts for DNA assembly) and intended throughput (simultaneous generation of >1000 PCR amplicons). To preserve part fidelity, Phusion polymerase is used for PCR due to its low error rate and high processivity. With the ability to analyze 96 samples in parallel and operate continuously, capillary electrophoresis instruments (e.g., Agilent Fragment Analyzer) facilitate a high-throughput workflow not possible with manual agarose gel electrophoresis while providing quantitative data including amplicon size, concentration, and purity [3]. As much as possible, the workflow is intended to be worklist driven and done on robots with barcoded source and destination plates. Location tracking of all reaction components is essential, and software-assisted tracking and worklist generation is greatly beneficial. This is particularly important for steps that require moving from 384-well plates to lower density 96-well plates.

2 Materials

This protocol uses reaction volumes of 30 μL in 384-well PCR plates (*see* **Note 1**). Material volumes required will depend on the desired number of reactions and the required product volume.

2.1 PCR

1. 50 μM desalted synthetic DNA oligonucleotide primers
2. Template DNA in the form of synthetic DNA fragments, extracted plasmid DNA, or isolated genomic DNA, 1–100 ng/μL.
3. Thermo-Fisher 2× Phusion Hot Start II High-Fidelity PCR Mastermix.
4. Molecular grade sterile water.
5. 384-well PCR plates
6. Adhesive PCR plate seals.
7. Echo Qualified 384-Well Polypropylene Microplates.
8. Low profile single well 96 diamond bottom reservoir.
9. (optional) Genomic DNA extraction kit (Gentra Puregene Cell Kit)

2.2 Verification by Capillary Electrophoresis

1. 96-well PCR plates
2. 1× TE Buffer: 10 mM Tris–HCl pH 8.0, 1 mM EDTA.

2.3 Reaction Cleanup and Preparation for Further Use

1. DpnI 20,000 units/mL.
2. Echo Qualified 384-Well Polypropylene Microplates.
3. Sterile water.

4. (optional) SPRI Magnetic beads and magnet plate: Beckman-Coulter Agencourt AMPure XP or Axygen Axyprep Mag PCR cleanup kit and Alpaqua.

2.4 Equipment

1. Multiplexed capillary electrophoresis instrument and reagent kit (AATI ZAG system with 33 cm 96-capillary array and accompanying ZAG-130 dsDNA Reagent Kit) (*see* **Note 2**).
2. Thermocyclers with 384-well heating blocks.
3. Automated robotic liquid handler with 96 well multichannel head and 20 μL tips (Beckman Coulter Biomek FX).
4. Acoustic liquid handler (Labcyte Echo 525).
5. (Optional) Qubit Fluorometer and DNA assay kit

3 Methods

3.1 PCR

1. Design and acquire primers to amplify each DNA part required for the assembly method from a commercial oligo supplier. This protocol requires 100 nL of each primer per reaction as desalted oligos at a concentration of 50 μM. If possible, these should be ordered in Echo qualified microplates for direct use on the acoustic liquid handler.
2. Acquire template DNA in the form of miniprepped plasmid, synthetic DNA fragments, or isolated genomic DNA (*see* **Note 3**). Material should be aliquoted into Echo qualified microplates.
3. Equilibrate template and primer plates to room temperature and centrifuge at 1200 × *g* for 2 min to remove any bubbles.
4. Use the Echo set to the 384PP_AQ_BP liquid type to transfer 100 nL each of the forward and reverse primer to the intended reaction well (Table 1).
5. Use the Echo set to the 384PP_AQ_BP liquid type to transfer 1–10 ng of template DNA to the intended reaction well (Table 1). Minimize the transfer volume to avoid diluting the mastermix and minimize transfer time (*see* **Note 4**).
6. Thaw and dilute the 2× Phusion Hot Start II High-Fidelity PCR Mastermix in half with sterile molecular grade water and store on ice until used.
7. Add the Phusion mastermix to all reaction plates using the liquid handler configured to add 30 μL to each reaction well from a central source reservoir (Table 1). Use fresh tips for each transfer and mix the reaction well by aspirating and dispensing at least three times.
8. Briefly spin plates to remove bubbles and seal the plates with thermal PCR seals.

Table 1
Concentration of components in PCR reaction

Component	Final concentration per 30 μL reaction
Thermo-Fisher 1× Phusion hot start II high-fidelity PCR mastermix	1×
Template DNA	1–10 ng
Forward primer (50 μM)	0.167 μM
Reverse primer (50 μM)	0.167 μM

9. Calculate the extension time for each plate based on the largest amplicon on the plate. It is advised to use 15 s per kilobase, or 30 s per kilobase if using genomic DNA as template, with a minimum extension time of 15 s.
10. Load plates into thermocyclers for the following conditions: Initial denaturation at 98 °C for 120 s; 10 cycles of denaturation at 98 °C for 12 s, annealing at 60 °C for 30 s, and extension at 72 °C for calculated extension time; 30 cycles of denaturation at 98 °C for 12 s, annealing at 55 °C for 30 s, and extension at 72 °C for the calculated extension time; final extension at 72 °C for 5 min; hold at 4 °C (*see* **Note 5**).

3.2 Verification by Capillary Electrophoresis

1. Prepare capillary electrophoresis instrument for use according to manufacturer's recommendations. A single ladder run can be used to export a calibration file for processing all subsequent sample runs (*see* **Note 6**).
2. Use the liquid handler to dilute 2 μL of the reaction volume into a 96-well PCR plate filled with 158 μL of 1× TE and mix well. Each reaction plate will require 4 96-well plates, one for each quadrant. Briefly spin down the plates to remove all bubbles before loading the plates into the instrument and running the appropriate method for the kit.
3. Process run data using the software accompanying the capillary electrophoresis instrument for association with an appropriate ladder calibration.
4. Assess the quality of the reaction by three metrics: the concentration of the amplicon should be above 5 ng/μL, the ratio of the size of observed amplicon to the expected size should be within 10%, and the molar concentration of the desired amplicon in comparison to the total molar concentration should be above 50% (*see* **Note 7**).

3.3 Reaction Cleanup and Preparation for Further Use

1. It may be desired to remove any plasmid templates used to amplify DNA parts that may interfere with downstream cloning steps by digesting with DpnI. Use the Echo set to the 384PP_AQ_GP liquid type to add 1200 μL to each required reaction well. Seal plates with PCR seals, briefly spin, and use a thermocycler to incubate at 37 °C for 30 min and 65 °C for 20 min to fully degrade all methylated plasmid DNA.
2. Unexpected amplification or excessive primer dimer concentration may necessitate cleanup of the PCR product by SPRI magnetic bead purification to remove any residual reaction components or contaminants from the PCR (*see* **Note 8**). Follow the manufacturer's instructions (*see* **Note 9**).
3. Transfer resulting DNA parts from PCR plates to Echo qualified plates for future use. Centrifuge for 2 min at 1200 × *g* to remove any bubbles.

4 Notes

1. This reaction volume may be further reduced depending on the downstream applications. In this protocol, 30 μL is used to provide enough material per PCR for use in multiple DNA assembly reactions. For applications such as amplicon sequencing or colony PCR where little actual product is needed, the reaction volume can be comfortably reduced to 10 μL or less, provided other reagent volumes are scaled accordingly. Note that evaporation may become an issue as reaction volumes decrease.
2. This configuration allows the resolution of fragments between 75 and 20,000 base pairs during a 30 min separation run. The kit includes the 75 and 20,000 base pair markers and the appropriately sized calibration ladder, plus the necessary separation gel, intercalating dye, and buffers required for instrument operation. Additional kits and configurations are available for smaller or more specific size ranges.
3. Genomic DNA does not always transfer well on the acoustic liquid handler, as long polymers are known to be problematic for acoustic liquid transfer [4]. Concentrations above 5 ng/μL may encounter transfer exceptions or the appearance of successful liquid transfers without any actual DNA being transferred in the droplet. For best chances of success, use a concentration near 1 ng/μL. Fluorometric methods of DNA quantification, such as Qubit, have been found to be more accurate and reliable than spectrophotometric methods, such as NanoDrop, for this purpose. Additionally, DNA extraction methods have influence on transferability as well; favorable

results have been observed with the Gentra Puregene Cell kit. Always test any new preparation by visual transfer confirmation and test reactions. If the volume of source material is limited or the transfer is difficult to visualize in the destination plate well, the source plate can be sealed with a transparent plate seal prior to running a test transfer out of the well. The droplet will be caught on the underside of the seal and even transfers under 100 nL may be visually confirmed by droplet formation on the seal. The plate can then be centrifuged to return the droplet to the well and the seal removed.

4. Synthetic DNA fragments and plasmid DNA commonly used in this process is approximately 30 ng/μL and 300 nL is typically used. When using genomic DNA as a template it must be fairly dilute to ensure transfer (*see* **Note 3**) and more volume will be needed; typically 1000 nL is used. Template volume may be adjusted based on the template source concentration and reaction volume.
5. The initial ten cycles of this program serve as a minimal touchdown PCR to increase primer binding specificity and reduce the likelihood of primer dimer formation. Exact annealing temperatures will depend on the melting temperature of the primers being used.
6. One full ladder plate can be created by diluting 200 μL of the ladder provided by the kit into 2800 μL of 1× TE buffer and aliquoting 30 μL into each well of a 96-well PCR plate.
7. These metrics have been found to indicate a successful PCR and serve as a good filter for processing a large number of reactions. These are also the minimum requirements for successful multipart DNA assembly via yeast homologous recombination [2] in a high-throughput process. The most common cause for failing these criteria is primer dimerization lowering the comparative molar concentration of the intended amplicon. This can be addressed by lowering the concentration of primer in the reaction (e.g., 2× reduction) or by magnetic bead cleanup steps that remove smaller DNA fragments. Also be aware of run-to-run "drift" in which there is a shifted retention time across all wells on the plate, most often resulting in a consistently larger observed amplicon size than expected. Consider analyzing these metrics not only per reaction but also for each plate as a whole to capture process variations such as this. Inaccurate sizing can be avoided by using fresh ladder and markers coupled with regular maintenance of the instrument and capillary array.
8. PCR cleanup may not be necessary depending on the sensitivity of downstream processes. Data obtained during verification will inform the extent of off-target amplification and primer

dimerization (*see* Subheading 3.2, **step 4**). Cleanup steps may be skipped and the reaction can be diluted in half with sterile water when transferring from PCR plates to Echo qualified plates (*see* Subheading 3.3, **step 3**). If the reaction is diluted rather than cleaned up with a kit, the buffer will need to be taken into account during future transfers on the Echo and the 384PP_AQ_SP liquid type should be used.

9. PCR product volume will need to be transferred to a plate with a larger available volume for this process, assuming a PCR volume of 30 μL. Round bottom plates have been found to work well in effectively forming a ring of beads when placed on the magnet plate to minimize their transfer during the final collection step. Modifications to the ratio of magnetic beads to reaction volume can shift the size range that is retained. A volume of beads equal to 70% of the reaction volume has been found to effectively remove primer dimers while retaining amplicons above 250 nucleotides.

References

1. Kanigowska P, Shen Y, Zheng Y et al (2016) Smart DNA fabrication using sound waves: applying acoustic dispensing technologies to synthetic biology. J Lab Autom 21:49–56
2. Chandran S, Shapland E (2017) Efficient assembly of DNA using yeast homologous recombination (YHR). In: Hughes R (ed) Synthetic DNA, Methods in molecular biology, vol 1472. Humana Press, New York, NY
3. Dharmadi Y, Patel K, Shapland E et al (2014) High-throughput, cost-effective verification of structural DNA assembly. Nucleic Acids Res 42:e22
4. Shapland EB, Holmes V, Reeves CD et al (2015) Low-cost, high-throughput sequencing of DNA assemblies using a highly multiplexed Nextera process. ACS Synth Biol 4:860–866

Chapter 5

High-Throughput DNA Assembly Using Yeast Homologous Recombination

Kristy Ip, Ron Yadin, and Kevin W. George

Abstract

Yeast homologous recombination is a reliable, low-cost, and efficient method for DNA assembly. Using homology regions as short as 24 base pairs, constructs of up to 12 unique parts can be assembled into a diverse range of vectors. The simplicity and robustness of this protocol make it amenable to laboratory automation and high-throughput operations. Here we describe a high-throughput protocol to generate DNA parts through PCR, assemble them into a vector via yeast transformation, and "shuttle" the resulting plasmid constructs into *E. coli* for storage and propagation. Though this protocol is intended for high-throughput workflows, it can be easily adapted for bench-scale DNA assembly.

Key words DNA assembly, Yeast homologous recombination, High-throughput, Automation, Synthetic biology

1 Introduction

Homologous recombination is a natural phenomenon in eukaryotes used to repair broken DNA at overlapping common sequences [1]. In synthetic biology, homologous recombination has been co-opted for DNA assembly and multi-part plasmid construction. *Saccharomyces cerevisiae* is typically the host of choice for this method due to its unmatched ability to facilitate homologous recombination. The only requirement for this method is sequence homology between adjacent pieces of DNA: when flanked with 20 to 40 base pairs of homologous sequences, DNA parts can be assembled in a linearized vector backbone with high efficiency (Fig. 1) [2]. The simplicity and flexibility of this protocol make it an attractive "BioFoundry" platform. With minor adjustments, this protocol can accommodate standardization (e.g., standardized homologous "linkers") and other application-specific modifications. For example, a commonly used variation of this protocol employs flanking restriction sites in the vector (e.g., PmeI) to liberate an assembled "payload" for homology-mediated genome

Sunil Chandran and Kevin W. George (eds.), *DNA Cloning and Assembly: Methods and Protocols*, Methods in Molecular Biology, vol. 2205, https://doi.org/10.1007/978-1-0716-0908-8_5, © Springer Science+Business Media, LLC, part of Springer Nature 2020

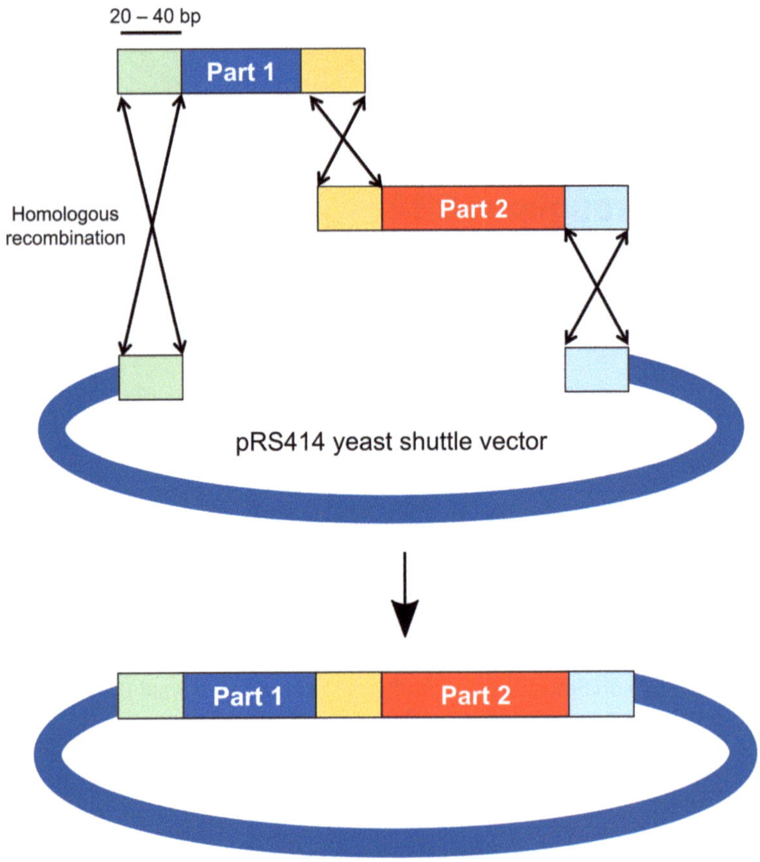

Fig. 1 Homologous recombination of a two-part assembly. (**a**) Parts flanked with 20–40 bp homology regions to adjacent DNA parts are generated by PCR. DNA parts are combined with a linearized yeast shuttle vector such as pRS414. (**b**) Yeast successfully assembles the DNA construct through homologous recombination. Assembled construct can be liberated from the vector via restriction digestion

integration. Similar modifications can be applied depending on the desired application.

The efficiency of homologous recombination in *Saccharomyces cerevisiae* makes it highly scalable for high-throughput, automated platforms. To maximize flexibility, the protocol described here uses PCR to generate compatible DNA parts. Homologous sequence "overhangs" of 20–40 bp are embedded in primer tails, effectively generating assembly-ready DNA in high throughput PCR reactions. Following verification and optional purification, these parts are transformed into *Saccharomyces cerevisiae* to be assembled in a "shuttle" vector containing selectable markers and replication sequences compatible with both yeast and *E. coli* [3]. After a 36-h "outgrowth" in selective media, assembled plasmids are isolated and transformed into *E. coli* that is subsequently plated on solid

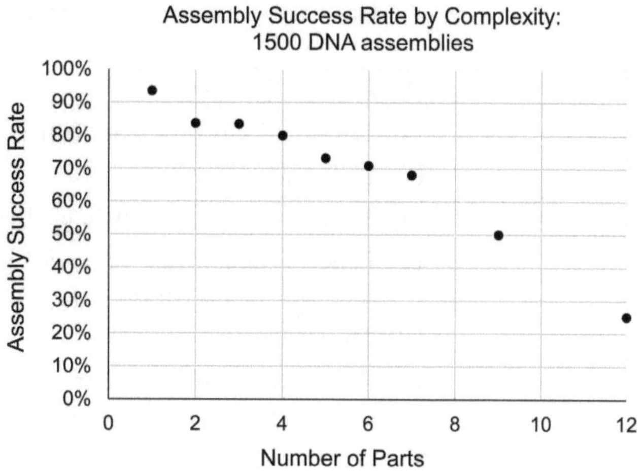

Fig. 2 Observed assembly rates for 1500 constructs of varying complexity, achieved with high-throughput operations. These assembly rates reflect individual replicates, i.e., assume only 1 colony is screened for each DNA construct. Screening 4 or more colonies for each construct results in success rates >90% for most complexities. The assemblies shown here used 24 bp homology sequences between compatible DNA parts. Note that DNA part concentrations in each assembly reaction were *not* normalized to facilitate high-throughput operations. If higher success rates are required, users can increase the homology length between adjacent DNA pieces and carefully normalize each reaction.

media. These resulting *E. coli* colonies are then screened via sequencing or restriction analysis for successful assemblies. Positive colonies can be banked as glycerol stocks or used for high copy plasmid isolation.

This protocol has been validated for the simultaneous assembly of 1500 DNA constructs of up to 12 parts each. A sample of observed success rates is shown in (Fig. 2). While this protocol is written primarily in 384- or 96-well format and is intended to be used with modern liquid handlers, volumes can be scaled up to perform homologous recombination at the bench-scale.

2 Materials

2.1 DNA Part Generation

1. 50 μM synthetic DNA oligonucleotide primers.
2. Synthetic DNA fragments or genomic DNA templates 10–100 ng/μL.
3. Phusion Hot-Start Flex DNA Polymerase with 5X HF buffer.
4. dNTP Solution Mix, 10 mM per nucleotide.
5. DpnI, 20,000 units/mL.
6. Molecular Grade or sterile water.

7. Capillary electrophoresis reagents and equipment, or agarose gel, loading dye, and ladder.
8. 96-well PCR plates.
9. 384-well PCR plates (optional).
10. PCR plate seals.
11. Liquid handling robot (e.g., Biomek FX).
12. Labcyte Echo 525 acoustic liquid handler and compatible 384-well plates (optional).
13. Thermocycler.
14. Fragment Analyzer or Zero Agarose Gel (ZAG) DNA Analyzer (Agilent) capillary electrophoresis system (optional).

2.2 Prepare Competent Yeast Cells

1. Yeast host strain: *Saccharomyces cerevisiae* strain CEN.PK2-1c, tryptophan auxotroph (*MATa ura3-52 trp1-289 leu2-3112 his3Δ1 MAL2-8C SUC2*) (*see* **Note 1**).
2. 100 mM Lithium Acetate (LiAC): Make 1 M LiAC by dissolving 102.02 g of Lithium Acetate dehydrate in 1 L of deionized water. Filter-sterilize with a 0.22 μm filter. Dilute 1 M LiAC with deionized water.
3. YPD + Kan: 1% yeast extract, 2% peptone, 2% dextrose, 50 mg kanamycin. Dissolve 10 g yeast extract and 20 g peptone into 700 mL deionized water. Raise volume to 900 mL with deionized water. Autoclave for 20 min at 15 psi on liquid cycle. Cool to 50 °C. Add 100 mL of 20% glucose and 1 mL of 50 mg/mL Kanamycin (*see* **Note 2**).
4. Molecular grade or sterile water.
5. Centrifuge with compatible rotor and bottles.
6. 15 mL culture tube.
7. Shaker flasks.
8. 30 °C shakers.

2.3 Yeast Transformation

1. pRS414 Yeast shuttle entry vector (*see* **Note 3**).
2. Salmon sperm DNA (ssDNA), 10 mg/L.
3. 50% PEG Solution: Weigh out 50% by volume of PEG and add to deionized water. Filter-sterilize.
4. 1 M Lithium Acetate (LiAC): Dissolve 102.02 g of lithium acetate dehydrate in 1 L of deionized water. Filter-sterilize with a 0.22 μm filter.
5. CSM -W (complete yeast synthetic medium without tryptophan): To a 2 L beaker add 700 mL of dH_2O and a stir bar. Add 0.74 g of CSM without tryptophan (114511012 MP Biomedicals or similar) and 43.7 g DOB (114025012 MP Biomedicals or similar) and stir until dissolved. Transfer the solution to

a glass media bottle. Loosely cap and sterilize by autoclaving for 20 min at 15 psi (1.05 kg/cm^2) on liquid cycle (*see* **Note 4**).

6. Molecular grade or sterile water.
7. 96-well PCR plates.
8. PCR plate seals.
9. 2.2 mL 96-well culture plates.
10. Gas-permeable plate seals.
11. Thermocycler.
12. Liquid handling robot (e.g., Biomek FX).
13. Labcyte Echo 525 acoustic liquid handler and compatible 384-well plates (optional).
14. Spectrophotometer.
15. Centrifuge with compatible 96-well plate rotor.
16. 30 °C shakers.

2.4 E. coli *Shuttling*

1. Column-based 96-well yeast miniprep kit (e.g., Zymoprep-96 Yeast Plasmid Miniprep).
2. 1× TE buffer: 10 mM Tris–HCl, pH 8.0, 1 mM EDTA.
3. Competent *E. coli* (e.g., DH5α).
4. SOC Medium: 0.5% yeast extract, 2% tryptone, 10 mM NaCl, 20 mM glucose. Dissolve 20 g of Tryptone, 5 g of Yeast Extract, and 0.5 g of NaCl in 0.5 L of deionized water. Raise volume up to 1 L. Autoclave for 20 min at 15 psi on liquid cycle. Cool to 50 °C. Add 20 mL of sterile 1 M Glucose and mix by swirling.
5. LB + Carb broth: Dissolve 25 g of powdered LB Broth into 900 mL of deionized water. Raise volume to 1 L. Autoclave for 30 min at 15 psi on liquid cycle. Cool to 50 °C. Add 1 mL of 50 mg/mL of carbenicillin.
6. LB + Carb agar plates: Dissolve 25 g of powdered LB Broth into 900 mL of deionized water. Raise volume to 1 L. Add 15 g of agar. Autoclave for 30 min at 15 psi on liquid cycle. Cool to 50 °C. Add 1 mL of 50 mg/mL of carbenicillin. Pour into plates. Allow to solidify and dry overnight. To use liquid handlers for plating, use compatible plate types such as Qtrays (Molecular Devices).
7. 96-well PCR plates.
8. Liquid handling robot (e.g., Biomek FX).
9. Centrifuge with compatible 96-well plate rotor.
10. 37 °C incubator.
11. Thermocycler.

3 Methods

3.1 DNA Part Generation

Unless otherwise stated, perform all liquid transfers above 1 μL on a traditional liquid handler (e.g. Biomek FX), and all transfers 1 μL or below on an acoustic liquid handler (e.g., Labcyte Echo 525). If only using a traditional liquid handler, adjust volumes and concentrations to minimize transfer error (*see* **Note 5**).

1. Design primers to amplify each DNA part. Primers should be both specific to the part and flanked with desired homology. Include 20–30 bp that anneal to the DNA part (T_m ~ 60 °C) and 20–40 bp that anneal to the adjacent DNA part. Note that the first and last DNA part will require homology to the vector backbone. Order these sequences using synthetic DNA ordering service at 25 nmol, desalted, and at a final concentration of 50 μM. For high throughput set-ups, it may be preferable to place order in 96-well or 384-well plates (some vendors will also deliver primers directly in 384-well Echo source plates, facilitating reaction setup).
2. Set up the PCR reactions. Rearray 100 nL of each primer at 50 μM and 10–100 ng of template DNA into each well on a 96- or 384-well PCR plate (*see* **Notes 6** and 7).
3. Prepare a PCR master mix of 5× HF buffer, dNTPs, molecular grade water, and Phusion DNA polymerase by hand on ice. Each well should contain: 6.67 μL of 5X HF Buffer, 22 μL of water, 0.67 μL of dNTPs, and 0.67 μL of Phusion polymerase. Use a liquid handler to dispense and mix 30 μL of premade master mix into each well containing re-arrayed primers and templates. Table 1 shows suggested recipes for each plate format (*see* **Note** 7).
4. Calculate PCR extension time based on the largest amplicon on a given plate. In general, 15–30 s/kilobase is used, with a minimum extension time of 16 s.

Table 1
Sample master mix volumes with safety factor implemented

Reagent	96-well plate (mL)	384-well plate (mL)
Molecular grade water	7.60	10.14
5X HF buffer	2.304	3.072
dNTPS (10 mM/nt)	0.2304	0.3072
Phusion DNA polymerase	0.2304	0.3072
Total	10.4	13.8

5. Place plate in the thermocycler and run the following PCR program: Initial denaturation at 98 °C for 2 min; 10 cycles of 98 °C for 12 s, 60 °C for 30 s, and 72 °C for calculated extension time; 30 cycles of 98 °C for 12 s, 55 °C for 30 s, and 72 °C for calculated extension time; final extension of 72 °C for 5 min.
6. Verify amplicons by running a small volume on an agarose gel or use capillary electrophoresis (*see* **Note 8**).
7. Reactions containing templates with plasmid backbones need to be digested before assembly. Add 20 U of DpnI to each reaction. Incubate at 37 °C for 30 min, 80 °C for 20 min to inactivate (*see* **Note 9**).

3.2 Prepare Competent Yeast Cells

1. Inoculate a yeast starter culture from frozen glycerol stock or streaked colony in 5 mL of YPD + Kan in a 15 mL culture tube. Grow overnight, shaking at 30 °C.
2. Check the OD600 of the starter culture (*see* **Note 10**).
3. Inoculate a flask with appropriate amount of starter into 0.5–1 L of YPD + Kan and grow overnight to OD of 0.7 (*see* **Note 11**).
4. Transfer culture to centrifuge bottle. Pellet the cells by centrifuging for 4000 × *g* for 4 min. Decant supernatant.
5. Completely re-suspend pellet with water. Use at least half of the original volume of culture to ensure thorough washing. Centrifuge at 4000 × *g* for 4 min. Decant supernatant.
6. Repeat **step 5** of this section with 100 mM LiAC.
7. After decanting, re-suspend the cell pellet with residual LiAC. Measure the volume. Add 100 mM LiAC to reach final desired OD-mL/well of 1.5 (*see* **Note 12**).

3.3 Yeast Transformation

1. Re-array 1 μL of each DNA part into 96-well PCR plates. Each well on the final assembly plate should represent a single assembly (*see* **Note 13**).
2. Prepare a mixture of 50% PEG, 1 M LiAC, and salmon sperm DNA (ssDNA) by hand. ssDNA should be boiled on a heat block at 95 °C for 10 min before use. Each transformation well should contain 42 μL of 50% PEG, 6.3 μL of 1 M LiAC, and 1.7 μL of ssDNA.
3. To the mixture prepared in the last step, add cells and entry vector. Each transformation should contain 10 μL of yeast cells and 2 ng of entry vector.
4. Transfer 62 μL of the final mixture into each well of 96-well plates containing mixed DNA parts from **step 1** of this section. Swirl the mixture before pouring into a reservoir. This will

homogenize the solution and prevent it from leaking out of the pipette tips during transfer.

5. Heat-shock using the thermocycler at 42 °C for 45 min.
6. Fill 2.2 mL 96-well culture plates with 1 mL of Molecular Grade or sterile water per well.
7. After heat shock is complete, transfer the entire transformation mixture to the pre-filled 2.2 mL plates. Wash by aspirating and dispensing several times.
8. Centrifuge to pellet the cells at 4000 × *g* for 3 min. Remove supernatant by inverting and flicking each plate over a waste container.
9. Resuspend pellets in 1 mL per well of CSM -W medium.
10. Seal each plate with a gas-permeable seal. Outgrow plate at 30 °C, shaking, for at least 36 h and up to 60 h.

3.4 *E. coli Shuttling*

1. Miniprep DNA from yeast using a column-based yeast miniprep kit following manufacturer's protocol. Elute in 30 μL of 1× TE buffer.
2. Thaw frozen competent *E. coli* cells on ice. Aliquot 50 μL per well into a 96-well plate.
3. Prepare for plating by drying LB + Carb agar plates under laminar flow hood until surface looks opaque. Do not over-dry.
4. Transfer and mix 10 μL of eluted DNA to 96-well plates containing 50 μL of competent cells. Work quickly to keep cells cold.
5. Keep plates containing cells with DNA on ice for 20 min.
6. Heat-shock plate at 42 °C for 45 s. This can be done by placing the plate, lightly covered, on a thermocycler set to 42 °C with the lid open.
7. Remove and immediately return to ice for at least 5 min.
8. Dilute cells in SOC (*see* **Note 14**).
9. Plate diluted cells on pre-dried agar plates. If using a liquid handler, calibrate the dispense height with the agar level. If plating multiple transformations on one plate, use appropriate droplet sizes to leave adequate spacing between wells to prevent contamination.
10. Let plates dry before handling. Cover, place upside down in an incubator, and grow overnight at 37 °C for about 18 h to obtain single isolate colonies.
11. Pick colonies into 1 mL/well of LB + Carb broth (*see* **Note 15**). After overnight incubation at 37 °C, isolate the plasmids via standard miniprep protocols and analyze via sequencing or restriction digest analysis. Alternatively, rolling circle amplification (RCA) may be used to generate sequence-ready DNA directly from boiled *E. coli* culture [4].

4 Notes

1. The protocol provided here is written for a tryptophan auxotroph. Auxotrophy can be achieved through point mutations or a complete knockout of *TRP1*. Other auxotrophic yeast strains or selectable markers can be used—users will need to adjust vector choice and growth media (i.e., replace CSM -W) accordingly.
2. Kanamycin is added to YPD to inhibit bacterial contamination that may occur during high-throughput operations.
3. pRS414 is an example shuttle vector compatible with this protocol. The key features of this shuttle vector are (1) low copy CEN/ARS sequence for replication in yeast, (2) higher copy pMB1 ori for replication in *E. coli*, (3) *TRP1* marker for auxotrophic selection in yeast, and (4) ampicillin resistance marker for selection in *E. coli*. The shuttle entry vector should be provided as a linearized DNA product.
4. Media used for yeast outgrowth should be compatible with the selectable marker of the shuttle vector. In this case, a shuttle vector containing the *TRP1* marker is paired with selection media without tryptophan.
5. In a high throughput setting, it is recommended to avoid liquid transfers <2 μL on standard liquid handling robotics such as the Biomek FX. Users can adjust volumes and stock concentrations accordingly to avoid low volume transfers.
6. Template DNA is frequently miniprepped plasmid DNA, synthetic DNA (e.g., gene fragments), or purified genomic DNA. If using an acoustic liquid handler, transfer of genomic DNA can be problematic. Users should dilute their genomic DNA stock to about 5 ng/μL and visually confirm transfer before proceeding.
7. When using liquid handlers to dispense reagents, the dead volume of the instrument should always be considered. Typically, a 10–20 mL dead volume should be added to all bulk-reagent transfers on the Biomek. To ease calculations for master mix preparation, a safety factor of 1.3×–1.5× can be multiplied for each reagent. On the Labcyte Echo, the dead volume of a polypropylene 384 square-well plate is 20 μL, with a maximum volume of 60 μL. Users should consult manufacturing instructions when changing labware.
8. For high-throughput verification via capillary electrophoresis, each DNA fragment should be within 10% of its expected size with a concentration greater than 5 ng/μL. We have found that as little as 5 ng of a DNA part can successfully assemble, though higher concentrations are preferable. Importantly, the molar

purity must be >50%, with limited nonspecific amplification. Primer dimers are often the most common nonspecific products, particularly for large amplicons. If primer dimers become problematic, users should consider decreasing the primer concentration by two-fold or purifying the PCR products via column or bead-based methods (e.g., AMPure XP).

9. Plasmid DNA from *E. coli* contains methylated DNA. DpnI cleaves methylated DNA and prevents any residual template from being carried over in the recombination process. Residual plasmid template due to ineffective DpnI digestion (typically due to a liquid transfer error) is a common failure mode in this process, though it is easily identified (i.e., picked *E. coli* colonies contain plasmid template rather than assembled DNA) and remedied.

10. To take an accurate OD of the starter culture, perform a 20-fold dilution before reading. For example, combine 50 μL of starter with 950 μL of media in a spectrophotometer cuvette. Then, multiply the reading by 20 to get the OD of the starter culture. Typically, a 5 mL overnight starter culture should reach an OD of 7 after 24 h.

11. Calculate the number of doublings by using this equation. 95 min is the expected doubling time. Total incubation time is calculated based on time between inoculation and harvest.

$$\text{num.of doublings} = \frac{\text{Total Incubation Time}}{95\ \text{min}}$$

Calculate the start inoculation OD by using this equation:

$$\text{Start inocul.OD} = \frac{0.7}{(\text{num.of doublings})^2}$$

Finally, calculate the start inoculation volume by using this equation. The total volume to be inoculated is based on how many transformations are needed. Typically, a single 1000 mL flask is enough for perform five 96-well plates of transformations (with dead volume considered).

$$\text{Vol of starter to inoculate} = \frac{\text{Start inocul.OD}}{\text{OD of starter}} \times (\text{total vol to be inoculated})$$

12. Each transformation well should contain 10 μL of harvested cells at an OD-mL per well of 1.5.

13. For maximum assembly efficiencies, users should normalize the molar concentrations of each DNA part. The recommended composition of each assembly reaction is 150 fmol of each DNA part and 5 fmol of shuttle vector. For high-throughput workflows, simply adding 1 μL of each DNA part (provided it

meets the QC metrics described in **Note 8**) is sufficient, though the minor decrease in assembly success rate may necessitate picking additional colonies. Note that the assembly efficiency metrics provided in Fig. 2 were generated *without* normalization.

14. A dilution factor of 16 yields pickable colonies for most transformations. A more concentrated dilution can be plated alongside within the same Q-tray well to ensure pickable colonies. Colony densities and growth time will differ depending on the agar plate dimensions. Plating dilution and method should be optimized for each plate-type.
15. Picking four colonies per assembly typically results in at least one positive construct for assemblies up to 12 parts. Users should pick additional replicates if more complex assemblies are attempted.

References

1. Eckert-Boulet N, Rothstein R, Lisby M (2011) Cell biology of homologous recombination in yeast. In: Tsubouchi H (ed) DNA recombination, Methods in molecular biology, vol 745. Humana Press, New York
2. Oldenburg K, Kham T, Michaelis S et al (1997) Recombination-mediated PCR-directed plasmid construction *in vivo* in yeast. Nucleic Acids Res 25(2):451–452
3. Chandran S, Shapland E (2017) Efficient assembly of DNA using yeast homologous recombination (YHR). In: Hughes R (ed) Synthetic DNA, Methods in Molecular Biology, vol 1472. Humana Press, New York, NY
4. Shapland E, Holmes V, Reeves C et al (2015) Low-cost, high-throughput sequencing of DNA assemblies using a highly multiplexed Nextera process. ACS Synth Biol 4(7):860–866

Chapter 6

Highly Multiplexed, Semiautomated Nextera Next-Generation Sequencing (NGS) Library Preparation

William Christie, Ron Yadin, Kristy Ip, and Kevin W. George

Abstract

High-throughput, inexpensive DNA sequencing is an essential component of large-scale DNA assembly operations. Using traditional and acoustic liquid-handling robotics, Illumina's Nextera Tagmentation reactions can be miniaturized and paired with custom PCR index primers to produce highly multiplexed NGS libraries for pooled sequencing. This chapter describes a high-throughput protocol that enables the simultaneous sequencing of thousands of DNA constructs in a single sequencing run at a dramatically reduced cost compared to bench-top methods.

Key words Multiplexed next-generation sequencing, High-throughput, Miniaturization, Synthetic biology

1 Introduction

As synthetic biology grows, so too does the need for high-throughput methods to sequence engineered DNA. Advances in DNA library preparation, liquid handling robotics, and next-generation sequencing (NGS) platforms have significantly increased researchers' access to high quality sequencing data. Recently described methods have demonstrated the simultaneous pooled sequencing of thousands of samples at reduced cost and dramatically increased throughput [1]. These methods rely on the construction of an expanded set of "barcode" index primers (*see* **Note 1**) and reaction miniaturization enabled by acoustic liquid handling robotics.

This chapter is an update to the protocol for high-throughput sequencing of plasmid DNA assemblies using a highly multiplexed Nextera process, as originally described for the bench scale by Shapland et al. [1]. With increased focus on automation and throughput, this version relies heavily on conventional pipetting robots capable of transferring 2–200 μL volumes using plastic tips (e.g., Beckman Coulter Biomek FX), as well as acoustic liquid

Sunil Chandran and Kevin W. George (eds.), *DNA Cloning and Assembly: Methods and Protocols*, Methods in Molecular Biology, vol. 2205, https://doi.org/10.1007/978-1-0716-0908-8_6, © Springer Science+Business Media, LLC, part of Springer Nature 2020

handlers capable of transferring 25–1000 nL in droplets with sound waves (e.g., Labcyte Echo 525). These platforms provide the efficiency and accuracy required to complete the complex liquid transfers which enable this method's miniaturization and multiplexity.

This protocol implements Illumina's Nextera Tagmentation library preparation method alongside an expanded set of i5 and i7 NGS PCR index primers (*see* Table 1) [1]. The 192 included DNA "barcodes" (96 for i5 and i7, respectively), can combine pairwise to provide a theoretical maximum plexity of 9216 unique combinations. If necessary, additional NGS barcodes can be designed using various open source software applications.

In practice, the upper plexity limit for this method is defined by the capacity of the chosen sequencing reagent kit. It is vital to choose a reagent kit with enough sequencing capacity to deliver high quality, statistically meaningful sequencing data for the requisite number of samples. The protocol described here has been successfully validated to prepare and sequence over 5000 plasmid constructs, between 4 and 18 kb in length, using a single MiSeq V2 – 2 × 150 bp reagent kit.

First, sample plasmid is liberated from *E. coli* cells via boiling and amplified into long, linear DNA concatemers using Rolling Circle Amplification (RCA). These concatemers are subsequently diluted and tagmented via Illumina's Nextera process. In each tagmentation reaction, "charged" transposases randomly fragment the DNA concatemers and add short adapter sequences to each fragment end. Next, a unique pair of i5 and i7 NGS PCR index primers is added to each sample well with an acoustic liquid-transferring robot. Each index primer contains a unique eight-nucleotide "barcode" flanked by two adapter sequences, one designed to anneal to the tagmented DNA and the other to anneal to NGS PCR terminal primers. The sequences introduced by the terminal primers allow the samples to bind to the MiSeq flowcell, while the barcode sequences from the index primers enable post-run data demultiplexing. After the NGS PCR is complete, the barcoded library can be pooled, purified, and characterized for loading into the sequencer.

2 Materials

2.1 Library Amplification

1. Plasmid-bearing *E. coli* liquid culture.
2. Laboratory grade water.

Table 1
List of unique i5 and i7 8-base barcodes to use for adapter primers

i5 name	Barcode sequence	i7 name	Barcode sequence
I5-Amy_3	CCATGTTG	I7-Amy_1006	GCTGTGTT
I5-Amy_6	ACACCGGC	I7-Amy_1017	ATGGCGAC
I5-Amy_11	GTATCCTA	I7-Amy_1018	GGCGAGCA
I5-Amy_13	GGATGAGC	I7-Amy_1019	GCTGTCCG
I5-Amy_14	GAGACTAG	I7-Amy_1030	CGAGTGAA
I5-Amy_15	GGCCTCTA	I7-Amy_1033	CCATCACT
I5-Amy_17	CAATGATA	I7-Amy_1036	AGTACACC
I5-Amy_21	CCTATCCA	I7-Amy_1049	TCGCTGAT
I5-Amy_22	TTGATATA	I7-Amy_1052	GGCGGTAA
I5-Amy_23	AGCGATAT	I7-Amy_1056	CCGCCGAA
I5-Amy_26	CCTACAGT	I7-Amy_1057	GATTGCGA
I5-Amy_27	ATGACAGT	I7-Amy_1058	ACATTCTC
I5-Amy_30	AGTGTACA	I7-Amy_1065	CCACTGGT
I5-Amy_31	CTGGCACG	I7-Amy_1078	CCTGCCAA
I5-Amy_250	TGGCTTGA	I7-Amy_1080	ATACGTCC
I5-Amy_37	CGCCTAAC	I7-Amy_1091	TCAACTCT
I5-Amy_38	CTCGTCGT	I7-Amy_1095	ACCGCTAC
I5-Amy_40	TACAGACA	I7-Amy_1097	GCAATGCT
I5-Amy_41	CAGTACCA	I7-Amy_1100	GGACCGCG
I5-Amy_45	AAGGTATC	I7-Amy_1101	CCTACTTA
I5-Amy_46	AATTGAAT	I7-Amy_1102	GATGATCT
I5-Amy_47	CGCAAGAG	I7-Amy_1112	CAGTGGAA
I5-Amy_48	CTCGATAA	I7-Amy_1115	GTTGACAT
I5-Amy_49	TTGTTCTC	I7-Amy_1125	GCCATAGA
I5-Amy_50	TGACATCT	I7-Amy_1134	TCTGGAAT
I5-Amy_52	TTCTGTTC	I7-Amy_1160	TGCCGATC
I5-Amy_54	TCAGCACC	I7-Amy_1173	ATGTAGCA
I5-Amy_56	GTTATCAC	I7-Amy_1174	GTCACCAA
I5 Amy_57	ACGTGTCC	I7-Amy_1193	CTAAGAGT
I5-Amy_60	TGGCTCCT	I7-Amy_1195	CCTCTCTC
I5-Amy_61	ATGCGAAG	I7-Amy_1199	GCTAATGA

(continued)

Table 1
(continued)

i5 name	Barcode sequence	i7 name	Barcode sequence
I5-Amy_62	AACTACCT	I7-Amy_1203	CTGGTGAT
I5-Amy_65	TGGTCATA	I7-Amy_1209	CTGATAGC
I5-Amy_68	ACATAACA	I7-Amy_1214	AATGCCGG
I5-Amy_69	ACAGGCAT	I7-Amy_1230	GCACATTG
I5-Amy_72	ACCTCTCT	I7-Amy_1233	TGTTGCAC
I5-Amy_88	AGATGATT	I7-Amy_1234	GCCTATCG
I5-Amy_92	AGACTCTT	I7-Amy_1244	GCCTTCGG
I5-Amy_93	ACTAGCAG	I7-Amy_1249	TCGGTGTC
I5-Amy_95	ATACACGT	I7-Amy_1258	GCGACGTA
I5-Amy_96	ACAGCATT	I7-Amy_1269	GCACGATT
I5-Amy_102	ATAATTAG	I7-Amy_1272	CTGCTACT
I5-Amy_108	TCTAGACC	I7-Amy_1274	GCTTACAA
I5-Amy_109	CTACAGAC	I7-Amy_1276	TGAACAAC
I5-Amy_111	CTAGTTGC	I7-Amy_1281	ACTTGTAA
I5-Amy_115	CATTGTAC	I7-Amy_1310	TCTGCGAC
I5-Amy_116	AGTATGAT	I7-Amy_1311	GGCCGAGT
I5-Amy_117	AGAATCAA	I7-Amy_1312	CACATTAC
I5-Amy_118	TTCATTGA	I7-Amy_1321	GATTCCAG
I5-Amy_120	ATCACTTA	I7-Amy_1331	TCCGCGGT
I5-Amy_121	TCAATCAT	I7-Amy_1343	ACTGACGA
I5-Amy_125	AGATGTCA	I7-Amy_1351	GCACACAT
I5-Amy_127	GTAATATG	I7-Amy_1354	CCTAGGAT
I5-Amy_129	CCACAGCA	I7-Amy_1356	CTTAACGA
I5-Amy_131	CATCCACC	I7-Amy_1357	CCACCATC
I5-Amy_133	CAGACTCA	I7-Amy_1359	TGAGCCGC
I5-Amy_135	ACTTCATA	I7-Amy_1366	TACTCCAC
I5-Amy_137	CCAACGGA	I7-Amy_1369	GGCAGCCG
I5-Amy_141	ACCAATCC	I7-Amy_1375	TTCGACTC
I5-Amy_145	TAGCATAA	I7-Amy_1386	TACGAATA
I5-Amy_146	TGACAGGA	I7-Amy_1392	AGGTCCTT
I5-Amy_147	CATGAAGT	I7-Amy_1397	CAGCGAGG

(continued)

Table 1
(continued)

i5 name	Barcode sequence	i7 name	Barcode sequence
I5-Amy_150	TATAGTAG	I7-Amy_1398	GACCTCAG
I5-Amy_152	ACCACATC	I7-Amy_1408	GCTAGGCG
I5-Amy_153	AATTATAG	I7-Amy_1414	TGCACGGA
I5-Amy_156	TTCCACAT	I7-Amy_1427	TAACGACC
I5-Amy_158	CAGGCATA	I7-Amy_1436	AACGGTTC
I5-Amy_256	TCACAGTC	I7-Amy_1437	TGCAATGC
I5-Amy_162	TAGTTAAC	I7-Amy_1439	ATTCGAGC
I5-Amy_163	CCGCATCT	I7-Amy_1440	TGCGTTCC
I5-Amy_164	ATGAATCT	I7-Amy_1447	ATGATCCA
I5-Amy_168	TGTGACTT	I7-Amy_1448	GGAACGAT
I5-Amy_169	AGGCTTAC	I7-Amy_1451	TCCGAAGC
I5-Amy_170	CTGTCCTG	I7-Amy_1453	CTGCCAAC
I5-Amy_171	GATACATT	I7-Amy_1462	AACCGCGG
I5-Amy_172	ACCGGAGT	I7-Amy_1466	AGAGCGAG
I5-Amy_173	TGACCTTC	I7-Amy_1470	TCGTATGT
I5-Amy_177	AGGACTAA	I7-Amy_1473	CTCGCTTC
I5-Amy_183	TCATTGAC	I7-Amy_1490	TGGAGCGC
I5-Amy_184	CAGGACAT	I7-Amy_1491	GTGGCCGT
I5-Amy_185	TAATACTC	I7-Amy_1493	TGGCCACC
I5-Amy_187	TATGCTTC	I7-Amy_1506	GCGCAGTT
I5-Amy_195	TTAGGAGA	I7-Amy_1507	TCTCCGTA
I5-Amy_199	GGCTAAGA	I7-Amy_1509	GCGTTGCG
I5-Amy_201	TAGTGAGT	I7-Amy_1511	GATAGCAT
I5-Amy_207	CCATCACT	I7-Amy_1512	AACCAGGT
I5-Amy_208	TTATAGTT	I7-Amy_1536	CATGACTA
I5-Amy_210	AGTACACC	I7-Amy_1541	GTCTCGGA
I5-Amy_211	CACTTGAG	I7-Amy_1543	CTCTAAGT
I5-Amy_213	AGTCCAAG	I7-Amy_1560	CATCGTGT
I5-Amy_215	TCACTACA	I7-Amy_1574	GCAACCTT
I5-Amy_216	AGAATTCC	I7-Amy_1577	GAGATTCT
I5-Amy_218	AATTAAGC	I7-Amy_1586	CACTGCTT

(continued)

Table 1
(continued)

i5 name	Barcode sequence	i7 name	Barcode sequence
I5-Amy_219	ACACCTAT	I7-Amy_1621	AGGTACGA
I5-Amy_221	ATTGCAAT	I7-Amy_1635	ACCGAGTC
I5-Amy_225	TGGATAAT	I7-Amy_1645	CACAAGTA

Each barcode is flanked on both ends by sequences specified by Illumina, as follows:
Index 1 Read 5′ CAAGCAGAAGACGGCATACGAGAT [i7 barcode] GTCTCGTGGGCTCGG
Index 2 Read 5′ AATGATACGGCGACCACCGAGATCTACAC [i5 barcode] TCGTCGGCAGCGTC
Underlined portions denote the terminal primers that are used in the second step of NGS PCR (*see* Subheading 3.3, **step 14**)

3. Rolling Circle Amplification (RCA) master mix containing phi29 polymerase, random hexamer primers, and reaction buffer (e.g., GE Healthcare Illustra™ TempliPhi).
4. 384-well PCR plates.
5. Plate vortex mixer (e.g., Eppendorf MixMate).
6. Thermocycler.
7. Centrifuge.
8. Liquid handler robot.

2.2 Library Quantification

1. Sterile or molecular grade water.
2. 1× TE Buffer: 10 mM Tris–HCl pH 8.0, 1 mM EDTA.
3. High sensitivity DNA quantification reagents and equipment (e.g., Qubit™ dsDNA HS Assay Kit).
4. Liquid handler robot (e.g., Biomek FX).
5. Echo Qualified 384-Well Polypropylene Microplates (Labcyte).
6. Labcyte Echo 525 (optional).

2.3 Next-Generation Sequencing (NGS) Library Preparation

1. 0.5% sodium dodecyl sulfate (SDS): Dissolve 0.5 mg of SDS in 70 mL of deionized water. Raise volume to 100 mL. Solution may precipitate and can be warmed at 65 °C before use.
2. i5 Terminal Primer: 100 μM (*see* Table 1).
3. i7 Terminal Primer: 100 μM (*see* Table 1).
4. i5 Adapter Primers: 100 μM (*see* Table 1).
5. i7 Adapter Primers: 100 μM (*see* Table 1).
6. Vent Polymerase, 2000 U/mL.
7. NexteraTagmentation Enzyme and Buffer (Illumina) (*see* **Note 2**).
8. 10× ThermoPol Reaction Buffer.

9. 100 mM $MgSO_4$.
10. dNTPs, 10 mM per nucleotide.
11. Echo Qualified 384-Well Polypropylene Microplates (Labcyte).
12. Echo Qualified Reservoir Plate (Labcyte).
13. 384-well PCR plates.
14. Liquid handler robot.
15. Labcyte Echo 525.
16. Thermocycler.
17. Centrifuge.

2.4 Library Cleanup and MiSeq Loading

1. Liquid handler robot (e.g., Biomek FX).
2. 96-well 1.1 mL round well plate.
3. 96-well round bottom plate compatible with Solid Phase Reversible Immobilization (SPRI) size selection (e.g., Costar Round Bottom).
4. Solid Phase Reversible Immobilization (SPRI) size selection Magnetic Beads (e.g., Beckman AMPure XP).
5. Solid Phase Reversible Immobilization (SPRI) size selection 96-well magnet plate.
6. Sterile water.
7. 1× TE Buffer: 10 mM Tris–HCl pH 8.0, 1 mM EDTA.
8. 70% ethanol.
9. 96-well PCR plate.
10. 12 × 200 μL multichannel pipette.
11. 25 mL sterile reservoir.
12. 15 mL sterile conical tube.
13. DNA purification and concentration columns and tubes (e.g., Amicon Ultra-0.5 mL Centrifugal Filters).
14. High sensitivity DNA quantification reagents and equipment (e.g., Qubit™ dsDNA HS Assay Kit).
15. DNA Fragment Analyzer reagents and equipment (e.g., Agilent BioAnalyzer & High Sensitivity dsDNA Assay Chip).
16. Illumina HT1 buffer.
17. PCR tubes.
18. Thermocycler.

3 Methods

All liquid transfers $\geq$2 μL should be performed on a liquid handler robot. Transfers $\leq$1 μL should be performed on a Labcyte Echo 525 (*see* **Note 3**).

3.1 Library Amplification

1. Determine the necessary number of 384-well PCR plates, and prefill all with 26 μL of lab grade water.
2. Transfer 4 μL of plasmid-bearing *E. coli* liquid culture to the prefilled, 384-well plates.
3. Centrifuge the plates at 500 × *g* for 1 min.
4. Liberate the sample plasmid from the *E. coli* cells by boiling the plates in the thermocycler as follows: 95 °C for 3 min, 4 °C until sample retrieval.
5. Transfer 4 μL of boiled material to a new, empty 384-well PCR plate.
6. Add 4 μL of RCA master mix to each well containing the 4 μL of boiled material.
7. Centrifuge the plates at 500 × *g* for 1 min.
8. Vortex the plates at 2000 rpm for 30 s.
9. Perform the RCA reaction in a thermocycler with the following program: 30 °C for 18 h, 65 °C for 10 min, incubate at 4 °C until sample retrieval (*see* **Note 4**).

3.2 Library Quantification

1. Measure the RCA-amplified DNA concentration using a plate-based quantification assay (e.g., PicoGreen) or select representative wells and measure the concentrations (*see* **Note 5**).
2. Dilute RCA-amplified material with 1× TE buffer or sterile water to a concentration of 1–5 ng/μL (*see* **Note 6**).

3.3 Next-Generation Sequencing (NGS) Library Preparation

1. Fill Echo Qualified 384-Well plates with 60 μL of diluted RCA-amplified DNA.
2. Transfer 200 nL of diluted DNA into 384-well PCR plates (*see* **Note 7**).
3. Prepare a tagmentation master mix in a 1:3 enzyme-to-buffer ratio by hand. *See* Table 2 for suggested volumes to use (*see* **Note 8**).
4. Fill an Echo Qualified Reservoir Plate with prepared tagmentation master mix. Centrifuge plate at 1000 × *g* for 1 min to remove bubbles.
5. Transfer 300 nL of tagmentation master mix to each well containing 200 nL of DNA.
6. Centrifuge the plates at 1000 × *g* for 2 min.

Table 2
Sample calculation for tagmentation master mix

	Volume needed for one 384-well plate (μL)	Dead volume per reservoir well (μL)
Tagmentation enzyme	19.2	62.6
Tagmentation buffer	57.6	187.7
Total volume	76.8	250.3

The volumes of tagmentation enzyme and buffer for one 384-well plate. These volumes can be scaled by multiplying by the number of plates being tagmented. The dead volume to add per Echo Reservoir well is also provided and can be scaled by the number of wells used. Each well can hold up to 2.8 mL of liquid

7. Place the plates in thermocyclers and run the following program to tagment the DNA: 55 °C for 5 min, 10 °C until sample retrieval.
8. Fill an Echo Qualified Reservoir Plate with 1 mL of 0.5% SDS solution. Centrifuge plate at 1000 × *g* for 1 min to remove bubbles.
9. After tagmentation is finished, remove the plates from the thermocyclers and transfer 125 nL of 0.5% SDS to each well (*see* **Note 9**).
10. Centrifuge the plates at 1000 × *g* for 2 min.
11. Fill an Echo Qualified 384-well plate with i5 and i7 adapter primers (*see* **Note 10**).
12. Transfer 125 nL of one unique i5 adapter primer and 125 nL of one unique i7 adapter primer to each well of the plates. Do not add the same combination of i5 and i7 index primers to more than 1 well (*see* **Note 10**).
13. Centrifuge the plates at 1000 × *g* for 2 min.
14. Prepare NGS PCR master mix on ice following these guidelines per well: 20.3 μL Sterile or Molecular Grade Water, 0.5 μL 10× ThermoPol buffer, 0.5 μL MgSO4, 0.5 μL dNTPs, 0.05 μL i5 terminal primer, 0.05 μL i7 terminal primer, 0.25 μL Vent Polymerase. *See* Table 3 for suggested volumes to use for a 384-well plate.
15. Transfer 24.5 μL of the NGS PCR master mix to each well containing tagmented DNA and adapter primers. Ensure samples are adequately mixed.
16. Centrifuge the NGS PCR plates at 500 × *g* for 1 min.
17. Place the NGS PCR plates in thermocyclers and run the following program: 72 °C for 3 min and then 12 cycles of 98 °C for 10 s, 63 °C for 30 s, and 72 °C for 30 s, incubate at 10 °C until sample retrieval.

Table 3
Sample NGS PCR master mix recipe with safety factor

Reagent	Volume needed for one 384-well plate with safety factor (1.2×) (mL)
Molecular grade water	9.34
10× ThermoPol buffer	1.15
100 mM $MgSO_4$	0.23
10 mM dNTPs	0.23
100 μM i5 terminal primer	0.02
100 μM i7 terminal primer	0.02
2000 U/mL vent polymerase	0.12
Total	10.96

The volumes needed to make NGS PCR master mix for one 384-well plate. These volumes include a safety factor of 1.2, which accounts for the volume lost when transferring liquids from one labware to another. If dispensing this master mix on a liquid handler, additional dead volume should be added

3.4 Library Pooling, Cleanup, and MiSeq Loading

1. Pool 10 μL from each well of all NGS PCR plates into a single 96-well 1.1 mL plate (many 384–to–one 96 transfer), then seal and vortex that plate for 1 min at 800 rpm (*see* **Note 11**).
2. Transfer 100 μL from each well of the 1.1 mL plate to a 96-well round bottom reaction plate that is compatible with 96-well Solid Phase Reversible Immobilization (SPRI) size selection PCR cleanup (e.g., Costar Round Bottom).
3. Mix SPRI magnetic beads (e.g., Beckman AMPure XP), and add 55 μL of beads to each well of the reaction plate, pipetting to mix well. Incubate the reaction plate at room temperature for 2–5 min.
4. Place the reaction plate onto a SPRI 96-well magnet plate for 2 min to separate the beads from the solution. The solution should appear clear before proceeding to the next step.
5. With the plate still seated on the magnet, carefully aspirate the clear solution from the plate and discard. Leave the last 5 μL of solution in the reaction plate to ensure that the beads are not accidentally drawn out.
6. With the plate still seated on the magnet, add 200 μL of 70% ethanol to each well and incubate at room temperature for 30 s. Aspirate out all the ethanol and discard.
7. Repeat **step 6** to complete a total of two washes.
8. Remove the reaction plate from the magnet plate, add 32 μL of TE buffer, and mix well by pipetting. Incubate the plate at room temperature for 2 min.
9. Place the reaction plate onto a SPRI 96-well magnet plate for 1 min to separate the beads from the solution.

10. Transfer 30 μL of eluent from the reaction plate to a clean 96-well elution plate, carefully avoiding the magnetic beads at the bottom of the wells.
11. Use a multichannel pipette to pool the entire volume of each well of the elution plate into a 25 mL sterile reservoir.
12. Transfer the pool to a 15 mL conical tube and vortex 5–10 s to mix well.
13. Load 400 μL of the pool into each of 2 DNA clean and concentrate columns (e.g., Amicon Ultra-0.5 mL Centrifugal Filters), and centrifuge for 12 min at 500 × *g*.
14. Discard flow-through, add 400 μL of TE buffer to each of the filters, and centrifuge for 12 min at 500 × *g*.
15. Discard flow-through, then invert columns into clean collection tubes. Centrifuge for 2 min at 1200 × *g* to collect the sample from the columns, then combine into one tube.
16. Measure the concentration of the pooled fragment library and dilute a small amount (<50 μL) to 5 ng/μL for characterization on the fragment analyzer.
17. Follow the manufacturer's instructions to analyze the 5 ng/μL dilution of the fragment pool on the fragment analyzer to obtain the average fragment size of the pool. Ideally, the average fragment size is between 900–1000 bp, ± 250 bp. Average fragment sizes outside this range might indicate improper tagmentation, with possibly detrimental effects for the sequencing run (*see* **Note 12**).
18. If within range, use the concentration and average fragment size measurements to make a 50 μL aliquot of the fragment pool diluted to ~4 nM in water (*see* **Note 13**).
19. Denature the fragment pool by combining 5 μL of the 4 nM pool with 5 μL of fresh 0.2 N NaOH in a 1.5 mL tube. Incubate at room temperature for 5 min.
20. Add 990 μL of prechilled Illumina HT1 buffer to further dilute the pool from 2 nM to 20 pM.
21. Combine 400 μL of the 20 pM pool with 400 μL of prechilled Illumina HT1 buffer, to get to the final load concentration of 10 pM.
22. Aliquot 190 μL of the diluted library into 5 PCR tubes and boil at 99 °C for 5 min in a thermocycler. Then immediately swirl tubes in ice water bath for 1 min to quickly reduce temperature.
23. Consolidate at least 800 μL of the pool into 1 tube. Keep on ice until ready to load sample onto the sequencer cartridge.

4 Notes

1. The terms *index primers* and *barcode primers* are used interchangeably in this chapter.
2. This protocol was developed using the Illumina Nextera Tagmentation kit, the bundled sale of which Illumina has since discontinued. However, it is still possible to buy the relevant reagents TD Tagment DNA Buffer and TDE1 Tagment DNA Enzyme from Illumina in bulk. Otherwise, the Nextera XT DNA Tagment kit can be used instead with some optimization of the reaction volumes and ratios.
3. This protocol is intended for high throughput environments. Reaction ratios can be scaled up at a benchtop scale.
4. The RCA reaction is very robust and will occur at room temperature. The primary benefit of using a thermocycler is the ability to heat inactivate the polymerase (Φ29 polymerase has exonuclease activity). The RCA reactions are incubated for 18 h to ensure the reaction goes to completion (i.e., dNTPs are fully exhausted). This has the added benefit of effectively "normalizing" the concentrations of amplified DNA.
5. The DNA concentration after RCA is important because it directly affects the tagmentation efficiency and thus the average DNA fragment size of the library. If the DNA is too concentrated, the library may be undertagmented, yielding a large average fragment size. Conversely, if the DNA is too dilute, the library will be overtagmented, yielding a small average fragment size. In both cases, poor quality sequencing data may be the result. It is recommended to use quantification methods that are specific for dsDNA (e.g., PicoGreen, which can be conveniently performed in a 96 or 384 well format).
6. It is important to dilute the RCA material to a DNA concentration of 1–5 ng/μL for 2 reasons. First, the Echo 525 does not robustly transfer RCA material (i.e., high molecular weight, linear concatemers) at high concentrations. This is similar to genomic DNA, which also must be diluted to ensure proper transfer. Second, this range of DNA concentration results in libraries with an ideal size distribution after tagmentation (with the reported reaction conditions). High-quality sequencing data has been repeatedly generated with these conditions for up to 5000 samples in one run.
7. If users measure the DNA concentration of every sample to be sequenced in Subheading 3.2 **step 1**, then it is possible to normalize the DNA going into the tagmentation reaction

(e.g., use the Echo to add additional volume for low concentration samples). Though normalization may lead to a more equal distribution of sequencing coverage, the observed difference is minor (and concentrations are often similar from sample to sample). Furthermore, normalization is not practical for high-throughput operations.

8. When making master mixes, a safety factor of at least 1.2 is recommended to account for volume lost when transferring from one labware to another. When using liquid handlers to dispense reagents, additional dead volume should be considered. Dead volume varies depending on the labware and instrument used. For example: In an Echo Qualified Reservoir, up to 2.55 mL can be dispensed from each well, with ~250 μL of dead volume. Users should consult the manufacturer for details when changing labware.
9. 0.5% SDS is used to remove the tagmentation enzyme from the DNA.
10. Each i5 and i7 adapter primer has its own unique sequence. These sequences are used to identify each piece of tagmented DNA individually and allows for multiplexing. Illumina provides 12 unique i7 adapter sequences and 8 unique i5 adapter sequences. This is enough for 96 samples to be sequenced, in order for us to process more than 96 samples per run we generated our own unique i5 and i7 adapter sequences (*see* Table 1).
11. At this stage, it is important to consider the coverage capacity of the chosen sequencing reagent kit, and pool the samples accordingly. If the sample number or size is too great to be run with a single reagent kit while still producing reliably robust data, it is possible to create two pools with your samples at this step, each to be run on its own sequencing run. We typically calculate our pooling strategy to ensure that we have a read depth of 10 or greater.
12. A fragment library with an average fragment size lower than the range accepted for sequencing (<650 bp) could result from tagmentation with too high a ratio of tagment enzyme to DNA, or from incubating the reaction too long. The opposite scenarios can lead to undertagmented DNA with average fragment sizes longer than the acceptable range (>1250 bp). Encountering a fragment library with an incorrect average fragment size should be remedied by retagmenting the original input DNA under corrected conditions.

13. To calculate the recipe for a 50 μL 4 nM dilution based on the concentration and average fragment size of the pool: Let C = initial pool concentration [ng/μL], S = avg. fragment size [bp], V = volume of initial pool needed [μL], and W = volume of water needed [μL]. Then:

$$V = (0.132)(S/C) \text{ and } W = 50 - V.$$

Reference

1. Shapland E, Holmes V, Reeves D et al (2015) Low-cost high-throughput sequencing of DNA assemblies using a highly multiplexed Nextera process. ACS Synth Biol 4(7):860–866. https://doi.org/10.1021/sb500362n

Part III

DNA Assembly Standards

Chapter 7

Generation of MoClo Standard Parts Using Golden Gate Cloning

Ramona Grützner and Sylvestre Marillonnet

Abstract

Availability of efficient DNA assembly methods is a basic requirement for synthetic biology. A variety of modular cloning systems have been developed, based on Golden Gate cloning for DNA assembly, to enable users to assemble multigene constructs from libraries of standard parts using a series of successive one-pot assembly reactions. Standard parts contain the DNA sequence coding for a genetic element of interest such as a promoter, coding sequence or terminator. Standard parts for the modular cloning system MoClo must be flanked by two BsaI restriction sites and should not contain internal sequences for two type IIS restriction sites, BsaI and BpiI, and optionally for a third type IIS enzyme, BsmBI. We provide here a detailed protocol for cloning of basic parts. This protocol requires the following steps (1) defining the type of basic part that needs to be cloned, (2) designing primers for amplification, (3) performing PCR amplification, (4) cloning of the fragments using Golden Gate cloning, and finally (5) sequencing of the part. For large basic parts, it is preferable to first clone subparts as intermediate level −1 constructs. These subparts are sequenced individually and are then further assembled to make the final level 0 module.

Key words Modular cloning, Synthetic biology, Biological parts, DNA assembly, Multigene constructs

1 Introduction

Golden Gate cloning is a DNA assembly method that enables users to assemble multiple DNA fragments (up to 24 fragments) using a one-pot DNA assembly reaction [1–3]. Golden Gate cloning is based on the use of type IIS enzymes such as BsaI, BpiI, or BsmBI (*see* **Note 1**). Modules flanked by two restriction sites for a type IIS enzyme are pipetted together with a recipient vector containing compatible type IIS enzyme restriction sites. Assembly is performed with a restriction–ligation using the type IIS restriction enzyme and ligase. Golden Gate cloning takes advantage of the fact that restriction sites for the type IIS enzyme used are not present in the ligated product of interest, but are still present in religated original vector or donor inserts plasmids. A restriction-ligation

Sunil Chandran and Kevin W. George (eds.), *DNA Cloning and Assembly: Methods and Protocols*, Methods in Molecular Biology, vol. 2205, https://doi.org/10.1007/978-1-0716-0908-8_7,

allows for repeated redigestion of the remaining or religated plasmids, making the fragments that they contain available for further rounds of ligation, until they are correctly assembled. Golden Gate cloning is used as the assembly method for several DNA cloning systems that have been developed for a variety of host organisms including bacteria, yeasts, fungi, plants, microalgae, cyanobacteria, and animal cells [4–19]. One of these systems is the Modular cloning system MoClo that we designed for DNA assembly for plants [18–21]. The MoClo system relies on libraries of standard parts. Each part consists of a DNA sequence containing a genetic element such as a promoter, coding sequence or terminator. Parts are flanked by two BsaI restriction sites in opposite orientations (Fig. 1). In addition, all parts should not contain any internal site for at least two, and optionally three, type IIS restriction enzymes that include BsaI, BpiI, and BsmBI. Generation of domesticated parts (lacking internal type IIS enzyme restriction sites) is the most labor-intensive part of cloning projects that use the MoClo system, since this process requires PCR amplification of DNA and sequencing of the resulting parts. In contrast, when cloned and sequenced parts are available, assembly of more complex constructs containing multiple parts is relatively easier since it only consists of performing one or several one-pot assembly reaction(s), and the resulting constructs do not necessarily need to be sequenced. One simple solution for generating basic parts is to order the sequence to a gene synthesis company. However, when doing this, it is important to be sure that the plasmid in which the part will be provided has an antibiotic selection marker that is compatible with the cloning system with which this part needs to be used. In addition, it is important to be sure that the sequence that one wants to order has been obtained from high quality sequencing data, as what is ordered is what the user will get. Therefore it is still useful to be able to clone basic parts in the laboratory from a source organism, to have the freedom to check and design the parts as needed. In this chapter, we provide a detailed protocol for cloning of MoClo level 0 basic parts. This protocol can also be used to clone parts for other modular cloning systems, provided that cloning vectors with appropriate antibiotic resistance markers are used. This protocol requires the following steps (1) defining the type of basic part that needs to be cloned, (2) designing primers for amplification, (3) performing PCR amplification, (4) cloning of the fragments using Golden Gate cloning, and finally (5) sequencing of the part. For large basic parts, it is preferable to first clone subparts as intermediate level −1 constructs. These subparts are sequenced individually and are then further assembled to make the final level 0 module.

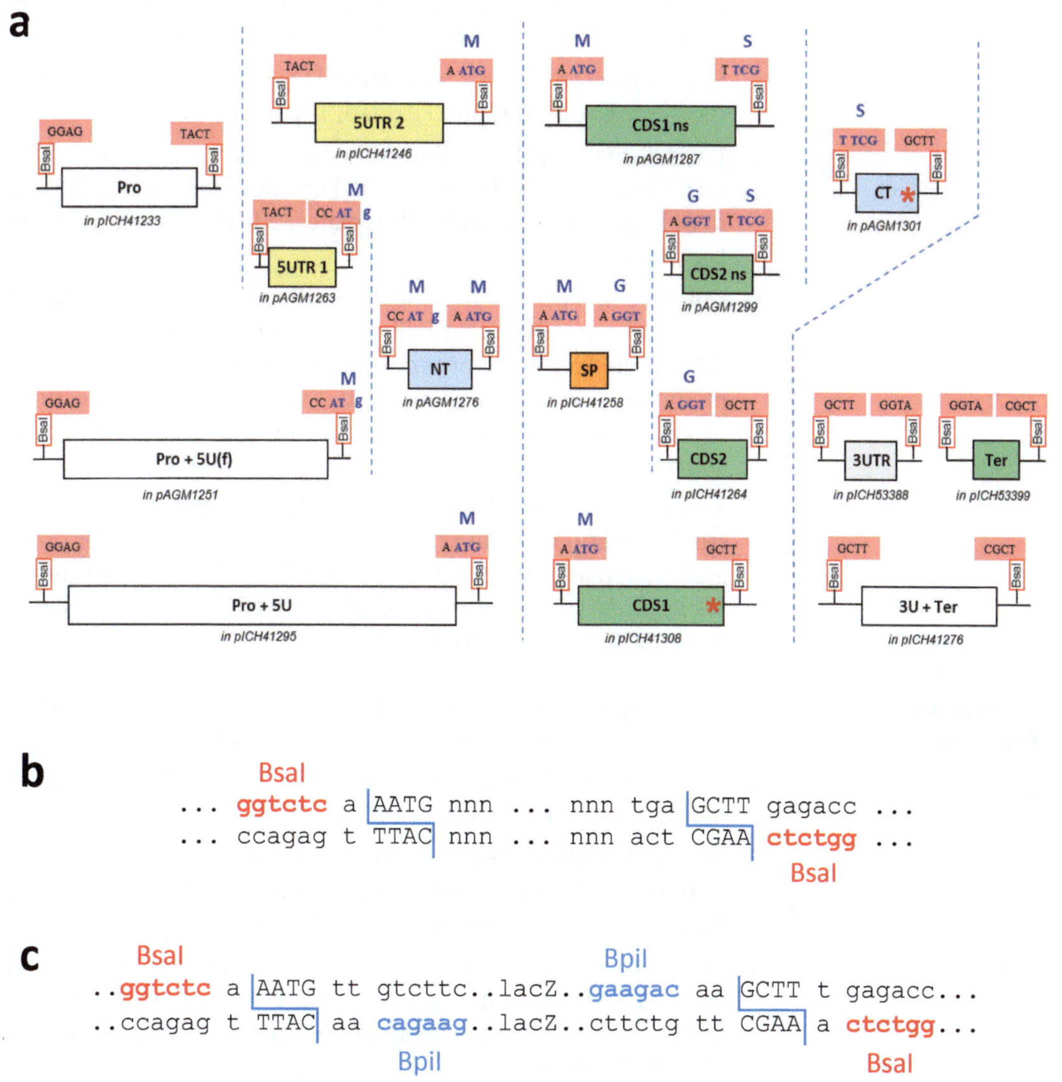

Fig. 1 Standard MoClo module types and cloning vectors. (**a**) Standard MoClo module types. Each module type has a special cloning vector, indicated below each module type. Pro: promoter, 5U/5UTR: 5′ untranslated leader, 3U/3UTR: 3′ untranslated sequence, NT: N-terminal fusion tag or sequence, CT: C-terminal fusion tag or sequence, SP: signal peptide, CDS: coding sequence, Ter: terminator. (**b**) Sequence detail of the CDS1 module (**b**) and its cloning vector (**c**) are given as examples

2 Materials

2.1 Cloning

1. 10 U/μL *Bsa*I (NEB, New England Biolabs Inc., Ipswich, MA, USA) (*see* **Note 2**).
2. 10 U/μL *Bpi*I (Thermo Fisher Scientific, Waltham, MA USA).
3. 3 U/μL T4 DNA Ligase or 20 U/μL T4 DNA Ligase (HC) (Promega Corporation, Madison, WI, USA), both

supplied with 10x ligation buffer (300 mM Tris–HCl pH 7.8, 100 mM $MgCl_2$, 100 mM DTT, 10 mM ATP).

4. Spectrophotometer for measuring DNA concentration such as the NanoDrop ND2000 (Peqlab, Erlangen).
5. Luria-Bertani (LB) medium: 1% (*w/v*) Bacto tryptone, 0.5% (*w/v*) yeast extract, 1% (*w/v*) NaCl in deionized water, adjusted to pH 7.0 with 5 N NaOH. For plates, 1.0% (*w/v*) agar is added (*see* **Note 3**).
6. kanamycin: filter-sterilized stock of 50 mg/mL in H_2O (stored in aliquots at −20 °C) are diluted 1:1000 (final concentration: 50 μg/mL) in an appropriate amount of medium after the medium has been autoclaved and cooled down. For spectinomycin, a stock of 40 mg/mL is made and is used at a final concentration of 100 μg/mL (dilution 1:400).
7. 5-bromo-4-chloro-3-indolyl-β-D-galactopyranoside (X-gal): stock solution of 20 mg/mL in dimethylformamide. For preparation of plates, the stock is diluted 1:500 (final concentration: 40 μg/mL) in an appropriate amount of LB agar after autoclaving/melting and cooling down.

2.2 Screening of Colonies

1. NucleoSpin® Plasmid Easy Pure (Macherey Nagel, Düren, Germany) for preparation of miniprep DNA.
2. DNA ladder: GeneRuler™ 1 kb DNA Ladder Plus is used as the marker for gel electrophoresis.
3. 50x TAE buffer: 242.0 g Tris, 57.1 mL of acetic acid, and 100 mL of 0.5 M EDTA, pH 8.0, in 1 L of deionized water. Running buffer for agarose gels is 1× TAE.
4. For preparation of gels for electrophoresis, agarose (0.7–1.5%) in 1× TAE is melted in a microwave oven and one drop of a 0.025% (w/v) ethidium bromide solution is added per 100 mL of melted agarose.
5. Gels are checked visually using a Fusion FX transilluminator (Vilber Lourmat, Eberhardzell, Germany), and pictures are taken by using the gel analysis software included.
6. DNA maps of plasmids are made by using the Vector NTIsoftware (Invitrogen).

3 Methods

3.1 Defining the Part Type

Cloning of standard parts requires to first define the part (or level 0 module) type. Parts can consist of promoters, coding sequences and terminators (Fig. 1a). Each part type is flanked by two BsaI sites in opposite orientations, as shown in Fig. 1b. The 4-nucleotide sequence of the cleavage sites, that we called fusion sites, is defined for each part type. For example, fusion sites for coding sequences are flanked by AATG at the start codon and GCTT (an arbitrary but fixed sequence) after the stop codon. Parts that a user may want to fuse to other sequences at its C-terminal end need to be cloned without a stop codon. In that case, the fusion site at the end of the sequence is defined as TTCG, with TCG coding for a serine (Fig. 1a). All possible part types are shown in Fig. 1a. For each part type, a specific cloning vector is available with fusion sites compatible with the part type. Alternatively a universal level 0 cloning vector pAGM9121 can be used for generation of parts with nonstandard sites that are sometimes needed for special applications. All cloning vectors are available at Addgene (https://www.addgene.org/cloning/moclo/marillonnet/).

Summary of actions:

1. Determine the sequence of the two fusion sites flanking the part

3.2 Domestication of the Part and Addition of Fusion Sites

Domestication consists of removing internal restriction sites for the type IIS enzymes that are used with a modular cloning system: BsaI, BpiI, and, optionally, BsmBI for the MoClo system. This step is first performed in silico. Removal of internal sites in coding sequences can always be done with a silent mutation. In contrast, removal of sites in noncoding sequences is more problematic since it is generally not known whether the site overlaps with an enhancer sequence important for promoter function. Therefore, after domesticating promoter sequences, it will be necessary to validate experimentally in the host organism that the promoter is working as intended. As an example for part domestication, we use here the Arabidopsis TT12 gene (GenBank accession NM_115765). This gene contains one BsaI site, three BpiI sites, and one BsmBI site (Fig. 2a). Therefore, five silent mutations are introduced to remove these sites (shown in green in Fig. 3). The next step consists of adding sequences for the standard fusion sites that will flank the part. Since this module will be a coding sequence with a stop codon (CDS1), we add one A before the start codon (resulting in sequence AATG) and the sequence GCTT after the stop codon (both fusion sites are highlighted in yellow in Fig. 3).

Summary of actions:

1. Introduce silent mutations in silico in the gene sequence.
2. Add to the sequence the two standard fusion sites flanking the part

3.3 Design of Primers

A pair of primers is made for each restriction site that needs to be removed (*see* **Note 4**). In addition, two primers are made at the beginning and the end of the coding sequence. At this step, we need to pay attention to how the amplified fragments will be

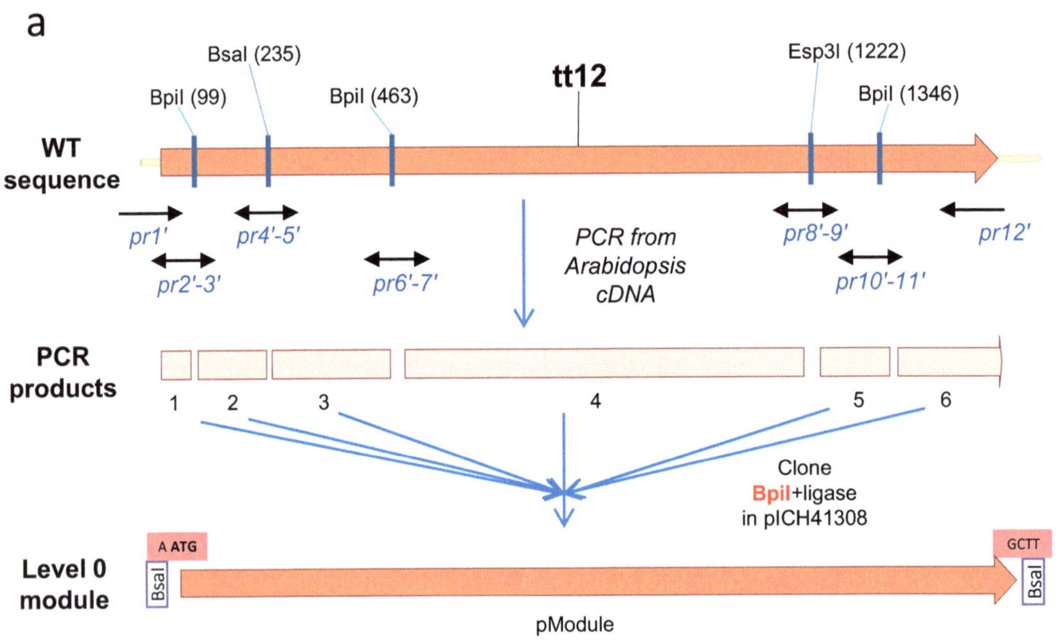

b

```
pr1'     tt gaagac aa aatg agctccacagagacatacgag
pr2'     tt gaagac aa cgat gaccgttcagttatctgagaatccg
pr3'     tt gaagac aa atcg ccagagatagaggagtttctc
pr4'     tt gaagac aa tacc actatagaggctccagagag
pr5'     tt gaagac aa ggta tctgttctgaattacatgctcagcttc
pr6'     tt gaagac aa gaac actgcagctgcaaggtgcaag
pr7'     tt gaagac aa gttc ctcacgttcctctactggtac
pr8'     tt gaagac aa cacg gctgctataacttctgcatcg
pr9'     tt gaagac aa cgtg tctgacctctttcctcttctcgc
pr10'    tt gaagac aa gtcc aatgacatagtacgtaacaagattcac
pr11'    tt gaagac aa ggac ttcctattggctgtgtccttggc
pr12'    tt gaagac aa aagc ttaaacacctgcgttagccatctcttg
```

Fig. 2 Cloning the Arabidopsis TT12 gene as an example of a CDS1 module. (**a**) The WT sequence contains five type IIS restriction sites that need to be removed. This requires the design of 12 primers (pr1′ to pr12′). Six PCR fragments are amplified and cloned using BpiI and ligase in level 0 cloning vector pICH41308. (**b**) sequence of the primers

assembled. Since assembly of the PCR fragments will be done using Golden Gate cloning, we need to define four-nucleotide sequences that will be used as fusion sites for assembly of the amplified gene fragments. These are nonstandard sites as they will be used only for assembly of the PCR products. In addition, we will also use the two standard fusion sites flanking the module (in this case AATG and

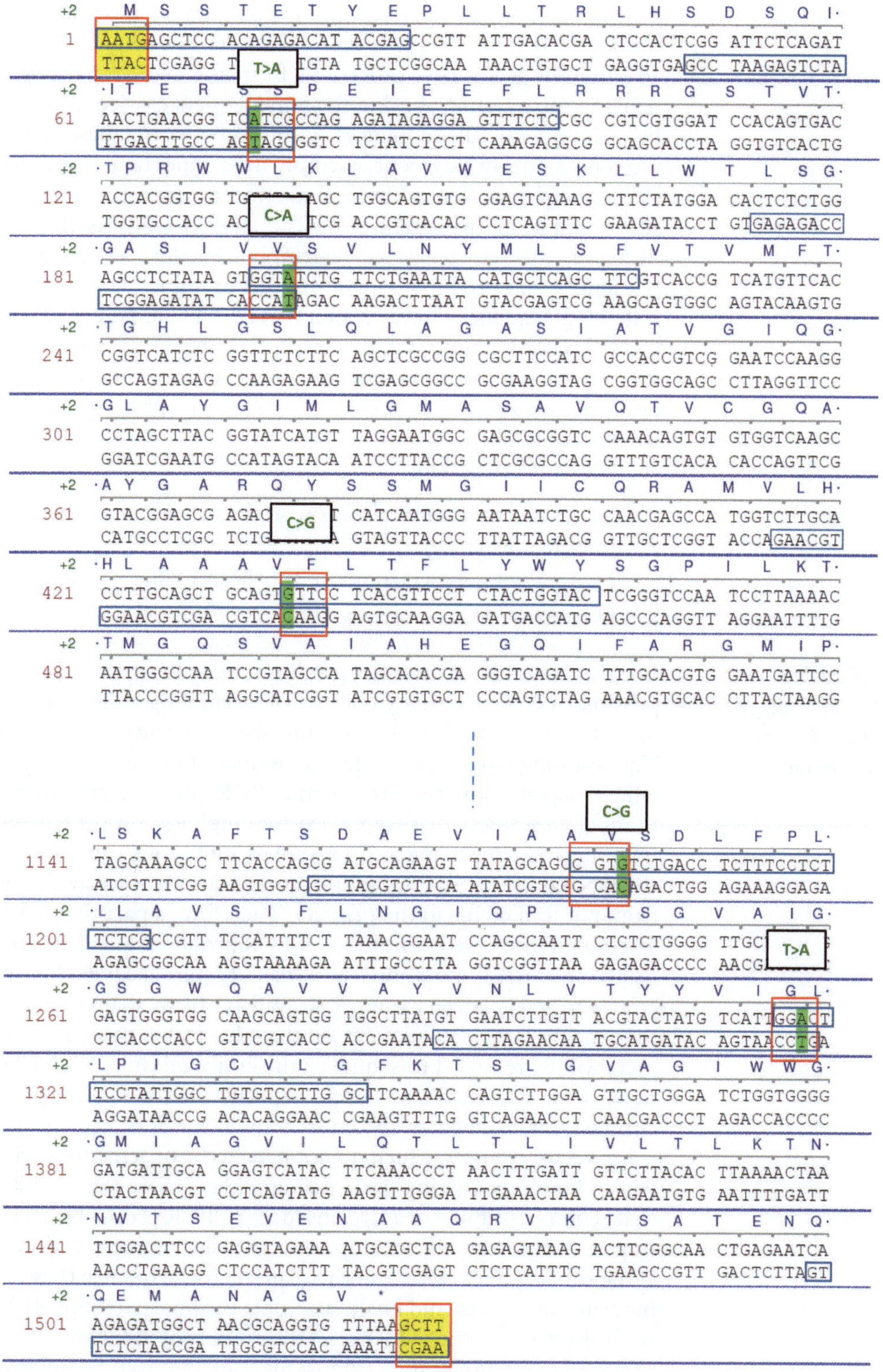

Fig. 3 Domestication of the TT12 gene. Nucleotide substitutions made to eliminate the type IIS enzymes restriction sites are highlighted in green. The 4-bp sequences chosen as fusion sites are boxed in red. Standard fusion sites are highlighted in yellow. Primer sequences selected for amplification of the fragments are boxed in black

GCTT). Therefore, for each introduced mutation, we define a 4-nucleotide sequence overlapping, or located near, each silent mutation. It is important to select 4-nucleotide sequences that are all different from each other and from the standard fusion sites flanking the module, as using the same sequence twice would result in deletion of all sequences located between these identical sequences during fragment assembly. It is important as well to not select a sequence that is identical to another sequence in the opposite orientation (for example GCTC and GAGC) as this would decrease the efficiency of assembly. Finally, all palindromic sequences should be avoided. In the example with TT12, the following sequences were chosen: AATG, ATCG, GGTA, GTTC, CGTC, GGAC, and GCTT (Fig. 3, red boxes). Primers are then defined starting with the sequence of each fusion site at the 5′ end (*see* **Note 5**). For the example of TT12, the 12 primers selected are shown in Fig. 3.

Summary of actions:

1. Select the nonstandard fusion sites that will be used for PCR fragment assembly.
2. Design primers starting at all fusion sites, standard and nonstandard

3.4 Defining a Cloning Strategy and Final Design of the Primer

For the assembly of the PCR products, two alternative options are possible. The first strategy consists of assembling all PCR products in a cloning vector and sequencing the resulting clone (Fig. 4a). The resulting clone may contain a number of unwanted mutations, which depends on the size of the PCR product and from the number of primers that were used for amplification (of the several fragments needed). Indeed, unwanted mutations can come from the DNA polymerase during PCR amplification, but may also come from random errors in the primers. Therefore, a part that has been assembled from a larger number of PCR products will likely contain more unwanted mutations and may require sequencing more clones to get a correct one. A second problem that occurs when sequencing large modules (more than 1.2 kb) is that internal primers will be required to sequence the entire module. Such primers need to be designed and ordered specifically for this purpose. In contrast, for small parts, the entire sequence will be easily obtained using vector primers, which do not need to be ordered for each new part. Therefore, cloning parts that are large (more than 1.2 kb) or that require assembly of a large number of PCR products in a single step is not a very efficient option. The alternative solution is to clone subfragments in an intermediate vector, sequence these intermediate constructs, and finally assemble the final part (Fig. 4b). For both alternatives, the sequence of the primers used for amplification will need to be different, as extensions containing a type IIS enzyme restriction site (for BsaI or BpiI) need to be added to the 5′ end of

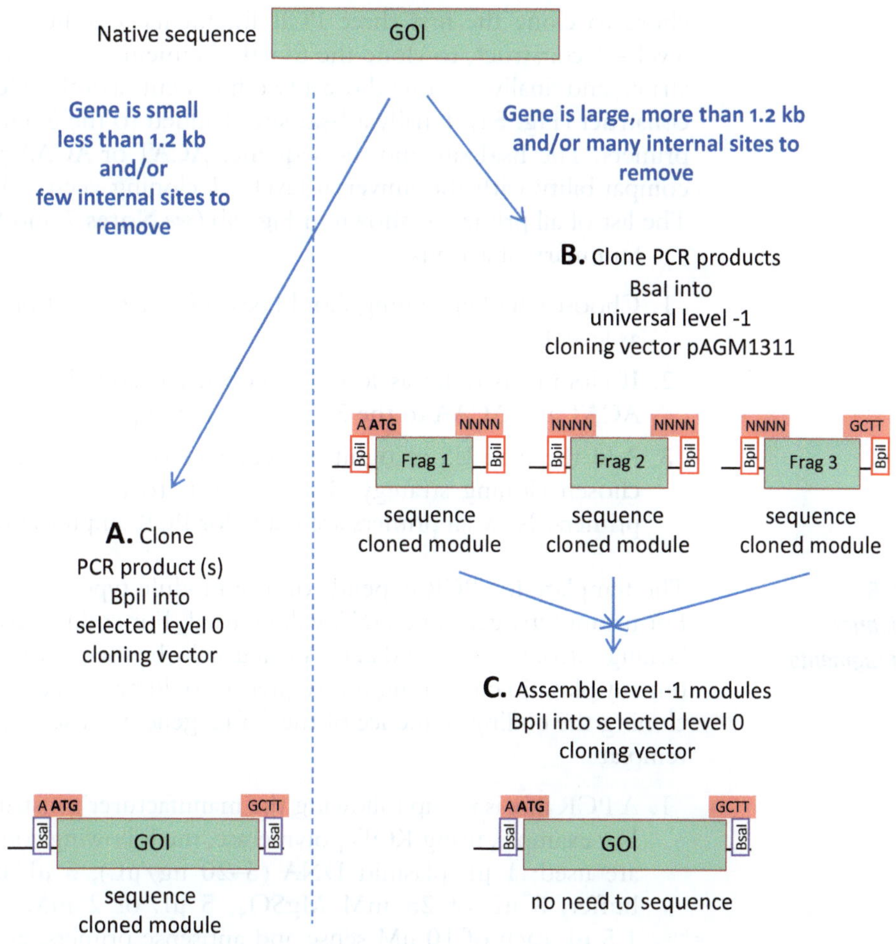

Fig. 4 Alternative strategies for cloning level 0 modules

the primers for providing compatibility to the cloning vector. Choosing the alternative of cloning all PCR products in a level 0 cloning vector requires adding a BpiI site at the 5′ end of the primers. With the example of TT12, the final sequence of the primers is shown in Fig. 2b. These primers provide compatibility with the level 0 cloning vector for CDS1 shown in Fig. 1c (*see* **Note 6**).

In the Case of TT12, it is, however, preferable to first clone several subfragments as intermediate level −1 constructs. Here again, several options are possible. The user could clone all fragments individually and sequence them. However, a more preferable alternative consists of subcloning groups of PCR products together to reduce the number of constructs that will need to be made and sequenced. The choice and number of which fragments should be cloned together can be determined by adding the sequence ACAA and ACAT at the 5′ and 3′ end of the fragments to provide compatibility to the universal level −1 cloning vector. For TT12 we chose to clone the first three PCR fragments together in a first level −1 construct, to clone the fourth fragment as a second construct, and finally to clone the last two fragments as a third level −1 construct (Fig. 5a). Finally a BsaI site is added to the 5′ end of all primers. The BsaI site and the sequence ACAT or ACAA provide compatibility with the universal level −1 cloning vector (Fig. 6). The list of all primers is shown in Fig. 5b (*see* **Notes 7** and **8**).

Summary of actions:

1. Choose whether cloning should be done via level −1 or directly as level 0.
2. If cloning is done as level −1 construct, add the sequences ACAT and ACAA to the 5′ end of selected primers.
3. Add the type IIS recognition sequence corresponding to the chosen cloning strategy (BsaI or BpiI) to the 5′ end of the primers. Now all primers are ready for PCR amplification.

3.5 PCR Amplification of the Fragments

The template for PCR depends on the module type that is made. For promoters, genomic DNA will be used. For coding sequences lacking introns, first strand cDNA will usually be used as a template, but any plasmid containing this sequence could be used as well. For cloning the coding sequence of the TT12 gene, we use cDNA as a template.

1. A PCR mix is set up following the manufacturer's instructions. For example, using KOD polymerase, the following conditions are used: 1 μL plasmid DNA (5–20 ng/μL), 5 μL of 10× buffer, 3 μL of 25 mM $MgSO_4$, 5 μL of 2 mM dNTPs, 1.5 μL each of 10 μM sense and antisense primers, and 1 μL of KOD Hot Start DNA polymerase (10 U/μL, final concentration 0.02 U/μL) in a total reaction volume of 50 μL.
2. PCR is performed using the following cycling conditions: (1) incubation at 95 °C for 2 min for polymerase activation, (2) denaturation at 95 °C for 20 s, (3) annealing at 58 °C for 10 s (the temperature for the annealing step can be adjusted for specific primers, but the temperature of 58 °C usually works well for most primers), (4) extension at 70 °C, the duration depends on the length of the expected fragment (from 10 s/kb for fragments smaller than 500 bp up to 25 s/kb for fragments larger than 3 kb, see the manufacturer's instructions); **steps 2–4** are repeated 35 times and are followed by a final extension step at 70 °C for 20 s–2 min (depending on fragment length). The reaction is then kept at 12 °C until taken out of the thermal cycler.
3. Of the PCR product obtained, 5 μL is analyzed by gel electrophoresis to make sure that a product of the correct size has been amplified.

a

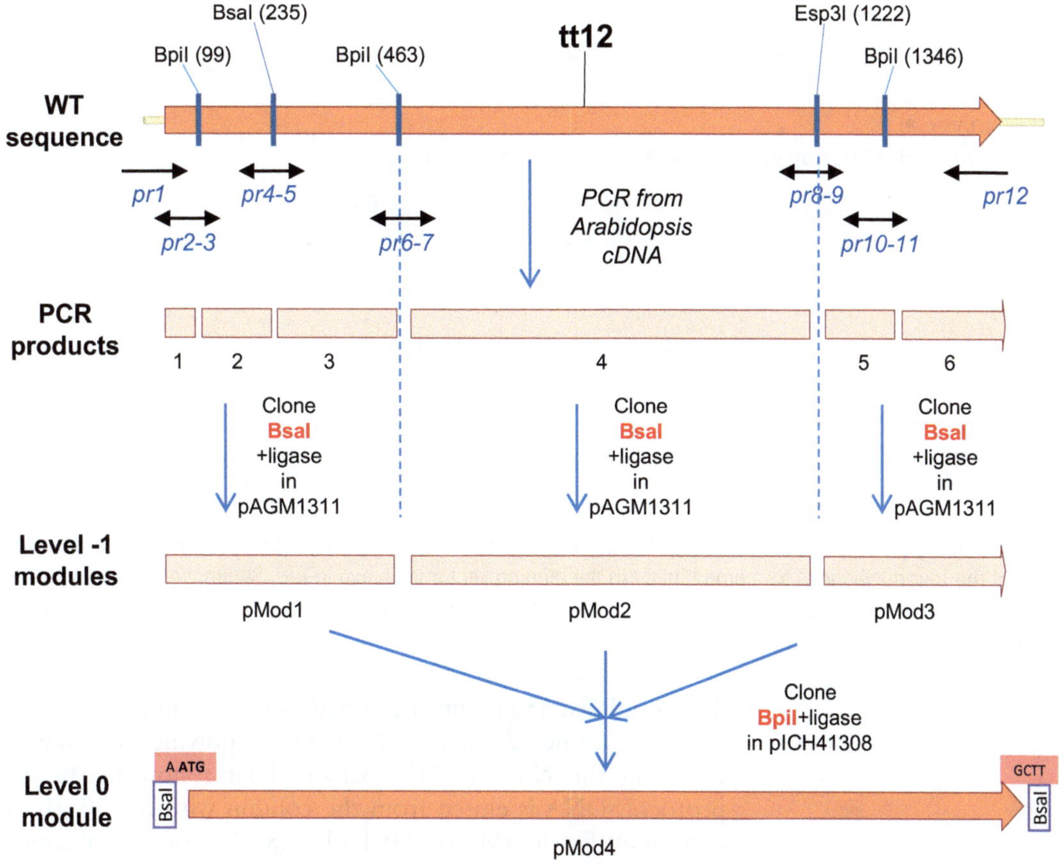

b

pr1	tt	ggtctc	a	ACAT	aatg	agctccacagagacatacgag
pr2	tt	ggtctc	a		cgat	gaccgttcagttatctgagaatccg
pr3	tt	ggtctc	a		atcg	ccagagatagaggagtttctc
pr4	tt	ggtctc	a		tacc	actatagaggctccagagag
pr5	tt	ggtctc	a		ggta	tctgttctgaattacatgctcagcttc
pr6	tt	ggtctc	a	ACAA	gaac	actgcagctgcaaggtgcaag
pr7	tt	ggtctc	a	ACAT	gttc	ctcacgttcctctactggtac
pr8	tt	ggtctc	a	ACAA	cacg	gctgctataacttctgcatcg
pr9	tt	ggtctc	a	ACAT	cgtg	tctgacctctttcctcttctcgc
pr10	tt	ggtctc	a		gtcc	aatgacatagtacgtaacaagattcac
pr11	tt	ggtctc	a		ggac	ttcctattggctgtgtccttggc
pr12	tt	ggtctc	a	ACAA	aagc	ttaaacacctgcgttagccatctcttg

Fig. 5 Cloning the Arabidopsis TT12 gene via level -1 cloning as an example for cloning of large modules. (**a**) 6 PCR fragments amplified from WT cDNA are cloned by groups of 3, 1, and 2 PCR products in the universal level -1 cloning vector pAGM1311. These clones are sequenced and then assembled in the level 0 cloning vector pICH41308. (**b**) Sequence of the primers used in (**a**)

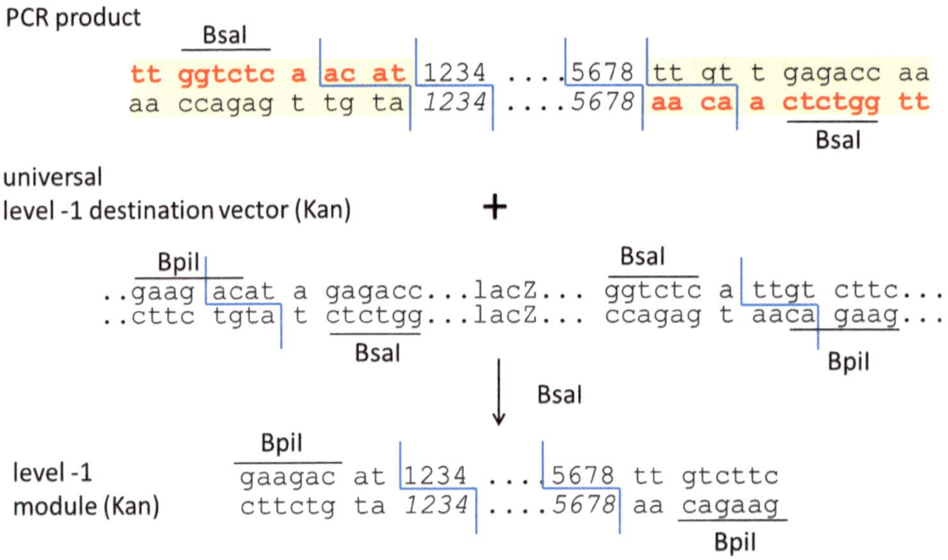

Fig. 6 Cloning of level -1 modules using the universal level -1 cloning vector. The sequence of the BsaI site and of the four nucleotides for compatibility to the cloning vector is shown in red. Sequences 1234 and 5678 correspond to the fusion sites specific for each module type, such as AATG and GCTT for CDS1 modules as an example

4. The amplified fragment is purified from remaining primers, potential primer dimers, and remaining polymerase enzyme by using the NucleoSpin® Extract II kit following the kit protocol. DNA is eluted from the column with 30–50 μL of elution buffer (5 mM Tris–HCl, pH 8.5). In case several bands were amplified rather than only the expected fragment, the same kit can also be used to cut and extract the appropriate DNA fragment from an agarose gel.

3.6 Cloning of Level −1 Constructs

1. Set up a restriction–ligation by pipetting into a tube 20 fmol (approximately 50 ng; *see* **Note 9**) of each purified PCR product and of the level −1 vector pAGM1311, 1.5 μL 10× ligation buffer, 5 U (0.5 μL) of BsaI, and 3 U (1 μL) of ligase (final volume of 15 μL).
2. Incubate the restriction–ligation mix in a thermal cycler for 1–4 h at 37 °C.
3. Restriction–ligation is followed by a digestion step (5 min at 50 °C) and then by heat inactivation for 10 min at 80 °C. The final incubation step at 80 °C is very important and is needed to inactivate the ligase at the end of the restriction–ligation. Omitting this step would lead to religation of some of the insert and plasmid backbone fragments when the reaction vessel is taken out of the thermal cycler, and would lead to a higher proportion of colonies containing incorrect constructs.

4. Transform the entire ligation into chemically compe- tentDH10B cells (*see* **Note 10**).
5. Thaw frozen chemically competentcells (50 μL per tube) on ice.
6. Add the entire ligation to the cells, and incubate on ice for 30 min.
7. Incubate for 90 s at 42 °C in a water bath.
8. Let the cells recover on ice for 2 min.
9. Add 0.5 mL of LB medium to the cells and incubate the tube at 37 °C in a shaker incubator (150 rpm) for 45 min to 1 h.
10. After incubation, plate 10–30 μL of the transformation on LB agar plates containing kanamycin and X-gal.
11. Incubate the plates overnight at 37 °C. Many white and very few blue colonies should be obtained.
12. Pick a few white colonies from the plate and transfer to liquid LB medium containing Kanamycin.
13. Grow bacterial culture over night at 37 °C and use for isolation of plasmid DNA.
14. Check plasmids by restriction digest using *Bpi*I (*see* **Note 11**).
15. Sequence the clones with vector primers moclof (agcgag- gaagcggaagagcg) and moclor (gccacctgacgtctaagaaacc) if clon- ing was done in the MoClo level −1 vector pAGM1311 (*see* **Notes 12** and **13**).

3.7 Assembly of Level 0 Modules

The Golden Gate reaction is virtually the same as described in Subheading 3.6 except that the enzyme BpiI is used instead of BsaI.

1. Set up a restriction–ligation by pipetting into a tube 20 fmol of each level −1 module and of the selected level 0 cloning vector (in the present example pICH41308), 1.5 μL 10× ligation buffer, 5 U (0.5 μL) of BpiI, and 3 U (1 μL) of ligase (final volume of 15 μL).
2. Incubate the restriction–ligation mix in a thermal cycler. For assembly of two to four level −1 modules, incubation for 1–4 h at 37 °C is sufficient. If more modules are ligated together, ligation is done in a volume of 25 μL (2.5 μL 10× ligation buffer, 1 μL of 10 U/μL BpiI, and 1.5 μL of 3 U/μL ligase) and a cycling program is used as following: 2 min at 37 °C followed by 5 min at 16 °C, both repeated 50 times.
3. Restriction–ligation is followed by a digestion step (5 min at 37 °C) and then by heat inactivation for 10 min at 80 °C.
4. Transformation of the reaction mix into *E. coli*. Cells are spread on LB plates containing spectinomycin and X-gal. Colonies containing correct constructs should be white, while colonies

containing uncut or religated vector should be blue. Plasmids should be isolated from a few colonies (usually two colonies are enough) and should be analyzed by restriction digest with BsaI. The final clones can be sequenced, but since PCR was not involved, this is not absolutely required.

Level 0 modules are then ready for assembly into level 1 constructs (usually containing complete transcription units) and then to level 2 [22], M or P [23] for assembly of multigene constructs.

4 Notes

1. The most commonly used type IIS enzymes used are BsaI, BpiI, and BsmBI or their isoschizomers. All of these have a 6 bp recognition site sequence and have a 4 nt cleavage site 1 or 2 bp away from the recognition site. SapI (isoschizomer, LguI) and AarI have a 7 bp recognition site sequence, which can be useful, as these enzymes will cleave DNA on average every 16 kb (for DNA with an average GC content of 50%) versus 4 kb for 6 bp cutters.
2. BsaI is sold by NEB, but isoschizomers of BsaI (with the same recognition site sequence) are also available from other companies, as for example Eco31I from Thermo Fisher Scientific. These enzymes do not necessarily have optimal cleavage efficiency in the same buffer. With the conditions that we are using, we have obtained better cloning efficiency with BsaI than with Eco31I, but it is possible that other restriction ligation buffers may work well for Eco31I as well. NEB also has several variants of BsaI, including BsaI-HF (High Fidelity) and BsaI-HFv2 (version 2). BsaI-HF did not work for us as well as regular BsaI for Golden Gate cloning. Fortunately, NEB has now replaced this enzyme with an optimized version (BsaI-HFv2) that seems to work quite well, as it was reported that 24 fragments could be assembled with good efficiency using this enzyme [3].
3. LB medium can be easily made using commercially available ready mixes. For example, we prepare liquid LB by adding 25 g of LB Broth High salt mix (from Duchefa) to 1 L of water. For LB solid medium, add 35 g of LB Agar High salt (Duchefa) to 1 L of water. Media is then autoclaved.
4. Two mutations that need to be introduced in close proximity can be eliminated with the same pair of primers. For example mutations that need to be introduced 40 bp apart may be eliminated by selecting a fusion site in the middle of the sequence between these sites. The two primers selected for amplification in different direction will each introduce a

mutation. It is important that make primers long enough that enough sequence is able to bind the template sequence downstream of the mutation that need to be introduced. Note that long primers up to 100 nt can easily be synthesized by companies. We usually select for the lower price available (no other purification other than salt free). Such primers will work just fine.

5. We usually design primers containing 12 G or C and with a length of at least 20 nt. For very AT-rich sequences, a longer primer might work with fewer G or C, but with a minimum of at least nine G or C. Alternatively, the user can use one of many online programs to select a primer with a desired melting temperature. It is important to calculate the T_m by considering only the sequence located downstream of the last introduced mutation, since what counts is annealing of the primer to its nonmutated template at the first amplification cycle.
6. Cloning a part with nonstandard fusion sites can be done using the universal level 0 vector pAGM9121. To use this vector sequences CTCA and CTCG need to be added to the primers that provide compatibility with the vector. In the example with TT12, prl' should be tt gaagac aa CTCA aatg agctccacagagacatacgag and primer 12′ should be tt gaagac aa CTCG aagc ttaaacacctgcgttagccatctcttg. All other primers should be as shown in Fig. 2b.
7. Cloning a part with nonstandard fusion sites using the level −1 universal cloning vector and then the level 0 universal cloning vector is possible. To do that, sequences CTCA and CTCG need to be added to the primers that provide compatibility with the universal level 0 vector. In the example with TT12, prl should be tt ggtctc a ACAT CTCA aatg agctccacagagacatacgag and primer 12 should be tt ggtctc a ACAA CTCG aagc ttaaacacctgcgttagccatctcttg. All other primers should be as shown in Fig. 5b.
8. When cloning PCR fragments in the level −1 universal vector, the fusion sites ACAT and TTGT (complementary of ACAA) will be used for cloning. It is important to check that the nonstandard fusion sites used for assembling the PCR fragments are all different from ACAT and TTGT.
9. In practice, if all module plasmids and the vector have approximately the same size (4–5 kb), simply adding 50 ng of DNA of each module and of the vector will work relatively well. However, when plasmids with widely different sizes are used, calculating an equimolar amount should provide a higher cloning efficiency. We have previously recommended using 40 fmol of each fragment and of the vector for cloning. It, however, appears that using less DNA leads to higher cloning efficiency.

We therefore now recommend using 20 fmol for each insert and vector. The following formula (from the NEB catalog) can be used: 1 μg of a 1000 bp DNA fragment corresponds to 1.52 pmol. Therefore, the volume of DNA to pipet (in μL) to have 20 fmol is given by the equation: 20 (fmol) x size (bp) of the DNA fragment/(concentration (ng/μL) × 1520). Frequently, this formula will lead to volumes <1 μL with standard miniprep DNA. In these cases, we recommend diluting the DNA in order to avoid pipetting errors.

10. Commercially produced commercial cells can be used, but usually lab-made competent are sufficient.
11. White colonies may contain inserts smaller than the expected size. This is because PCR amplification often leads to at least some amount of primer dimer production. Primer dimers should be removed by column purification of the products. However, it is frequently the case that some primer dimers remain in the purified products. The dimers are flanked by BsaI restriction sites and can therefore be cloned as efficiently as the full length products. Digestion of the clones with BpiI allows for the identification of these incorrect clones, and allows for sending for sequencing only clones with full-length inserts.
12. Usually, sequencing one clone may be enough. However, when amplifying sequences from DNA or cDNA of an organism of interest, it is quite common to find sequence mismatches that differ from the expected sequence. If the polymorphism leads to an amino-acid change, it is important to know whether we are dealing with a polymorphism or a PCR mutation. Therefore, a second clone needs to be sequenced. If the same polymorphism is seen twice, one can assume that we are dealing with a polymorphism and not a PCR mutation.
13. A second problem might be encountered when sequencing level −1 clones (or alternatively level 0 clones) that are assembled from more than one fragment, and when the gene of interest is part of a gene family. It is possible that more than one target sequence is amplified at the same time if the primers bind to more than one gene. In that case, chimeric sequences derived from several genes (in several fragments in a successive order) might be cloned together. If that is the case, it might be useful to go back one step in the design of the primers, to clone all fragments separately instead of by groups of several fragments. After sequencing, the correct fragments can then be selectively chosen and further assembled at the next step of cloning (level 0 cloning).

References

1. Engler C, Gruetzner R, Kandzia R et al (2009) Golden gate shuffling: a one-pot DNA shuffling method based on type IIs restriction enzymes. PLoS One 4:e5553
2. Engler C, Kandzia R, Marillonnet S (2008) A one pot, one step, precision cloning method with high throughput capability. PLoS One 3: e3647
3. Potapov V, Ong JL, Kucera RB et al (2018) Comprehensive profiling of four base overhang ligation fidelity by T4 DNA ligase and application to DNA assembly. ACS Synth Biol 7:2665–2674
4. Agmon N, Mitchell LA, Cai Y et al (2015) Yeast Golden Gate (yGG) for the efficient assembly of S. cerevisiae transcription units. ACS Synth Biol 4:853–859
5. Binder A, Lambert J, Morbitzer R et al (2014) A modular plasmid assembly kit for multigene expression, gene silencing and silencing rescue in plants. PLoS One 9:e88218
6. Crozet P, Navarro FJ, Willmund F et al (2018) Birth of a photosynthetic chassis: a MoClo toolkit enabling synthetic biology in the microalga Chlamydomonas reinhardtii. ACS Synth Biol 7:2074–2086
7. Hernanz-Koers M, Gandia M, Garrigues S et al (2018) FungalBraid: a GoldenBraid-based modular cloning platform for the assembly and exchange of DNA elements tailored to fungal synthetic biology. Fungal Genet Biol 116:51–61
8. Iverson SV, Haddock TL, Beal J et al (2016) CIDAR MoClo: improved MoClo assembly standard and new E. coli part library enable rapid combinatorial design for synthetic and traditional biology. ACS Synth Biol 5:99–103
9. Kowarschik K, Hoehenwarter W, Marillonnet S et al (2018) UbiGate: a synthetic biology toolbox to analyse ubiquitination. New Phytol 217:1749–1763
10. Lampropoulos A, Sutikovic Z, Wenzl C et al (2013) GreenGate—a novel, versatile, and efficient cloning system for plant transgenesis. PLoS One 8:e83043
11. Lee ME, Deloache WC, Cervantes B et al (2015) A highly characterized yeast toolkit for modular, multipart assembly. ACS Synth Biol 4:975–986
12. Moore SJ, Lai HE, Kelwick RJ et al (2016) EcoFlex: a multifunctional MoClo kit for E. coli synthetic biology. ACS Synth Biol 5:1059–1069
13. Occhialini A, Piatek AA, Pfotenhauer AC et al (2019) MoChlo: a versatile, modular cloning toolbox for chloroplast biotechnology. Plant Physiol 179:943–957
14. Prielhofer R, Barrero JJ, Steuer S et al (2017) GoldenPiCS: a Golden Gate-derived modular cloning system for applied synthetic biology in the yeast Pichia pastoris. BMC Syst Biol 11:123
15. Sarrion-Perdigones A, Vazquez-Vilar M, Palaci J et al (2013) GoldenBraid 2.0: a comprehensive DNA assembly framework for plant synthetic biology. Plant Physiol 162:1618–1631
16. Terfruchte M, Joehnk B, Fajardo-Somera R et al (2014) Establishing a versatile Golden Gate cloning system for genetic engineering in fungi. Fungal Genet Biol 62:1–10
17. Vasudevan R, Gale GR, Schiavon AA et al (2019) CyanoGate: a modular cloning suite for engineering cyanobacteria based on the plant MoClo syntax. Plant Physiol 180 (1):39–55
18. Weber E, Engler C, Gruetzner R et al (2011) A modular cloning system for standardized assembly of multigene constructs. PLoS One 6:e16765
19. Werner S, Engler C, Weber E et al (2012) Fast track assembly of multigene constructs using Golden Gate cloning and the MoClo system. Bioengineered Bugs 3:38–43
20. Engler C, Youles M, Gruetzner R et al (2014) A golden gate modular cloning toolbox for plants. ACS Synth Biol 3:839–843
21. Gantner J, Ordon J, Ilse T et al (2018) Peripheral infrastructure vectors and an extended set of plant parts for the modular cloning system. PLoS One 13:e0197185
22. Marillonnet S, Werner S (2019) Assembly of complex pathways using type IIs restriction enzymes. Methods Mol Biol 1927:93–109
23. Marillonnet S, Werner S (2020) Assembly of multigene constructs using the modular cloning system MoClo. In: Chandran S, George K (eds) DNA cloning and assembly: methods and protocols. Springer, New York

Chapter 8

Assembly of Multigene Constructs Using the Modular Cloning System MoClo

Sylvestre Marillonnet and Stefan Werner

Abstract

Modular cloning systems that rely on type IIS enzymes for DNA assembly have many advantages for complex pathway engineering. These systems are simple to use, efficient, and allow users to assemble multigene constructs by performing a series of one-pot assembly steps, starting from libraries of cloned and sequenced parts. The efficiency of these systems also facilitates the generation of libraries of construct variants. We describe here a protocol for assembly of multigene constructs using the Modular Cloning system MoClo. Making constructs using the MoClo system requires users to first define the structure of the final construct to identify all basic parts and vectors required for the construction strategy. The assembly strategy is then defined following a set of standard rules. Multigene constructs are then assembled using a series of one-pot assembly steps with the set of identified parts and vectors.

Key words Modular cloning, Synthetic biology, Biological parts, DNA assembly, Multigene constructs

1 Introduction

For many years, constructs were made using protocols that required a succession of labor-intensive DNA manipulation steps that included, for example, DNA digestion using one or two restriction enzymes, partial or complete filling of DNA ends, dephosphorylation of vector ends, gel extraction, ligation, transformation of *E.coli* cells and screening of colonies. These protocols worked, and are still used today, but are relatively time-consuming and inefficient. Fortunately, they can now be replaced by more efficient strategies that include homology-based assembly methods and type IIS-based assembly methods. Type IIS-based methods rely on Golden Gate cloning, a method that allows assembly of multiple DNA fragments using a simple one-pot, one-step assembly reaction. Golden Gate cloning combines the use of a type IIS enzyme with a restriction–ligation reaction [3]. Type IIS enzymes, such as BsaI or BpiI, are characterized by the fact that the restriction

Sunil Chandran and Kevin W. George (eds.), *DNA Cloning and Assembly: Methods and Protocols*, Methods in Molecular Biology, vol. 2205, https://doi.org/10.1007/978-1-0716-0908-8_8, © Springer Science+Business Media, LLC, part of Springer Nature 2020

enzyme cleavage site lies outside of the enzyme DNA recognition site. This allows DNA digestion and ligation to be performed simultaneously as restriction sites for the type IIS enzyme used for cloning are eliminated during the cloning reaction. Golden Gate has been used as the base of several related but not identical modular cloning systems for plants, yeast, mammalian cells and fungi [1, 2, 6–9, 11–21]. These cloning systems differ by the specific type IIS enzymes used for cloning at the various assembly steps, by the sequences of the type IIS cleavage sites that are used for cleavage and assembly, or by the antibiotic selection markers present on the various vector backbones. We present here a protocol for the modular system MoClo that was first designed to be used in plants [19], but can easily be adapted for use in other organisms such as yeast and bacteria. The MoClo system is a hierarchical assembly system by which transcription units are assembled from a set of level 0 modules that contain basic genetic elements such as promoters, coding sequences and terminators. The resulting level 1 constructs are then assembled in multigene constructs using two alternative strategies: (i) assembly in level 2 constructs by successive addition of sets of up to 6 level 1 transcription units to a growing construct, or (ii) assembly of several intermediate level M multigene constructs, which are then assembled in a level P construct (Fig. 1) (*see* **Note 1**). This chapter describes how to make multigene constructs using level M and P constructs.

2 Materials

2.1 Cloning

1. 10 U/μL BsaI (NEB, New England Biolabs Inc., Ipswich, MA, USA) (*see* **Note 2**).
2. 10 U/μL BpiI (Thermo Fisher Scientific, Dreieich, Germany).
3. 3 U/μL T4 DNA Ligase or 20 U/μL T4 DNA Ligase (HC), both supplied with 10× ligation buffer (300 mM Tris-HCl pH 7.8, 100 mM $MgCl_2$, 100 mM DTT, 10 mM ATP).
4. A spectrophotometer for DNA quantification, for example the NanoDrop ND2000 (Peqlab, Erlangen).
5. Luria-Bertani (LB) medium: 1% (*w/v*) Bacto tryptone, 0.5% (*w/v*) yeast extract, 1% (*w/v*) NaCl in deionized water, adjusted to pH 7.0 with 5 N NaOH. For plates, 1.5% (*w/v*) agar is added.
6. Antibiotics carbenicillin (used instead of ampicillin) and kanamycin: filter-sterilized stocks of 50 mg/mL in H_2O (stored in aliquots at −20 °C) are diluted 1:1000 (final concentration: 50 μg/mL) in an appropriate amount of medium after the medium has been autoclaved and cooled down. For spectinomycin, a stock of 40 mg/mL is made and is used at a final concentration of 100 μg/mL (dilution 1:400).

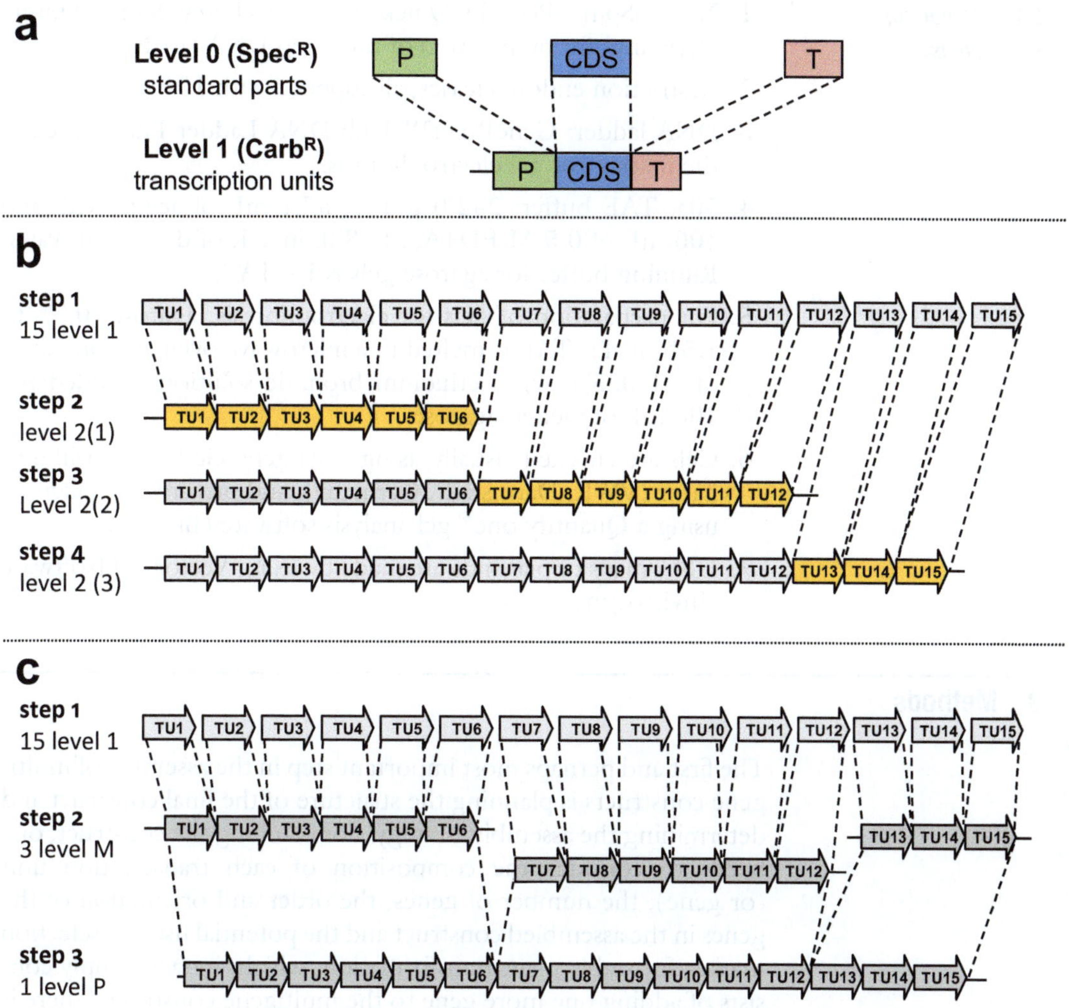

Fig. 1 Comparison of multigene assembly using level 2 versus level M and P. (**a**) level 1 constructs are assembled from standard parts (level 0 modules) such as promoters (P), coding sequences (CDS), and terminators (T) using a first Golden Gate cloning reaction, resulting in constructs containing a transcription unit (TU). To make a construct containing 15 TU (as an example), 15 level 1 constructs (represented as arrows in **b** and **c**) are assembled in parallel from selected level 0 modules. (**b**) assembly of multigene constructs using level 2 constructs. Up to six level 1 constructs are assembled at each step into a level 2 construct that grows larger at each step. For example, step 2 to 4 shows the successive addition of 6, 6 and 3 transcription units to a level 2 construct (transcription units cloned directly from level 1 constructs are shown in yellow). (**c**) assembly of multigene constructs using level M and P constructs. Groups of up to six transcription units are preassembled in level M constructs. The level M constructs are then assembled in a level P construct

7. 5-bromo-4-chloro-3-indolyl-β-D-galactopyranoside (X-gal): stock solution of 20 mg/mL in dimethylformamide. For preparation of plates, the stock is diluted 1:500 (final concentration: 40 μg/mL) in an appropriate amount of LB agar after autoclaving/melting and cooling down.

2.2 Screening of Colonies

1. NucleoSpin® Plasmid Quick Pure (Macherey-Nagel, Düren, Germany) for preparation of miniprep DNA (*see* **Note 3**).
2. Restriction endonucleases, all supplied with 10× buffer.
3. DNA ladder: GeneRuler™ 1 kb DNA Ladder Plus is used as the marker for gel electrophoresis.
4. 50× TAE buffer: 242.0 g Tris, 57.1 mL of acetic acid, and 100 mL of 0.5 M EDTA, pH 8.0, in 1 L of deionized water. Running buffer for agarose gels is 1× TAE.
5. For preparation of gels for electrophoresis, agarose (0.7% to 1.5%) in 1× TAE is melted in a microwave oven and one drop of a 0.025% (w/v) ethidium bromide solution is added per 100 mL of melted agarose.
6. Gels are checked visually using a Syngene GelVue transilluminator (VWR, Darmstadt, Germany), and pictures are taken by using a Quantity one® gel analysis software (Biorad).
7. DNA maps of plasmids are made by using Vector NTIsoftware (Invitrogen).

3 Methods

The first and perhaps most important step in the assembly of multigene constructs is planning the structure of the final construct and determining the assembly strategy. For planning the construct, one needs to consider the composition of each transcription unit (or gene), the number of genes, the order and orientation of the genes in the assembled construct and the potential use of a selection marker for stable plant transformation, which, in fact, simply consists of adding one more gene to the multigene construct. There is no theoretical limit to the number of genes that can be assembled; however, the limit is set by the size and the stability of the plasmid which has to be cloned and maintained in *E. coli*. The process of selecting standard or individualized parts for the transcription units, the choice of destination vectors and their assembly into higher order constructs is described below.

3.1 Selection of Standard Parts and Vectors

3.1.1 Selection of Standard Parts

Assembly of multigene constructs starts with the assembly of transcription units (genes) from basic parts (Fig. 1a). MoClo basic parts, also called level 0 modules or standard parts, consist of a DNA fragment of interest flanked by two BsaI restriction sites in opposite orientations cloned in a plasmid with a spectinomycin resistance cassette (Fig. 2a). The various types of level 0 modules comprise, but are not limited to, promoters, 5'untranslated regions (5'UTRs), coding sequences, 3'UTRs, and terminators (*see* **Note 4**). The 4-bp sequences of the BsaI cleavage sites (the 'fusion sites') are specific for each type of module and define which parts can be fused together. For example, a promoter module (with

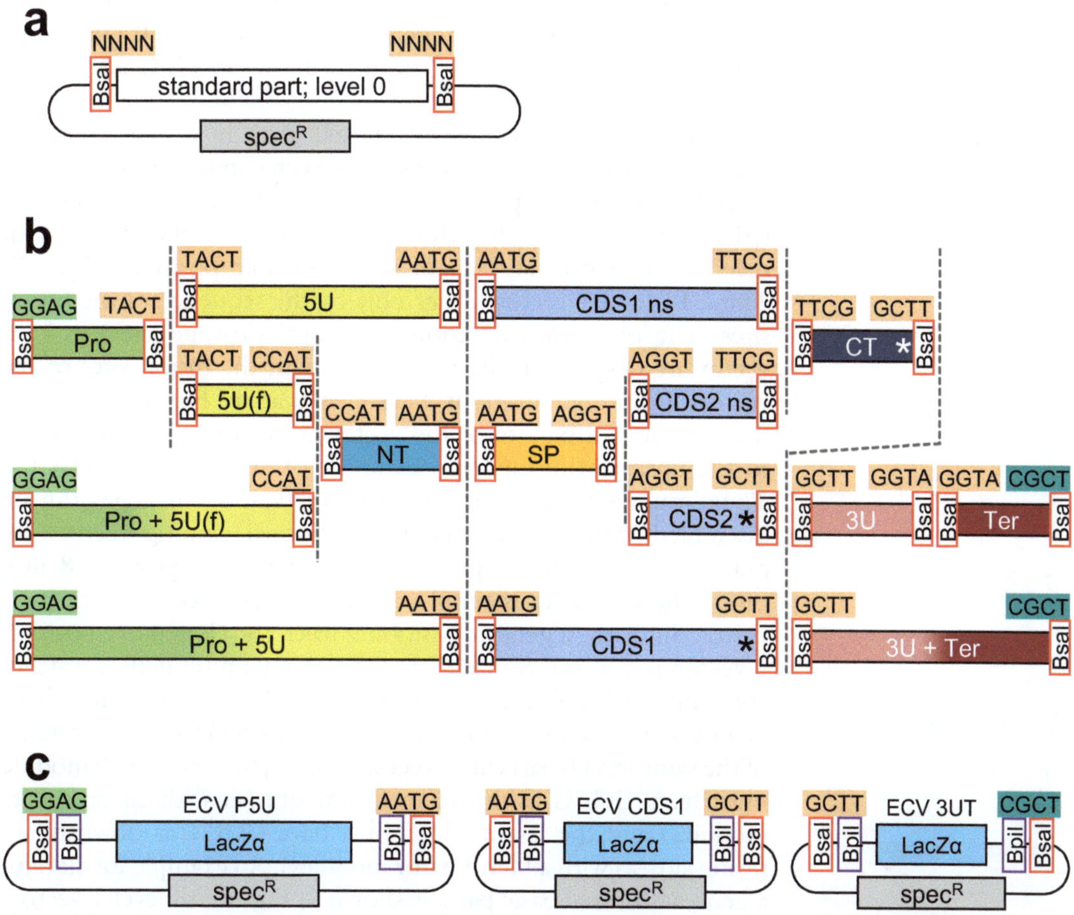

Fig. 2 Level 0 modules (standard parts) and cloning vectors. (**a**) General structure of a level 0 plasmid. The level 0 module is flanked by two inverted BsaI sites. **(b)** Different types of standard parts with their corresponding fusion sites. Promoter (Pro), 5′- and 3′-UTR (5U, 3U), N- and C-terminal tag (NT, CT), signal peptide (SP), coding sequence (CDS) with stop codon (*) or without (ns), terminator (Ter). **(c)** Level 0 cloning vectors. Only three vectors are shown as example, but vectors for each module type are available

TACT as 3′ fusion site) can be fused only to an untranslated leader sequence with the same fusion site at the 5′ end. The different types of standard parts and their fusion sites are shown in Fig. 2b.

Making a multigene construct first requires identification of all basic parts needed. Several parts may already be available; for example, the promoters or terminators may have already been cloned by the same or another laboratory. For example, a selection of 96 standard biological parts (including plant promoters and terminators, tags, reporter genes, selectable markers) can be obtained from Addgene (https://www.addgene.org/cloning/MoClo/) [4]. In addition, parts are being cloned and characterized by research labs using modular cloning systems [5]. Parts that are not already available can easily be cloned by users into corresponding level

0 cloning vectors (Fig. 2c; level 0 vectors and all other MoClo vectors are also available at Addgene). Cloning of level 0 modules is described in detail in Chap. 7 of this volume (*see* **Note 5**).

3.1.2 Selection of Level 1 Destination Vectors

The position and orientation of each gene in a final construct determine which level 1 destination vector must be chosen for assembly of a transcription unit. A total of 14 vectors are available (pL1F1-7, pL1R1-7) for cloning of genes in either forward or reverse orientation at each of seven possible positions (Fig. 3). These 14 level 1 vectors differ only by the sequence of the fusion sites. The two external fusion sites (BpiI cleavage sites) of each vector are designed to be compatible with the fusion sites of the vectors from the positions before and after. Furthermore, the downstream fusion site of the vector at position 7 is compatible to the upstream fusion site of the vector at position 1 (TGCC). Thus, multigene constructs with more than seven genes can be assembled without the need for additional cloning vectors. In practice, a gene that is planned to be cloned at position 8 in a multigene construct is simply cloned in a position 1 destination vector. Similarly, a gene at position 9 has to be cloned in a position 2 vector and so on. An example for making a multigene construct containing 18 genes is shown in Fig. 4. All level 1 vectors have identical internal fusion sites (BsaI cleavage sites) to allow assembly of the same level 0 parts in all vectors, that is, the first level 0 module must have GGAG as upstream fusion site (as with all promoter modules) and the last module must have CGCT as downstream fusion site (as with all terminator modules). An example for cloning a gene with three basic parts is shown in Fig. 5. All level 1 vectors are binary plasmids that can be used for *Agrobacterium* mediated transient expression in plants. These vectors are nevertheless suitable for making constructs for other organisms, as level 1 constructs are usually used only as intermediate constructs, and the only feature that is required for this purpose is replication and selection in *E.coli*. Other vectors (e.g., nonbinary vectors) could also be used as long as they contain the features outlined in Fig. 3 and an appropriate selectable marker (an ampicillin/carbenicillin resistance gene).

3.2 Assembly of Transcription Units

Once the level 0 modules and the destination vector are made or selected from previously made constructs, assembling the gene only requires pipetting all components into a reaction mix and incubating the reaction vessel in a thermal cycler.

1. Set up a restriction–ligation by pipetting into a tube 20 fmol (approximately 50 ng; *see* **Note 6**) of each level 0 module (*see* **Note 7**) and of the vector, 2 μL 10× ligation buffer, 10 U (1 μL) of BsaI, and either 3 U (1 μL) of ligase for assembly of two to four modules or 20 U (1 μL) HC ligase for assembly of more than four modules (final volume of 20 μL; *see* **Note 8**).

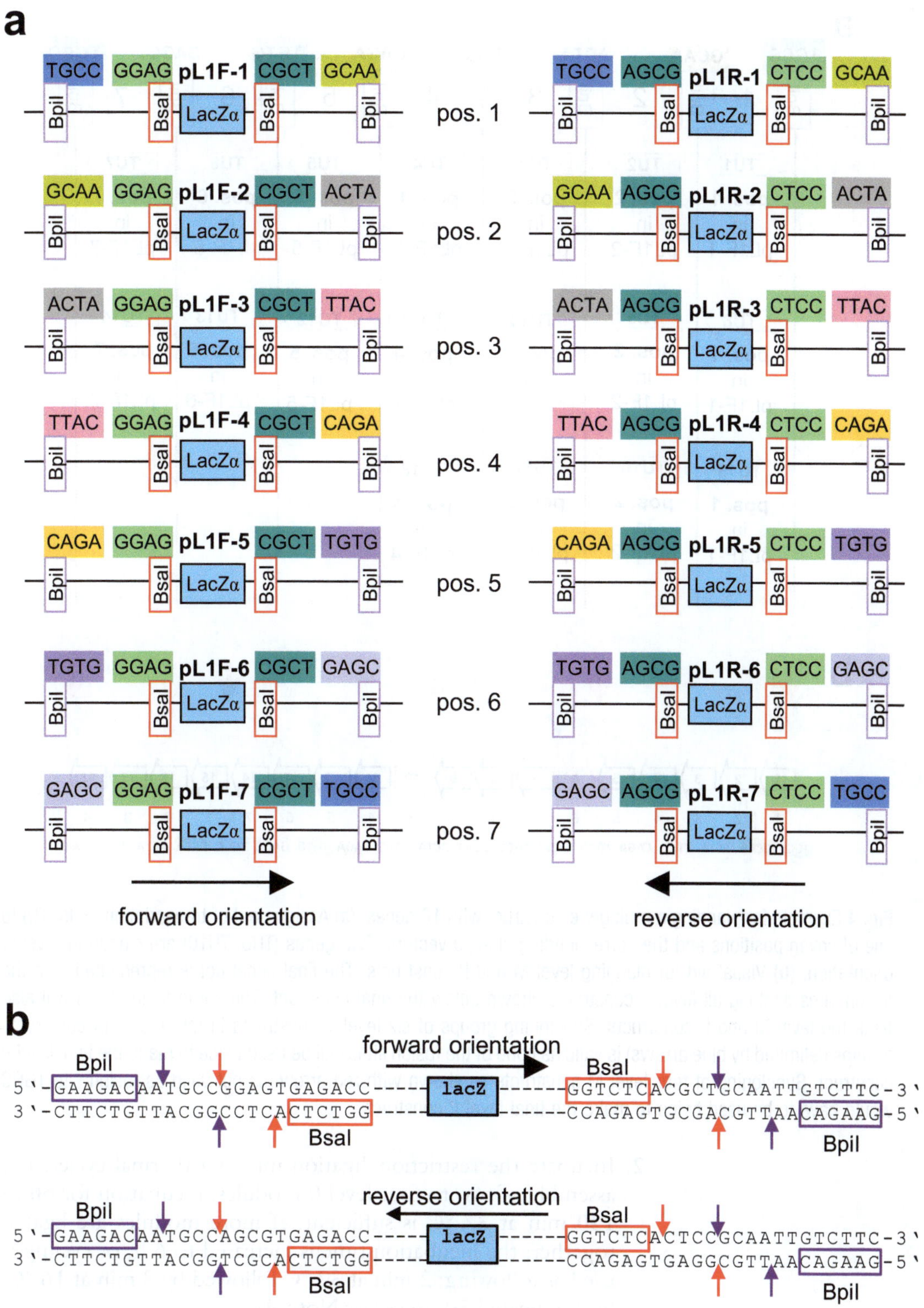

Fig. 3 Level 1 destination vectors. (**a**) Genes can be cloned at one of seven positions in forward (pL1F-vectors) or reverse (pL1R-vectors) orientation. (**b**) Sequence of the type IIS recognition (rectangles) and cleavage sites (arrows) of vectors pL1F-1 and pL1R-1

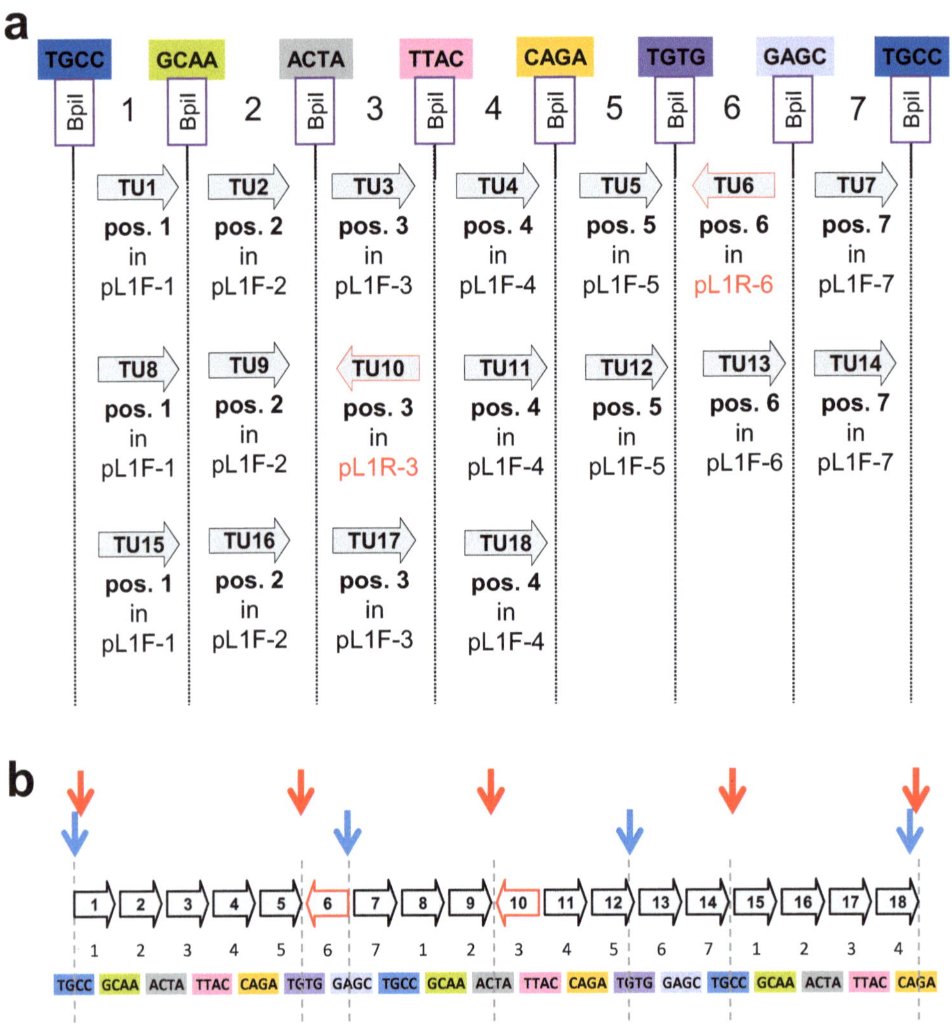

Fig. 4 Example for planning a multigene construct with 18 genes. (**a**) Assignment of transcription units (TU) to one of seven positions and the corresponding cloning vectors. Two genes (TU6, TU10) are cloned in reverse orientation. **(b)** Visual aid for planning level M and P constructs. The final construct is represented with the fusion sites flanking all level 1 constructs shown below the final construct. The same fusion sites will also flank the level M and P constructs. Subcloning groups of six level 1 constructs in three level M constructs (groups delimited by blue arrows) is valid as none of the fusion sites will be used twice to make the final level P construct. Subcloning of four level M constructs as shown with red arrows is not valid as fusion site TGCC would have to be used twice to make the final level P construct

2. Incubate the restriction–ligation mix in a thermal cycler. For assembly of two to four level 0 modules, incubation for 60 to 120 min at 37 °C is sufficient. If more modules are ligated together, the incubation time is increased to 6 h, or cycling is used as following: 2 min at 37 °C followed by 3 min at 16 °C, both repeated 50 times (*see* **Note 9**).

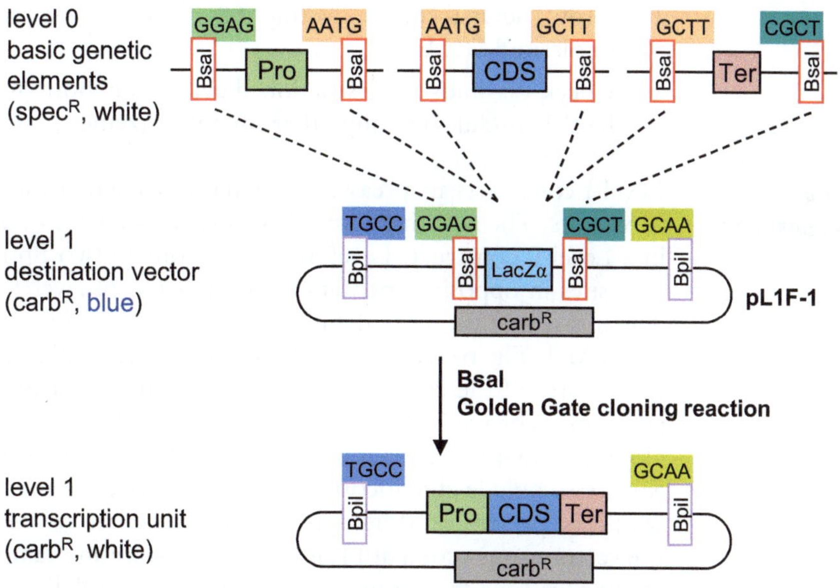

Fig. 5 Example of cloning a transcription unit consisting of three basic parts. Correct clones are identified by the antibiotic resistance marker ($carb^R$) and the color of the colonies on plates with X-gal (white)

3. Restriction–ligation is followed by a digestion step (5 min at 50 °C) and then by heat inactivation for 10 min at 80 °C. The final incubation step at 80 °C is very important and is needed to inactivate the ligase at the end of the restriction–ligation. Omitting this step would lead to religation of some of the insert and plasmid backbone fragments when the reaction vessel is taken out of the thermal cycler, and would lead to a higher proportion of colonies containing incorrect constructs.
4. Transform the entire ligation into chemically competent DH10B cells (*see* **Note 10**).
5. Thaw frozen chemically competent cells (100 μL per tube) on ice.
6. Add the entire ligation to the cells, and incubate on ice for 30 min.
7. Incubate for 90 s at 42 °C in a water bath.
8. Let the cells recover on ice for 5 min.
9. Add 1 mL of LB medium to the cells, and incubate the tube at 37 °C in a shaker incubator (150 rpm) for 45 min to 1 h.
10. After incubation, plate 25–100 μL of the transformation on LB agar plates containing carbenicillin and X-gal.
11. Incubate the plates overnight at 37 °C. Many white and very few blue colonies should be obtained.
12. Pick a few white colonies from the plate and transfer to liquid LB medium containing carbenicillin (*see* **Note 11** and **12**).

13. Grow bacterial culture overnight at 37 °C and use for isolation of plasmid DNA.
14. Check plasmids by restriction digest using BpiI (for MoClo level 1 modules) or any other suitable enzyme.

3.3 Assembly of Level M Constructs

Level 1 constructs are preassembled in groups of up to six into level M vectors. The resulting level M constructs will later be assembled in a Level P construct. Level M vectors contain two BpiI recognition sites in opposite orientation for insertion of level 1 modules. The upstream fusion site defined by the first BpiI site (e.g., TGCC for pLM-1; Fig. 6) is compatible with fusion sites flanking upstream fusion sites of the level 1 vectors, whereas the downstream fusion site consists of a universal sequence (GGGA). This design allows for cloning of two to six genes in the same vector (using more genes in this step would lead to incorrect clones because the same fusion site would be present on different modules). The last gene is fused to the vector using a compatible linker. We use here an example for the cloning of four transcription units in a final level P vector, with cloning of transcription units by groups of 2 in level M constructs (this small number of genes could be cloned directly in a final level M construct, but is chosen here for the sake of space in the figures). Cloning of two transcription units in the first level M construct will require using vector pLM1 and end-linker pELM-2 (see compatibility of the fusion sites in Figs. 6 and 7). Cloning of the next two transcription units will require vector pLM3 and end-linker pELM-4.

After cloning, level M constructs become flanked by two BsaI sites, one coming from the level M vector, and the other coming from the level M end-linker (Figs. 6 and 7). The fusion sites flanking these level M vectors are derived from those used previously in level 1 constructs. It is important to check, before starting cloning of level M constructs, that all level M constructs that are made will be compatible for assembly in level P vectors. We can take an example for defining a strategy for making a construct containing 18 genes (*see* Fig. 4b). One possibility is to clone the transcription units in level M vectors in three groups of six genes. This strategy is valid as the fusion sites that will flank the level M constructs (TGCC, GAGC, TGTG, and CAGA) are all different (Fig. 4b, blue arrows). Alternatively, one could decide to clone the level 1 constructs in groups of no more than 5 transcription units for higher cloning efficiency, for example in 4 groups of 5, 4, 5, and 4 transcription units. However, this strategy is not valid as the fusion sites that would flank the resulting level M constructs are TGCC, TGTG, ACTA, TGCC, and CAGA, with TGCC present twice (Fig. 4b, red arrows). An alternative would be to clone groups of 5, 4, 4, and 5 transcription units, with fusion sites TGCC, TGTG, ACTA, GAGC, and CAGA flanking the resulting level M constructs.

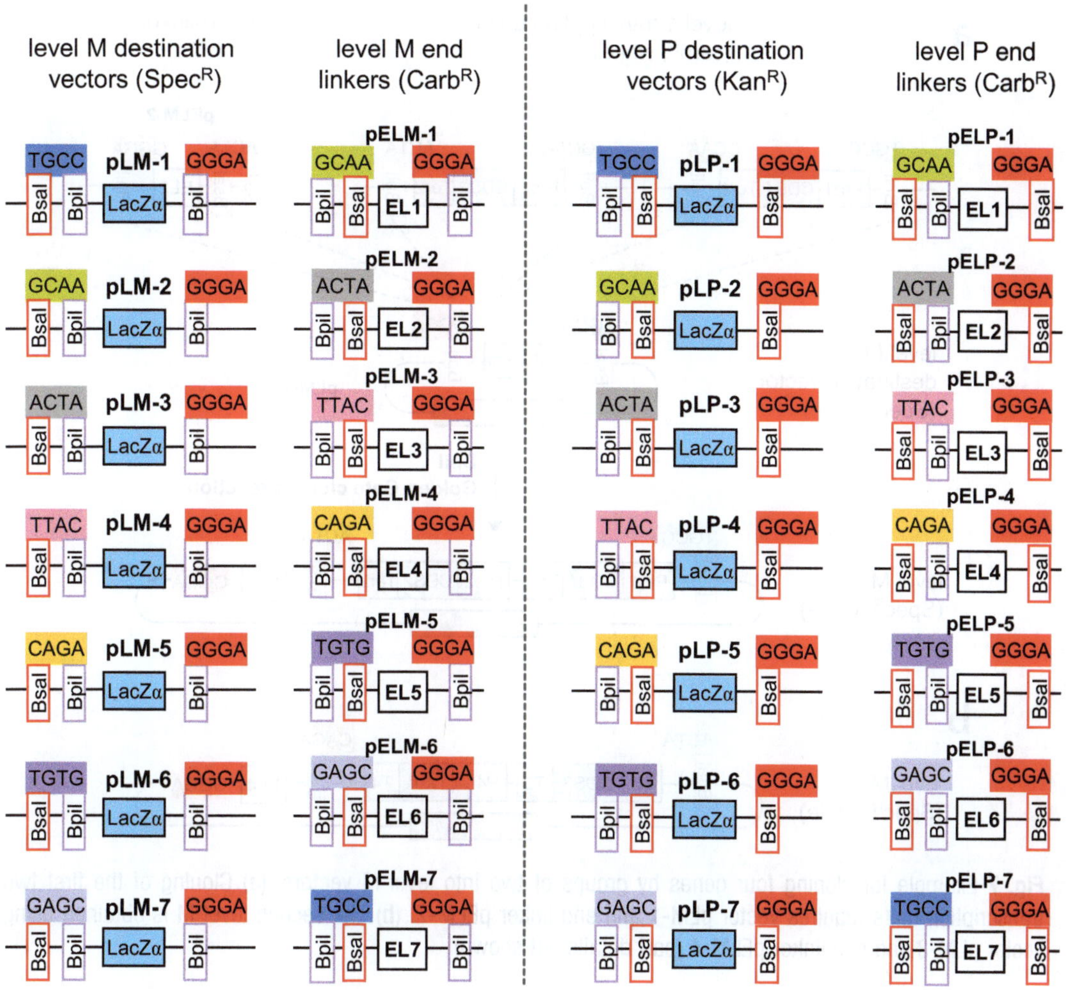

Fig. 6 Level M and P destination vectors and end-linkers

The Golden Gate reaction for making level M constructs is virtually the same as described in Subheading 3.2 except that the enzyme BpiI is used instead of BsaI.

1. Set up a restriction–ligation by pipetting into a tube 20 fmol of each level 1 module, of the end-linker and of the level M vector, 2 μL 10× ligation buffer, 10 U (1 μL) of BpiI, and either 3 U (1 μL) of ligase for assembly of two to four modules or 20 U (1 μL) HC ligase for assembly of more than four modules (final volume of 20 μL).
2. Incubate the restriction–ligation mix in a thermal cycler. For assembly of two to four level 1 modules, incubation for 60–120 min at 37 °C is sufficient. If more modules are ligated together, the incubation time is increased to 6 h, or cycling is used as following: 2 min at 37 °C followed by 3 min at 16 °C, both repeated 50 times (*see* **Note 9**).

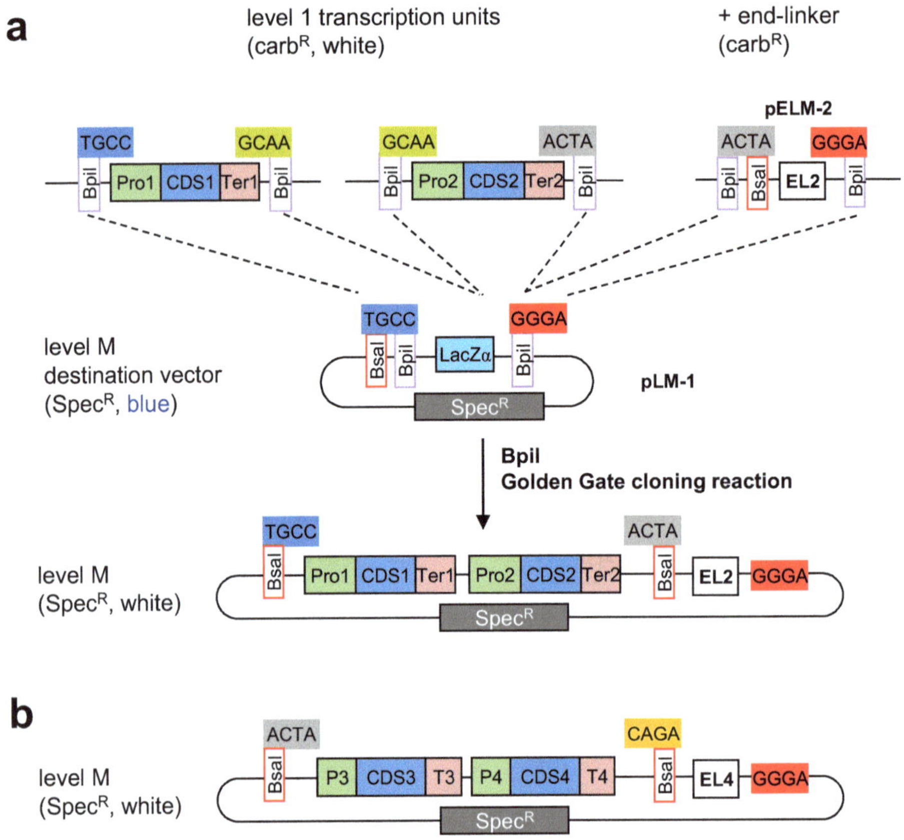

Fig. 7 Example for cloning four genes by groups of two into level M vectors. (**a**) Cloning of the first two transcription units requires vector pLM-1 and end linker pELM-2. (**b**) The second level M is obtained using vector pLM-3 and end linker pELM-4 (part details not shown)

3. Restriction–ligation is followed by a digestion step (5 min at 37 °C; *see* **Note 13**) and then by heat inactivation for 10 min at 80 °C.
4. Transformation of the reaction mix into *E. coli* is performed as described in Subheading 3.2 and cells are spread on LB plates with spectinomycin and X-gal. Colonies containing correct constructs should be white whereas colonies containing uncut or religated vector should be blue. Plasmids should be isolated from a few colonies and should be analyzed by restriction digest with a suitable enzyme.

3.4 Assembly of Level P Constructs

Level P vectors and end-linkers are similar to level M, but all BpiI sites and BsaI sites are reciprocally exchanged. In addition, level P vectors have a different antibiotic selectable marker: Kan^R instead of $Spec^R$ (Fig. 6). As for level M constructs, cloning in level P vectors requires selecting a suitable level P vector and an end-linker. In the example where 2 level M constructs consisting of two transcription

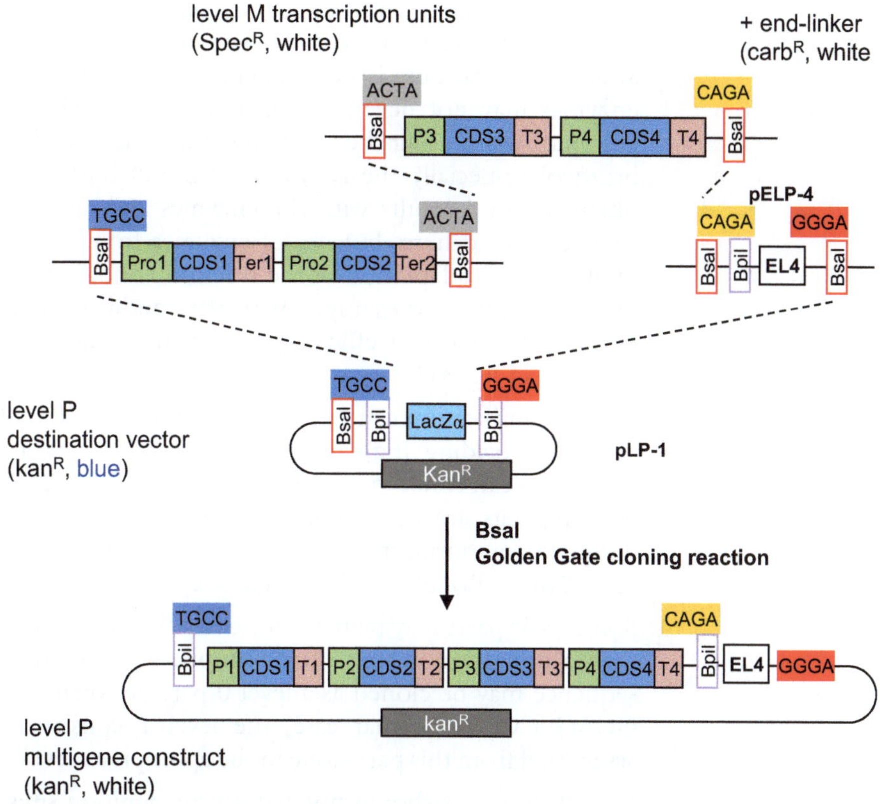

Fig. 8 Example of level P cloning from 2 level M constructs

units are cloned, vector pLP-1 and end-linker pELP-4 are used (Fig. 8). The Golden Gate reaction is the same as described in Subheading 3.2. Level P constructs are flanked by BpiI restriction sites and have therefore a similar structure as level 1 constructs (except for the selectable marker). As level 1 constructs, several level P constructs can be assembled into level M vectors. As discussed with level M constructs, the user needs to make sure that all level P vectors that need to be assembled are compatible with respect of their fusion sites. The suitability of the fusion sites has to be determined at the beginning of the project, before starting making level M constructs.

4 Notes

1. Several level P constructs can be assembled in Level M vectors to make larger constructs. This process can be repeated indefinitely by using alternatively level M and level P vectors. The limitation will, of course, be the size of the final construct and its ability to be transformed and to replicate in *E.coli*.

2. Type IIS restriction enzymes and ligase can be purchased from several commercial suppliers. Users should nevertheless be aware that various isoschizomers of type IIS restriction enzymes may not necessarily all cleave DNA with the same efficiency when using the reaction conditions described in this protocol (especially the restriction–ligation buffer). We have obtained good results with the enzymes and reaction conditions described here, but use of enzymes from other sources might require adaptation of the reaction conditions to obtain optimal results. For example, with the conditions that we are using, better cloning efficiency is obtained when using BsaI rather than Eco31I.
3. Plasmid purification can be done with any commercially available kit providing high purity DNA. Cloning of multigene constructs can result in very large plasmids for which standard miniprep kits are not suited. In these cases, the use of kits designed for purification of large plasmids is recommended (e.g., NucleoBond® PC from Macherey-Nagel).
4. Some basic parts may be made for constructing something else than a transcription unit. For example, a recombination site sequence may be cloned as a level 0 part between fusion sites GGAG CGCT. In that case, the level 1 construct will be assembled from this part alone in the appropriate level 1 vector.
5. Level 0 modules should not contain recognition sites for the type IIS restriction enzymes used during the different cloning steps (BsaI, BpiI, and optionally BsmBI for use of the modules with level 2 vectors [10] or for other applications). If recognition sites for these enzymes are present, they will have to be removed using a process called domestication. Domestication is usually done by PCR amplification of the sequences of interest using primers designed to introduce point mutations (silent mutations for sites in coding sequences) in the restriction site sequences. Another possibility is to have the sequence lacking these restriction sites synthesized by a commercial gene synthesis provider. This solution provides the added benefit of allowing codon-optimization (in case of coding sequences), which might be useful for expression in nonnative host species.
6. In practice, if all module plasmids and the vector have approximately the same size (4–5 kb), simply adding 50 ng of DNA of each module and of the vector will work relatively well. However, when plasmids with widely different sizes are used, calculating an equimolar amount should provide a higher cloning efficiency. We have previously recommended using 40 fmol of each fragment and of the vector for cloning. It however appears that using less DNA leads to higher cloning efficiency. We therefore now recommend using 20 fmol for each insert and

vector. The following formula (from the NEB catalog) can be used: 1 μg of a 1000 bp DNA fragment corresponds to 1.52 pmol. Therefore, the volume of DNA to pipet (in μL) to have 20 fmol is given by the equation: 20 (fmol) × size (bp) of the DNA fragment/(concentration (ng/μL) × 1520). Frequently, this formula will lead to volumes <1 μL with standard miniprep DNA. In these cases, we recommend diluting the DNA in order to avoid pipetting errors.

7. The Golden Gate cloning reaction is quite efficient and leads to a large proportion of colonies containing the desired construct. This high efficiency makes it possible to construct libraries of constructs variants rather than individual constructs. Construct libraries are made by using several modules of the same type (flanked by the same fusion sites) in the ligation mix rather than a single module of each type. In that case, if *n* variant modules replace a single module, the amount of each variant is divided by *n*, but the total amount of fragments for this module type remains the same, that is, 20 fmol. Constructs libraries can be made in vectors containing a CCDB marker for counter-selection to avoid growing empty vectors, but can also be made in normal constructs containing a standard LacZ marker if the presence of a few empty plasmid vectors in the resulting library is acceptable for the project. Libraries can be made for level 1 constructs, but also for constructs of all successive levels M and P.
8. The total volume of enzymes in the reaction should not exceed 10% (*v/v*). Otherwise, glycerol from the storage buffer of the enzymes can lead to decreased efficiencies.
9. Both the temperature conditions and ligase concentration are complementary parameters. For simple constructs requiring ligation of only one or two inserts, any program should work efficiently, even with ligase at regular concentration, and in most cases continuous incubation at 37 °C should be sufficient. The cycling program may improve the cloning efficiency especially when more modules are used and when plasmids become larger. For some applications that require high cloning efficiencies (e.g., the cloning of promoter libraries), use of high-concentration ligase and cycling is recommended.
10. Any other *E. coli* strain can also be used as long as it provides complementation for the lacZα fragment to allow blue–white selection. However, different *E. coli* strains show different transformation efficiencies which can be important for certain applications or when transforming large plasmids. In some cases and with some strains, electroporation can provide better results. In this case, DNA from the restriction–ligation mix should first be ethanol-precipitated and resuspended in 10 μL of water.

11. Two colonies should be sufficient in most cases, since the majority of colonies should contain the correct fragment. Clones with a correct restriction pattern do not necessarily need to be sequenced since constructs are normally assembled only from sequenced level 0 modules.
12. When libraries are made rather than individual constructs, an aliquot of the transformation is plated to evaluate the efficiency of cloning. In this case, there is usually no need to screen these colonies. The rest of the transformation is grown as a single culture. A miniprep made from this culture can be used in the same way as a miniprep from a single construct for the next stage of cloning. It can also be digested to check the quality of the library; however, the size of the fragments will of course depend on what is present in the library.
13. The optimal temperature for restriction with BpiI is 37 °C. The final digestion is thus carried out at 37 °C.

References

1. Andreou AI, Nakayama N (2018) Mobius assembly: a versatile Golden-Gate framework towards universal DNA assembly. PLoS One 13:e0189892
2. Binder A, Lambert J, Morbitzer R et al (2014) A modular plasmid assembly kit for multigene expression, gene silencing and silencing rescue in plants. PLoS One 9:e88218
3. Engler C, Kandzia R, Marillonnet S (2008) A one pot, one step, precision cloning method with high throughput capability. PLoS One 3: e3647
4. Engler C, Youles M, Gruetzner R et al (2014) A golden gate modular cloning toolbox for plants. ACS Synth Biol 3:839–843
5. Feike D, Korolev AV, Soumpourou E et al (2019) Characterizing standard genetic parts and establishing common principles for engineering legume and cereal roots. Plant Biotechnol J 17(12):2234–2245
6. Iverson SV, Haddock TL, Beal J et al (2016) CIDAR MoClo: improved MoClo assembly standard and new E. coli part library enable rapid combinatorial design for synthetic and traditional biology. ACS Synth Biol 5:99–103
7. Larroude M, Park YK, Soudier P et al (2019) A modular Golden Gate toolkit for Yarrowia lipolytica synthetic biology. Microbial Biotechnol. https://doi.org/10.1111/1751-7915.13427
8. Lee ME, Deloache WC, Cervantes B et al (2015) A highly characterized yeast toolkit for modular, multipart assembly. ACS Synth Biol 4:975–986
9. Lin D, O'callaghan CA (2018) MetClo: methylase-assisted hierarchical DNA assembly using a single type IIS restriction enzyme. Nucleic Acids Res 46:e113
10. Marillonnet S, Werner S (2015) Assembly of multigene constructs using Golden Gate cloning. Methods Mol Biol 1321:269–284
11. Moore SJ, Lai HE, Kelwick RJ et al (2016) EcoFlex: a multifunctional MoClo kit for E. coli synthetic biology. ACS Synth Biol 5:1059–1069
12. Occhialini A, Piatek AA, Pfotenhauer AC et al (2019) MoChlo: a versatile, modular cloning toolbox for chloroplast biotechnology. Plant Physiol 179:943–957
13. Pollak B, Cerda A, Delmans M et al (2019) Loop assembly: a simple and open system for recursive fabrication of DNA circuits. New Phytol 222:628–640
14. Prielhofer R, Barrero JJ, Steuer S et al (2017) GoldenPiCS: a Golden Gate-derived modular cloning system for applied synthetic biology in the yeast Pichia pastoris. BMC Syst Biol 11:123
15. Rajkumar AS, Varela JA, Juergens H et al (2019) Biological parts for Kluyveromyces marxianus synthetic biology. Front Bioeng Biotechnol 7:97
16. Sarrion-Perdigones A, Vazquez-Vilar M, Palaci J et al (2013) GoldenBraid 2.0: a comprehensive DNA assembly framework for plant synthetic biology. Plant Physiol 162:1618–1631
17. Taylor GM, Mordaka PM, Heap JT (2019) Start-stop assembly: a functionally scarless

DNA assembly system optimized for metabolic engineering. Nucleic Acids Res 47:e17

18. Terfruchte M, Joehnk B, Fajardo-Somera R et al (2014) Establishing a versatile Golden Gate cloning system for genetic engineering in fungi. Fungal Genet Biol 62:1–10
19. Weber E, Engler C, Gruetzner R et al (2011) A modular cloning system for standardized assembly of multigene constructs. PLoS One 6:e16765
20. Werner S, Engler C, Weber E et al (2012) Fast track assembly of multigene constructs using Golden Gate cloning and the MoClo system. Bioengineered Bugs 3:38–43
21. Wu D, Schandry N, Lahaye T (2018) A modular toolbox for Golden-Gate-based plasmid assembly streamlines the generation of Ralstonia solanacearum species complex knockout strains and multi-cassette complementation constructs. Mol Plant Pathol 19:1511–1522

Chapter 9

Hierarchical Modular DNA Assembly Using MetClo

Da Lin and Christopher A. O'Callaghan

Abstract

DNA assembly methods are essential for multiple applications including synthetic biology. We recently developed MetClo, a method that uses a single type IIS restriction enzyme for hierarchical modular DNA assembly. This offers great flexibility in the design of the assembly experiment and simplicity of execution. Here we describe a protocol for hierarchical assembly of large DNA constructs from modular DNA parts using the MetClo vector set, a set of assembly vectors designed for the MetClo method.

Key words MetClo, DNA assembly, Synthetic biology, Type IIS restriction enzyme, Modular assembly

1 Introduction

The advancement of synthetic biology requires efficient DNA assembly. Of the existing methods available for DNA assembly, type IIS restriction enzyme–based hierarchical DNA assembly methods such as MoClo have significant advantages for modular assembly of standardized DNA parts, including simple one-pot assembly reaction setup using circular plasmid DNA as an input [1, 2]. A major drawback of the assembly methods based on type IIS restriction enzymes is the requirement to remove internal type IIS restriction sites for the multiple restriction enzymes used in the hierarchical assembly process. For example, depending on the assembly topology, the standard MoClo assembly method requires the use of two or three different type IIS restriction enzymes [1–3], placing constraints on the DNA sequences that can be assembled.

To address this issue, we recently developed MetClo, an assembly method that uses only a single type IIS restriction enzyme for hierarchical assembly [4]. The method is based on a process we term "methylation-switching," whereby a sequence-specific DNA methylase methylates a type IIS restriction enzyme recognition site that has been engineered to overlap with the sequence motif that the methylase recognizes. This methylation of the type IIS

Sunil Chandran and Kevin W. George (eds.), *DNA Cloning and Assembly: Methods and Protocols*, Methods in Molecular Biology, vol. 2205, https://doi.org/10.1007/978-1-0716-0908-8_9,

restriction enzyme recognition site ensures that it is no longer recognized by the type IIS restriction enzyme and thus remains intact. The design of assembly vectors with methylase-switchable and normal nonswitchable type IIS restriction sites allows DNA fragments to be assembled into the assembly vectors in a one-pot reaction with a type IIS restriction enzyme, and then the assembled DNA can subsequently be released using the same restriction enzyme (Fig. 1). This assembled DNA can then be used as an insert for further rounds of DNA assembly. This allows the use of only a single type IIS restriction enzyme throughout the entire assembly process. Consequently, there are significantly fewer sequence constraints in modular part design, greater exchangeability of DNA parts between existing modular part libraries, and greater flexibility in the design of DNA assembly schemes.

To facilitate the use of the MetClo method for modular hierarchical DNA assembly, we designed a set of modular assembly vectors for MetClo-based hierarchical assembly using the type IIS restriction enzyme BsaI [4]. Here we describe the MetClo hierarchical DNA assembly process using this vector set, the principles guiding the choice of assembly vector in individual assembly reactions, and the experimental protocols for DNA assembly using MetClo.

2 Materials

2.1 Molecular Biology Reagents

1. 20 U/μL BsaI-HFv2 (NEB).
2. 30 U/μL T4 DNA Ligase (HC) (Thermo Fisher Scientific).
3. 10× T4 Ligase buffer (NEB): 500 mM Tris–HCl, 100 mM $MgCl_2$, 10 mM ATP, 100 mM DTT, pH 7.5. Use 1× buffer for DNA assembly reaction.
4. LB medium: 10 g tryptone, 5 g yeast extract, 10 g NaCl in 1 L water, autoclave. For agar plates, add 1.5% agar (*see* **Note 1**).
5. Low salt LB medium: 10 g tryptone, 5 g yeast extract, 5 g NaCl in 1 L water, autoclave. For agar plates add 1.5% agar. This medium is used for culturing of *E.coli* with zeocin selection.
6. Antibiotics: Ampicillin is prepared as stock solution of 100 mg/mL in water and used at a final concentration of 100 μg/mL. Kanamycin is prepared as stock solution of 30 mg/mL in water and used at a final concentration of 30 μg/mL. Chloramphenicol is prepared as stock solution of 25 mg/mL in ethanol and used at a final concentration of 25 μg/mL. Zeocin is prepared as stock solution of 100 mg/mL in water and used at a final concentration of 25 μg/mL.

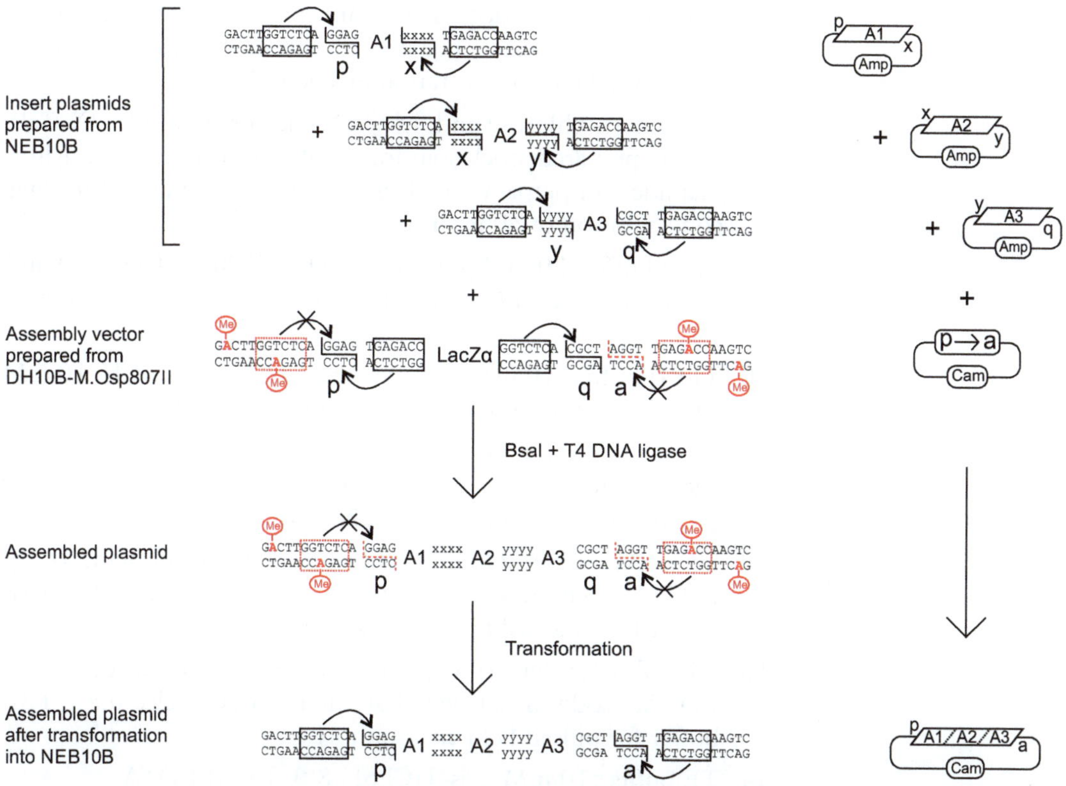

Fig. 1 MetClo DNA assembly using BsaI. The insert plasmids contain DNA fragments to be assembled (A1, A2 and A3). These fragments are flanked by BsaI sites (boxed with solid lines) that when cut generate adhesive ends compatible with each other (x and y) and with the assembly vector (p and q). The flanking BsaI sites overlap with the M.Osp807II methylase recognition sequence. Insert plasmids were prepared from a normal strain (NEB10B) that does not express the M.Osp807II switch methylase. As a result, the BsaI sites are not methylated and so the insert DNA fragments can be released by BsaI digestion. The assembly vector contains a LacZα selection marker flanked by head-to-head BsaI sites. The outer pair of BsaI sites (boxed in red dotted lines) closer to the vector backbone overlap with an M.Osp807II methylation sequence and so are methylation-switchable, whereas the inner pair of BsaI sites (boxed in solid black lines) do not overlap with a methylase recognition site and so are not methylation-switchable. Preparation of the assembly vector in the M.Osp807II switch methylase-expressing DH10B strain results in selective blocking of the outer pair of BsaI sites (methylated bases are shown in red). The LacZα fragment can be released by BsaI through cutting at the inner pair of BsaI sites (boxed in solid black lines), generating adhesive ends compatible with the insert fragments (p and q). Following a one-pot reaction using BsaI and T4 DNA ligase, ligation among compatible adhesive ends results in ordered assembly of DNA fragments into the assembly vector backbone. The assembled fragment in the assembled plasmid is flanked by methylated BsaI sites (boxed in red dotted lines), which are not cut by BsaI. Following transformation into a normal strain (NEB10B) that does not express the M.Osp807II switch methylase, methylation of the flanking restriction sites is lost. The assembled fragment can be released by BsaI for the next round of a multistage assembly. Once released by BsaI, the assembled fragment carries adhesive ends "p" and "a." The process is shown schematically on the right of the figure. The insert plasmids carry antibiotic selection marker (Amp, ampicillin) different from the assembly vector (Cam, chloramphenicol)

7. Isopropyl β-D-1-thiogalactopyranoside (IPTG): stock solution of 100 mM in water. For preparation of agar plates, IPTG is added at a final concentration of 100 μM.
8. 5-bromo-4-chloro-3-indolyl β-D-galactopyranoside (X-Gal): freshly prepared stock solution of 50 mg/mL in dimethylformamide. For preparation of agar plates, X-Gal is added at a final concentration of 50 μg/mL.
9. TFBI buffer: 30 mM potassium acetate, 100 mM KCl, 10 mM $CaCl_2$, 50 mM $MnCl_2$, 15% glycerol. Filter sterilize and store at 4 °C.
10. TFBII buffer: 10 mM MOPS free acid, 75 mM $CaCl_2$, 10 mM KCl, 15% glycerol. Filter sterilize and store at 4 °C.
11. DNA markers: 1 kb Plus DNA Ladder (NEB) for standard agarose gel electrophoresis, and MidRange PFG Marker (NEB) for pulsed field electrophoresis.
12. 50× TAE buffer: 242 g Tris base, 57.1 mL acetic acid, 37.2 g EDTA disodium salt dihydrate in 1 L water. Use 1× TAE buffer for standard gel electrophoresis.
13. 10× TBE buffer: 108 g Tris base, 55 g boric acid, 7.44 g EDTA disodium salt dehydrate in 1 L water. Use 0.5× TBE buffer for pulsed field electrophoresis.
14. TE buffer: 10 mM Tris–HCl pH 8.0, 1 mM EDTA.
15. GeneJET Plasmid Miniprep Kit (Thermo Fisher Scientific).
16. NucleoBond Xtra BAC Kit (Macherey-Nagel).
17. PhasePrep Bacterial Artificial Chromosomes DNA Kit (Sigma).

2.2 Strains

1. DH10B-MOsp807II for vector DNA preparation. We developed this strain to be resistant to 25 μg/mL zeocin and it is available from Addgene. (*see* **Note 2**).
2. NEB10B for insert DNA preparation. This strain is available as chemical-competent or electrocompetent cells from NEB. Alternatively DH10B (Thermo Fisher Scientific) can be used.

2.3 Plasmids

MetClo vector set (Addgene). This vector set that we developed is available as glycerol stocks from Addgene. The plasmids should be transformed into DH10B-MOsp807II to use as assembly vectors for MetClo assembly reaction.

2.4 Equipment

1. Microcentrifuge: 5424 Centrifuge (Eppendorf).
2. Benchtop centrifuge: Avanti J-15R Centrifuge (Beckman Coulter).
3. UV spectrometer: ND-1000 Spectrophotometer (NanoDrop Technologies).

4. Thermocycler: PTC-225 Peltier Thermal Cycler (MJ Research).
5. Gel imager: ChemiDoc MP Imaging System (Bio-Rad).
6. Electroporator: Gene Pulser Electroporation System (Bio-Rad).
7. Shaking incubator: Innova 44 Incubator (New Brunswick).
8. Bacteria incubator: Model M50C (Genlab).
9. Gel electrophoresis device: Sub-Cell GT (Bio-Rad).
10. Pulsed field electrophoresis device: CHEF-DR II Pulsed Field Electrophoresis System (Bio-Rad).

3 Methods

The MetClo vector set consists of 48 assembly vectors which each have one of 3 different antibiotic selection markers (chloramphenicol, kanamycin, or ampicillin), and one of 2 different replication origins (F or p15a) for hierarchical modular assembly by MetClo using BsaI (Table 1 and Fig. 2). The vector names are divided in two parts separated by a space. The first part represents the vector backbone and specifies the antibiotic selection marker and replication origin of the vector backbone. The second part represents the type of adaptor sequences carried by the assembly vector (Fig. 3). For example, pMXLC_pFa represents MetClo assembly vector with p15a replication origin and chloramphenicol selection marker. The fragment assembled into this vector will be in the forward orientation and carry adaptor sequence "p" (GGAG) at the start and adaptor sequence "a" (AGGT) at the end.

The MetClo vector set is designed to facilitate the assembly of large DNA constructs from an arbitrary number of DNA fragments in a given order. During each round of a MetClo assembly using the MetClo vector set, multiple DNA fragments are assembled into specific MetClo assembly vectors in a single pot reaction. The newly assembled fragments within these assembly vectors can then be used directly in the next round of assembly and this process can be repeated in multiple sequential rounds of assembly using the standard vector set. The choice of assembly vectors to use in any given round of assembly is based on the required order of the fragments in the final construct arising from that assembly reaction. The planning phase of an assembly experiment involves:

1. Generation or sourcing of suitable DNA fragments for assembly.
2. Determination of the positional order of individual fragments in the final construct.

Table 1
The MetClo vector set

Name	ID	Selection marker	Replication origin	Adaptor sequence
pMXLC_pFa[a]	POC12355	Chloramphenicol	p15a	pFa
pMXLC_aFb	POC12356	Chloramphenicol	p15a	aFb
pMXLC_bFc	POC12357	Chloramphenicol	p15a	bFc
pMXLC_cFd	POC12358	Chloramphenicol	p15a	cFd
pMXLC_dFe	POC12359	Chloramphenicol	p15a	dFe
pMXLC_aFq	POC12362	Chloramphenicol	p15a	aFq
pMXLC_bFq	POC12363	Chloramphenicol	p15a	bFq
pMXLC_cFq	POC12364	Chloramphenicol	p15a	cFq
pMXLC_dFq	POC12365	Chloramphenicol	p15a	dFq
pMXLC_eFq	POC12366	Chloramphenicol	p15a	eFq
pMXLK_pFa	POC12367	Kanamycin	p15a	pFa
pMXLK_aFb	POC12368	Kanamycin	p15a	aFb
pMXLK_bFc	POC12369	Kanamycin	p15a	bFc
pMXLK_cFd	POC12370	Kanamycin	p15a	cFd
pMXLK_dFe	POC12371	Kanamycin	p15a	dFe
pMXLK_aFq	POC12374	Kanamycin	p15a	aFq
pMXLK_bFq	POC12375	Kanamycin	p15a	bFq
pMXLK_cFq	POC12376	Kanamycin	p15a	cFq
pMXLK_dFq	POC12377	Kanamycin	p15a	dFq
pMXLK_eFq	POC12378	Kanamycin	p15a	eFq
pMXLP_pFa	POC12379	Ampicillin	p15a	pFa
pMXLP_aFb	POC12380	Ampicillin	p15a	aFb
pMXLP_bFc	POC12381	Ampicillin	p15a	bFc
pMXLP_cFd	POC12382	Ampicillin	p15a	cFd
pMXLP_dFe	POC12383	Ampicillin	p15a	dFe
pMXLP_aFq	POC12386	Ampicillin	p15a	aFq
pMXLP_bFq	POC12387	Ampicillin	p15a	bFq
pMXLP_cFq	POC12388	Ampicillin	p15a	cFq
pMXLP_dFq	POC12389	Ampicillin	p15a	dFq
pMXLP_eFq	POC12390	Ampicillin	p15a	eFq
pMXBC_pFa	POC12391	Chloramphenicol	F	pFa

(continued)

Table 1
(continued)

Name	ID	Selection marker	Replication origin	Adaptor sequence
pMXBC_aFb	POC12392	Chloramphenicol	F	aFb
pMXBC_bFc	POC12393	Chloramphenicol	F	bFc
pMXBC_aFq	POC12394	Chloramphenicol	F	aFq
pMXBC_bFq	POC12395	Chloramphenicol	F	bFq
pMXBC_cFq	POC12396	Chloramphenicol	F	cFq
pMXBK_pFa	POC12397	Kanamycin	F	pFa
pMXBK_aFb	POC12398	Kanamycin	F	aFb
pMXBK_bFc	POC12399	Kanamycin	F	bFc
pMXBK_aFq	POC12400	Kanamycin	F	aFq
pMXBK_bFq	POC12401	Kanamycin	F	bFq
pMXBK_cFq	POC12402	Kanamycin	F	cFq
pMXBP_pFa	POC12403	Ampicillin	F	pFa
pMXBP_aFb	POC12404	Ampicillin	F	aFb
pMXBP_bFc	POC12405	Ampicillin	F	bFc
pMXBP_aFq	POC12406	Ampicillin	F	aFq
pMXBP_bFq	POC12407	Ampicillin	F	bFq
pMXBP_cFq	POC12408	Ampicillin	F	cFq

[a]In the four letters representing the vector backbone, the first two letters "MX" are fixed, showing that the vectors are for MetClo ("M") using M.Osp807II-based BsaI assembly ("X"); the third letter represents the type of replication origin: "L" for p15a (Low copy) and "B" for F replication origin (BAC); the fourth letter represents the antibiotics selection marker: "P" for ampicillin (penicillin), "C" for chloramphenicol, and "K" for kanamycin. In the letters representing the adaptor sequence, the first letter and the last letter are in lower case and represent the front and end adaptor sequence carried by the fragment once assembled into this vector ("p," "q," "a," "b," "c," "d," "e"). The middle letter in upper case represent the orientation of the assembled fragment in the vector ("F" for forward orientation)

3. Design of the entire assembly process by determination of the number of rounds of hierarchical assembly and selection of the specific assembly vectors for each assembly reaction.

The first two steps specify the sequence of the final construct, and the last step specifies the assembly process based on the construct to be assembled.

3.1 Planning of an Assembly Experiment

In the first round of an assembly experiment, smaller fragments (termed "component" or "modular parts") are assembled into larger fragments. The resulting larger fragments are assembled into plasmids with flanking BsaI sites. These larger fragments we term "elementary units" and they form the substrate for MetClo

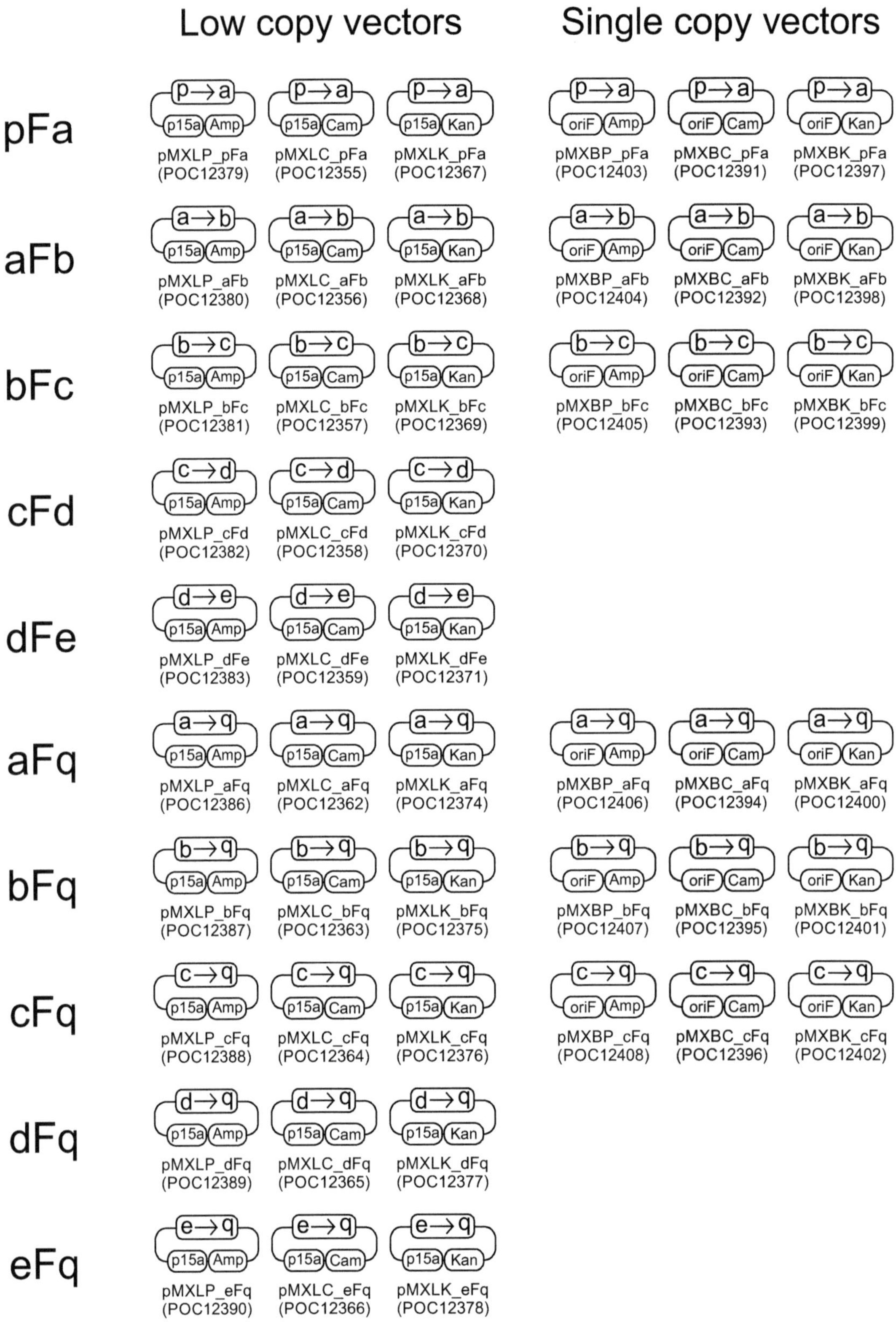

Fig. 2 The MetClo vector set. The vector set contains 30 low copy vectors with a p15a replication origin and either ampicillin (Amp), kanamycin (Kan) or chloramphenicol (Cam) selection markers, and a LacZα negative

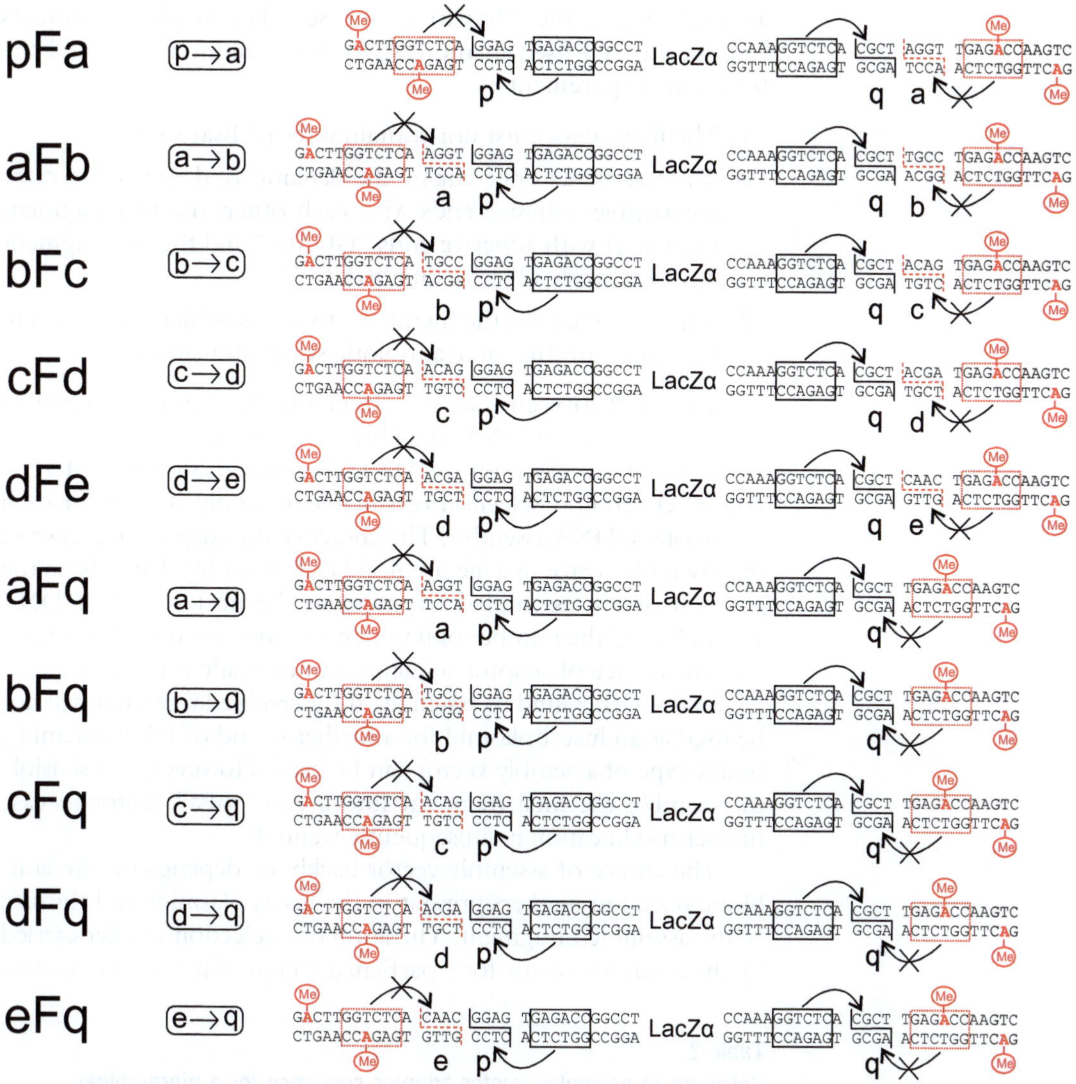

Fig. 3 Adaptor sequence design of the MetClo vector set. The MetClo vector set contains assembly vectors with 10 different types of adaptor sequence design. In each type a negative selection marker (LacZα) is flanked by two head-to-head BsaI sites. The outer pair of BsaI sites closer to the vector backbone (boxed in red dotted lines) overlap with an M.Osp807II methylation sequence and so are methylation-switchable, whereas the inner pair of BsaI sites (boxed in solid black lines) are not. Preparation of the assembly vector in a strain expressing the M.Osp807II switch methylase results in selective blocking of the outer pair of BsaI sites (methylated bases are shown in red). For each design, the inner pair of BsaI sites can be used to release the negative selection marker for assembly of the same set of insert fragments with compatible adhesive ends "p" and "q." The assembled fragment can be released from the assembly vector backbone using the outer pair of BsaI sites, and the adhesive ends carried by the assembled fragment depends on the design. For example, assembly into type "pFa" generates an assembled fragment with adhesive ends "p" and "a"

Fig. 2 (continued) selection marker flanked by one of 10 different adaptor sequence designs. The vector set also contains 18 low copy number vectors with an F replication origin (oriF), one of the three antibiotic selection markers, and a fragment containing a ColE1 replication origin and a LacZα negative selection marker flanked by one of six different adaptor sequence designs assembly using the MetClo vector set. The smaller fragments (modular parts) used in the first round assembly must fulfill the following requirements:

1. The fragments must not contain internal BsaI sites.
2. The fragments once cut by the flanking BsaI sites must carry compatible adhesive ends with each other, the first fragment must start with adhesive ends "GGAG," and the last fragment must end with adhesive end "CGCT."
3. (Optional) Ideally, the fragments to be assembled are carried in plasmids with the same antibiotic selection marker.

The MetClo vector set is designed to be directly compatible with several existing DNA part libraries [3, 5, 6].

Hierarchical assembly of DNA fragments using the MetClo vector set requires selection of suitable assembly vectors for each round of the DNA assembly. The choice of the adaptor sequence of the assembly vector in a hierarchical DNA assembly depends on the position of the assembled fragment in the next round of DNA assembly, and the number of fragments in the next round of assembly. The choice of adaptor sequence can be made using Table 2.

In the final round of assembly, the assembled fragment will not be used as an insert plasmid for a further round of DNA assembly, so any type of assembly vector can be used. However, it is sensible to assemble the final fragment into a type "pFa" vector in case further modification is subsequently wanted.

The choice of assembly vector backbone depends on the antibiotic selection markers present in the insert plasmids and the size of the assembled fragment. The antibiotic selection marker carried by the assembly vector for a particular round of the assembly must

Table 2
Selection of assembly vector adaptor sequence for a hierarchical assembly experiment

Number of fragments to assemble in the next round of assembly	Position of the assembled fragment in the next round of assembly					
	1	2	3	4	5	6
2	pFa[a]	aFq				
3	pFa	aFb	bFq			
4	pFa	aFb	bFc	cFq		
5	pFa	aFb	bFc	cFd	dFq	
6	pFa	aFb	bFc	cFd	dFe	eFq

[a]Adaptor sequence type of the assembly vector for the current round assembly

be different from those in each of the insert plasmids used in that round. Therefore, it is convenient for all the initial fragments to be in plasmids which have the same antibiotic selection marker. Similarly, it is sensible to ensure that in each round of the assembly the plasmids which are donating inserts all have the same antibiotic selection marker.

The size of the assembled fragment determines the choice of replication origin for the assembly vector into which it will be assembled. Vectors with the low copy number p15a origin are suitable for assembling fragments up to 30 kb, whereas larger fragments should be assembled into vectors with the single copy F replication origin (*see* **Note 3**).

A hierarchical assembly process can be planned using the following procedure. First, the larger fragments we term "elementary units" are assembled from smaller fragments into appropriate vectors. Next, the total number of elementary units and their order in the final assembled construct are noted.

If the total number of elementary units to be assembled is less than 6, then only two stages of assembly are required. In stage 1, modular parts are assembled into elementary units using assembly vectors with a p15a-based replication origin and with adaptor sequence type as indicated in Table 2 according to the position of the elementary unit in the final construct. In stage 2, the resulting elementary units are assembled into type "pFa" assembly vector with an F replication origin. As an example, a six-unit assembly can be undertaken using the assembly vectors listed in Table 3.

If the total number of elementary units to be assembled exceeds six, then the assembly requires multiple rounds. As an example, for a final construct of nine elementary units built from multiple modular parts, the assembly can be broken down into three stages: in stage 1 the modular parts are assembled into nine elementary units; in stage 2 the nine elementary units are assembled three per group

Table 3
Assembly of a six-unit construct using the MetClo vector set

	Insert plasmids	Assembly vector	Assembled plasmid
Stage 1	A1, A2...	pMXLC_pFa	pMXLC_A
	B1, B2...	pMXLC_aFb	pMXLC_B
	C1, C2...	pMXLC_bFc	pMXLC_C
	D1, D2...	pMXLC_cFd	pMXLC_D
	E1, E2...	pMXLC_dFe	pMXLC_E
	F1, F2...	pMXLC_eFf	pMXLC_F
Stage 2	pMXLC_A, pMXLC_B, pMXLC_C, pMXLC_D, pMXLC_E, pMXLC_F	pMXBK_pFa	pMXBK_ABCDEF

Table 4
Assembly of a nine-unit construct using the MetClo vector set

	Insert plasmids	Assembly vector	Assembled plasmid
Stage 1	A1, A2...	pMXLC_pFa	pMXLC_A
	B1, B2...	pMXLC_aFb	pMXLC_B
	C1, C2...	pMXLC_bFq	pMXLC_C
	D1, D2...	pMXLC_pFa	pMXLC_D
	E1, E2...	pMXLC_aFb	pMXLC_E
	F1, F2...	pMXLC_bFq	pMXLC_F
	G1, G2...	pMXLC_pFa	pMXLC_G
	H1, H2...	pMXLC_aFb	pMXLC_H
	I1, I2...	pMXLC_bFq	pMXLC_I
Stage 2	pMXLC_A, pMXLC_B, pMXLC_C	pMXBK_pFa	pMXBK_ABC
	pMXLC_D, pMXLC_E, pMXLC_F	pMXBK_aFb	pMXBK_DEF
	pMXLC_G, pMXLC_H, pMXLC_I	pMXBK_bFq	pMXBK_GHI
Stage 3	pMXBK_ABC, pMXBK_DEF, pMXBK_GHI	pMXBC_pFa	pMXBC_ABCDEFGHI

into three large fragments; and in stage 3 the three large fragments are assembled into the final construct. The selection of assembly vectors for this assembly is specified in Table 4.

Once a construct is assembled, the inherent flexibility of MetClo assembly, which arises from its use of only a single type IIS restriction enzyme, means that it is possible to add additional elementary units with minimal effort. By default, the final construct for a standard MetClo hierarchical assembly is assembled into a vector of adaptor sequence type "pFa" in the final stage of assembly (as illustrated for a single assembly in Fig. 1). Therefore, if between 1 and 5 elementary units are to be added to the 3′ end of an existing construct, these elementary units can be assembled into vector types corresponding to positions 2–6 in Table 2, and used as fragments 2–6 in an assembly reaction in which the original previously assembled construct fragment to be modified acts as the first fragment (type "pFa"). It is important to use an assembly vector with an antibiotic selection marker that is different from that in the plasmid carrying the fragment to be modified and the plasmid carrying the individual elementary units to be added. As long as all the elementary units carry the same antibiotic selection marker, it is always possible to choose one of the three antibiotic selection markers available in the MetClo vector set. Notably, any individual elementary units assembled into vector type "aFq" can be added one by one indefinitely at the end of any elementary units in a type "pFa" vector.

Once the assembly process is designed, the assembly experiment can be undertaken using the following protocols. Briefly, all the insert plasmids are prepared in NEB10B cells, the assembly vector in DH10B-MOsp807II cells, and the assembly reactions are carried out using a thermal cycler and transformed into NEB10B cells. Correctly assembled plasmids are then used as insert plasmids in the next round of assembly, until the final construct is made.

3.2 Preparation of Chemically Competent DH10B-MOsp807II

1. Streak glycerol stock of DH10B-MOsp807II onto Low salt LB agar plates with 25 μg/mL zeocin and incubate the plate overnight at 37 °C.
2. Pick a single colony into 5 mL Low salt LB medium with 25 μg/mL zeocin and shake overnight at 37 °C 220 rpm.
3. Transfer the 5 mL overnight culture into 250 mL LB medium and shake at 37 °C 220 rpm until OD600 = 0.6 (*see* **Note 2**).
4. Pellet cells at 1500 × *g* at 4 °C for 15 min.
5. Resuspend cells in 50 mL TFBI buffer, incubate on ice for 30 min.
6. Pellet cells at 1500 × *g* at 4 °C for 15 min.
7. Resuspend cells in 10 mL TFBII buffer, incubate on ice for 15 min.
8. Aliquot competent cells into 1.5 mL Eppendorf tubes, flash-freeze in liquid nitrogen and store at −80 °C.

3.3 Preparation of MetClo Assembly Vectors

The MetClo assembly vectors are prepared in DH10B-MOsp807II to generate vectors with selective blocking of BsaI sites by methylation switching.

1. Streak glycerol stock of MetClo assembly vector onto LB agar plates with appropriate antibiotics (ampicillin, kanamycin or chloramphenicol) and incubate the plate overnight at 37 °C (*see* **Note 2**).
2. Pick a single colony into 5 mL LB medium with appropriate antibiotics and shake overnight at 37 °C 220 rpm.
3. Purify the plasmid from the 5 mL LB overnight culture using the GeneJET Plasmid Miniprep Kit following the manufacturer's instructions.
4. Transform 1 μL plasmid into chemically competent DH10B-MOsp807II, and spread onto LB agar plates with appropriate antibiotics.
5. Pick a single colony into 100 mL LB medium with appropriate antibiotics (ampicillin, kanamycin or chloramphenicol) and shake overnight at 37 °C 220 rpm (*see* **Note 4**).
6. Purify the plasmid from the 100 mL overnight culture using the QIAGEN Plasmid Midiprep kit. The purified plasmid can be used as an assembly vector in MetClo assembly reactions (*see* **Note 4**).

3.4 DNA assembly by MetClo

1. Set up assembly reaction in a 0.2 mL PCR tube on ice:
   (a) 30 fmol of each insert plasmids prepared in NEB10B (*see* **Note 5**).
   (b) 30 fmol of assembly vector prepared in DH10B-MOsp807II (*see* **Note 5**).
   (c) 2 μL 10× T4 ligase buffer.
   (d) 0.5 μL T4 DNA ligase.
   (e) 0.5 μL BsaI-HFv2 (*see* **Note 6**).
   (f) Add water to top up the reaction volume to 20 μL.
2. Pipette up and down to mix well.
3. Run the assembly reaction on a thermal cycler using the following condition: 37 °C for 15 min, followed by 45 cycles of 37 °C for 2 min plus 16 °C for 5 min, then 37 °C for 20 min, and 80 °C for 5 min.
4. (Optional) For assembly of large fragments over 30 kb, add 1 μL BsaI-HFv2 to the assembly reaction, pipette up and down to mix well, and incubate at 37 °C for 45 min and then 80 °C for 5 min.
5. For assembly of small fragments less than 10 kb, transform the assembly reaction into chemically competent NEB10B cells: Add 10 μL assembly reaction to 50 μL competent cells in a 1.5 mL Eppendorf tube, incubate on ice for 15 min. Heat shock at 42 °C for 45 s, then incubate on ice for 2 min. Add 250 μL LB medium to the transformed cells, shake at 37 °C for 1 h at 220 rpm. Spread 100 μL cells onto LB plates with appropriate antibiotics plus 50 μg/mL X-Gal and 100 μM IPTG. Incubate at 37 °C overnight.
6. For assembly of large fragments greater than 10 kb, transform the assembly reaction into electrocompetent NEB10B cells: Add 50 mL water to a 10 cm petri dish. Place a piece of 0.05 μm Millipore filter on the surface of water. Transfer 20 μL assembly reaction on top of the filter, drop dialyze at room temperature for 1 h. Add 5 μL dialyzed assembly reaction to 25 μL electrocompetent NEB10B cells in an Eppendorf tube on ice, pipette up and down once to mix well. Transfer the cells to 0.1 cm electroporation cuvette prechilled on ice. Electroporate at 0.9 kV 100 Ω 25 μF. Add 1 mL LB medium to the cells in the cuvette. Transfer the cells to a 30 mL universal container, shake at 37 °C for 1 h at 220 rpm. Spread 100 μL cells onto LB plates with appropriate antibiotics plus 50 μg/mL X-Gal and 100 μM IPTG. Incubate at 37 °C overnight.

3.5 Analysis of the Assembled DNA

1. Pick single white clones into 5 mL LB medium with appropriate antibiotics, shake at 37 °C overnight at 220 rpm.
2. Transfer 0.5 mL overnight culture to 0.5 mL sterile 50% glycerol in a screw-capped 1.5 mL tube, invert the tubes several times to mix well and store at −80 °C as glycerol stock.
3. For constructs assembled into low copy vectors with p15a replication origin, purify the plasmid using GeneJET Plasmid Miniprep Kit and elute with 50 μL elution buffer.
4. For constructs assembled into low copy vectors with the F replication origin, purify the plasmid using the PhasePrep Bacterial Artificial Chromosomes DNA Kit following the Micro Scale Preparation protocol until the Nucleic Acid Preparation step, then resuspend the dried pellet in 40 μL TE buffer.
5. Analyze the assembled DNA by standard gel electrophoresis: Digest 2 μL assembled plasmid with p15a replication origin or 8 μL assembled plasmid with F replication origin in 10 μL reaction using appropriate restriction enzyme for 1 h (*see* **Note 7**). Prepare 0.5% agarose gel with Midori green in 1× TAE buffer. Load the restricted samples along with 1 kb plus DNA ladder onto the gel, run at 10 V/cm for 30 min. Visualize the gel under UV light for analysis.
6. Analyze the assembled DNA by pulsed field electrophoresis: Digest 2 μL assembled plasmid with p15a replication origin or 10 μL assembled plasmid with F replication origin in 20 μL reaction using appropriate restriction enzyme for 1 h. Prepare 1% agarose gel in 0.5× TBE buffer for pulsed field electrophoresis. Load the restricted samples along with 1 kb plus DNA ladder and MidRange PFG Marker onto the gel. Run pulsed electrophoresis using the CHEF-DR II system under the following settings: 6 V/cm for 16 h at 14 °C, with initial switch time 2 s, and final switch time 16 s. Stain the gel with 50 μL Midori green in 1 L 0.5× TBE buffer for 30 min at room temperature. Rinse the gel with water and visualize the gel under UV light for analysis.

3.6 Preparation of Assembled DNA as Insert Plasmids for Next Round of DNA Assembly

1. Streak glycerol stock onto LB plates with appropriate antibiotics, incubate overnight at 37 °C.
2. For plasmids with p15a replication origin, purify the plasmid from 100 mL overnight culture: Pick a single colony into 100 mL LB medium with appropriate antibiotics, shake at 37 °C overnight at 220 rpm. Purify the plasmid using Qiagen Plasmid Midi kit, dissolve the plasmid in 50 μL TE buffer.
3. For plasmids with F replication origin, purify the plasmid from 500 mL overnight culture: Pick a single colony into 500 mL LB medium with appropriate antibiotics, shake at 37 °C overnight at 220 rpm. Purify the plasmid using NucleoBond Xtra BAC Kit, dissolve the plasmid in 50 μL TE buffer.

4 Notes

1. For preparation of agar plates, antibiotics, X-gal, and IPTG should be added after the medium has cooled down to 50 °C.
2. DH10B-MOsp807II expresses the M.Osp807II methylase from the *arsB* locus of the *E. coli* chromosome. A zeocin selection cassette was integrated into the same locus during the generation of the strain, which renders the strain zeocin-resistant. It is not necessary to use zeocin selection to maintain selection pressure for M.Osp807II methylase expression. Therefore, zeocin selection is not necessary during the large volume culturing step in DH10B-MOsp807II competent cell preparation or during the preparation of assembly vectors in this strain.
3. There are cases where an identical set of insert fragments can be successfully assembled into vectors with the F origin, but not vectors with p15a replication origin, despite the assembled fragment being less than 10 kb. This may be due to toxicity of the assembled fragment at high copy number. It is therefore advised that an assembly vector with the F origin be used in cases where assembly with p15a-based assembly vectors has failed.
4. The assembly vectors with F replication origin carry a high copy ColE replication origin in the negative selection marker LacZα fragment. For these vectors, miniprep from 5 mL culture is sufficient to generate plasmids with enough purity and concentration for MetClo assembly.
5. 15 fmol of each insert plasmid and assembly vector is sufficient for successful MetClo DNA assembly.
6. BsaI from NEB can be used instead of BsaI-HFv2. The number of white colonies from assemblies using BsaI is fewer than assemblies made using BsaI-HFv2, possibly due to greater activity or stability of BsaI-HFv2 in the assembly buffer.
7. BsaI, NotI, and XhoI are common restriction enzymes used for analysis. The MetClo assembly vectors have flanking NotI sites in the vector backbone. Restriction with either BsaI or NotI will separate assembled fragments from the vector backbone for standard gel electrophoresis or pulsed field electrophoresis. XhoI or other 6 base cutters usually generate smaller fragments and are therefore suitable for standard gel electrophoresis.

References

1. Weber E, Engler C, Gruetzner R et al (2011) A modular cloning system for standardized assembly of multigene constructs. PLoS One 6(2): e16765
2. Werner S, Engler C, Weber E et al (2012) Fast track assembly of multigene constructs using Golden Gate cloning and the MoClo system. Bioeng Bugs 3(1):38–43
3. Sarrion-Perdigones A, Vazquez-Vilar M, Palací J et al (2013) GoldenBraid 2.0: a comprehensive DNA assembly framework for plant synthetic biology. Plant Physiol 162(3):1618–1631
4. Lin D, O'Callaghan CA (2018) MetClo: methylase-assisted hierarchical DNA assembly using a single type IIS restriction enzyme. Nucleic Acids Res 46(19):e113. https://doi.org/10.1093/nar/gky596
5. Duportet X, Wroblewska L, Guye P et al (2014) A platform for rapid prototyping of synthetic gene networks in mammalian cells. Nucleic Acids Res 42:13440–13451
6. Engler C, Youles M, Gruetzner R et al (2014) A golden gate modular cloning toolbox for plants. ACS Synth Biol 3(11):839–843

Chapter 10

PaperClip DNA Assembly: Reduce, Reuse, Recycle

Maryia Trubitsyna, Annegret Honsbein, Uma Jayachandran, Alistair Elfick, and Christopher E. French

Abstract

Creating DNA constructs is a basic and fundamental step in molecular and synthetic biology. While prices for gene synthesis are decreasing, it is still more economical in most cases to assemble constructs from a library of components (Parts). Many methods for DNA assembly are available, but most require either a fixed and inflexible format for the construct, with all Parts first being cloned in specific donor plasmids, or remaking Parts with new homology ends for each specific assembly reaction, requiring large numbers of single-use oligonucleotides. PaperClip assembly allows Parts stored in any format (linear PCR products or synthetic DNA, or cloned in any plasmid) to be used in totally flexible assembly reactions; up to 11 parts can be assembled in a single reaction, in any order, to give a linear or circular construct, and the oligonucleotides required in the assembly process can be reused in any subsequent assembly. In addition to constructing plasmids for bacterial transformation, PaperClip is also well suited to generate linear products for direct transfection of yeast, mammalian, or cyanobacterial cell lines. Thus, PaperClip offers a simple, flexible, and economical route to multipart assembly of constructs for a wide variety of purposes.

Key words Synthetic biology, Molecular cloning, DNA assembly, Single-pot assembly, Oligonucleotide annealing, Ligation, PAGE, PCR, Agarose gel electrophoresis, CPEC, Selection

1 Introduction

Despite the steadily decreasing costs of DNA synthesis, it is often most economical to assemble DNA constructs from a library of DNA components, which in the literature are often designated as Parts. This is particularly true when large numbers of constructs are made in a combinatorial way—with different arrangements of Parts, total synthesis of all of the desired combinations could be very costly. Thus, DNA assembly remains a vital technique in synthetic biology and genetic engineering. Modern DNA assembly methods can be divided in two main groups: restriction-based methods [1–4] and homology-based methods [5–7]. The former group includes hierarchical methods such as BioBrick assembly, which assemble Parts two at a time, as well as multi-part assembly systems such as Golden Gate and its numerous variants and

Sunil Chandran and Kevin W. George (eds.), *DNA Cloning and Assembly: Methods and Protocols*, Methods in Molecular Biology, vol. 2205, https://doi.org/10.1007/978-1-0716-0908-8_10,

descendants. The latter group includes widely used methods such as Gibson assembly, yeast in-vivo assembly, and circular polymerase extension cloning (CPEC).

All of the methods have their advantages and limitations. Restriction-based methods can assemble repetitive sequences, but the user must ensure that the DNA sequences (Parts) are free of the "forbidden" restriction sites used in assembly. This may require additional mutagenesis steps to delete these sites to prevent undesired cleavage. In addition, hierarchical systems such as BioBrick assembly require multiple assembly steps to assemble large constructs, whereas multi-part assembly systems such as Golden Gate and its variants require the assembly to have a fixed format, with certain types of Parts in certain locations. Both types require Parts to be initially cloned in suitable donor plasmids bearing the necessary restriction sites, which adds a layer of up-front complexity and cost, though the assembly procedures can be simple and rapid once the library has been prepared. By contrast, homology-based methods do not require deletion of unwanted sites, but since the order of assembly is based on end-homology between Parts, it is necessary to remake Parts bearing new homology ends for every new assembly. This normally involves amplification of the sequence of interest with primers carrying the homology area of the neighboring Parts, thus requiring single-use oligonucleotides and numerous PCR reactions prior to assembly. This is particularly inconvenient when many different combinations are to be assembled.

PaperClip DNA assembly [8] is a homology-based method that does not involve use of restriction enzymes and thus does not require the deletion of forbidden restriction sites. PaperClip allows construction of multi-Part plasmids or linear fragments in a single reaction. In contrast to other methods, it allows Parts to be assembled in any order without the requirement for cloning into specific donor plasmids or modification of Part ends. While PaperClip assembly is directed by oligonucleotides, these are Part-specific rather than assembly-specific, and the same oligonucleotides will be reused for any assembly involving a given Part, regardless of the neighboring Parts. Another key feature of PaperClip assembly is that Parts may be introduced into the assembly reaction in various forms, including linear PCR products, linearized plasmids containing the desired sequence, or circular plasmids bearing the desired Part. This provides speed and flexibility, since it is not necessary to prepare a library of Parts in a specific format prior to assembly.

PaperClip assembly requires four oligonucleotides to be obtained for each Part (Fig. 1). These oligonucleotides will be reused in every assembly which involves that Part. Each oligonucleotide is around 40 bases in length: two (forward and reverse) match the upstream end of the Part, and two (forward and reverse) match the downstream end of the Part. Forward and Reverse oligonucleotides are annealed in pairs to generate double stranded

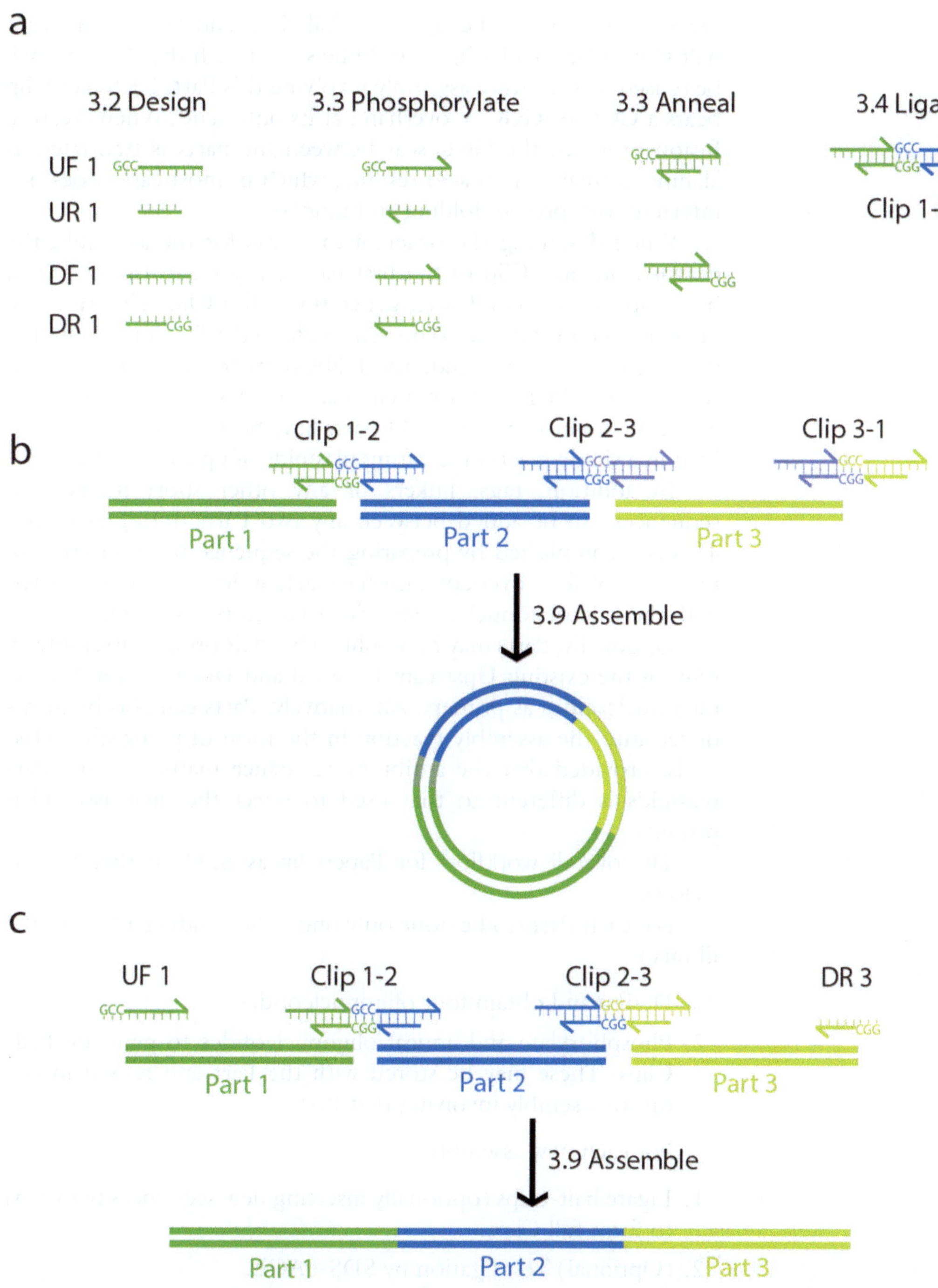

Fig. 1 Scheme of the PaperClip assembly method. (**a**) For each DNA part to be joined, four oligonucleotides ~40 bp should be obtained. These oligonucleotides are phosphorylated and annealed in pairs (forward with reverse) to create half-Clips. The upstream half-Clip contains a GCC sticky end, and the downstream half-Clip contains a compatible GGC sticky end. During the preparation for assembly downstream half-Clips are ligated with upstream half-Clips according to the desired order of the DNA parts in the final construct, creating full Clips. The ligase is then heat-inactivated to prevent interference during next steps. Clips guide the order of DNA parts during the assembly step to create circular (**b**) or linear products (**c**)

DNA "half-Clips"—the upstream half-Clip and the downstream half-Clip. These half-Clips should be stored with the Part and will be reused in any future assembly involving this Part. Each half-Clip bears a GCC or GGC 5′ overhang at its outer end. When creating fusion proteins, the GCC scar between the parts is translated as alanine, a small amino acid residue, which in most cases does not interfere with protein folding and function.

When designing the order of the Parts for the assembly, the downstream half-Clip of the first part is ligated to the upstream half-Clip of the second part, generating a full Clip (Fig. 1a). The ligase is then inactivated to prevent undesired religation during the next steps. The Clips guide the DNA parts to join in the desired order during PCR to form a circular (Fig. 1b) or linear (Fig. 1c) product. In our hands up to 11 parts have been successfully assembled in a single reaction, creating plasmids of up to 12.3 kb.

In addition, tags, linkers or any other short intervening sequences can be added between any two Parts during assembly. This is accomplished by preparing the sequence to be inserted in the form of four overlapping oligonucleotides, which are ligated with the Clip oligonucleotides prior to assembly (Fig. 2).

Optionally, Parts may be amplified by PCR prior to assembly by reusing the existing Upstream Forward and Downstream Reverse oligonucleotides as primers. Alternatively, Parts can also be introduced into the assembly reaction in the form of preexisting plasmids, provided that the antibiotic resistance marker on the Part plasmids is different to that used to select the final assembled product.

The overall workflow for PaperClip assembly is therefore as follows:

For each Part (to be done only once when adding a Part to the library):

1. Design and obtain four oligonucleotides.
2. Phosphorylate and anneal oligonucleotides to generate half-Clips. These may be stored with the Part and reused in any future assembly involving that Part.

For each new assembly:

1. Ligate half-Clips (optionally inserting new sequences between) to form full Clips.
2. (Optional) Test ligation by SDS-PAGE.
3. (Optional) Amplify Parts by PCR prior to assembly.
4. Conduct the assembly reaction using Clips and Parts by PCR (a modification of CPEC).
5. Transform competent cells and select for correct assembly.

The complete workflow, indicating essential and optional steps, is shown in Fig. 3.

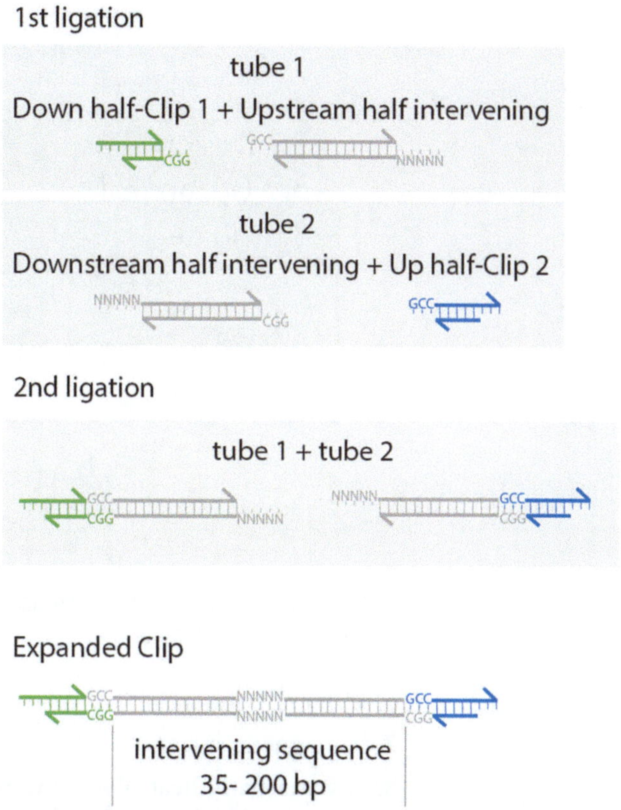

Fig. 2 Intervening sequences of 35–200 bp can be inserted between the DNA parts. These can be protein tags, linkers, protein domains, and so on. The intervening sequence should be obtained in the form of four oligonucleotides. After phosphorylation and annealing, the first ligation is performed: the Upstream half of the intervening sequence is ligated to the downstream half-Clip of the previous DNA part (tube 1) and the 3′ half of the intervening sequence is ligated to the upstream half-Clip of the next DNA part (tube 2). After 1 h incubation, the contents of the two tubes is mixed together and the intervening sequence is restored by 5 bp sticky end ligation during the second ligation step, thus creating an expanded Clip including the inserted sequence

2 Materials

All solutions should be prepared using ultrapure water at room temperature. All DNA should be kept on ice during use. DNA stocks should be stored at −20 °C.

2.1 General Materials Required

1. 1.5 mL microcentrifuge tubes.
2. 0.2 mL clear PCR tubes; a microcentrifuge.
3. Ice.
4. Pipettes and tips.
5. Nuclease free water (Qiagen).

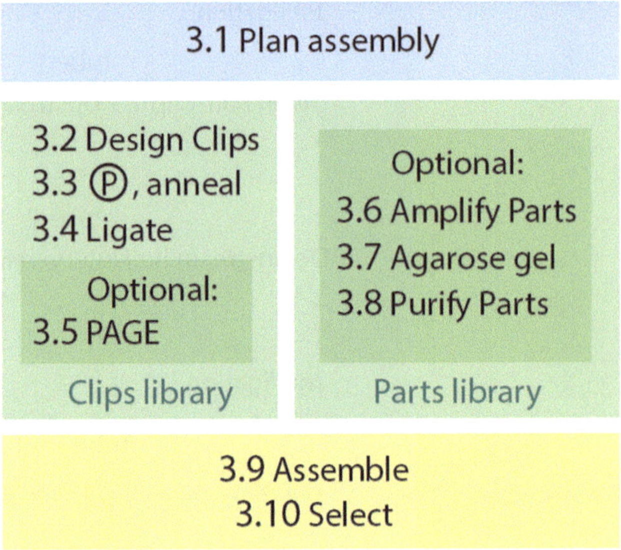

Fig. 3 Overall workflow of PaperClip assembly. Numbers refer to protocols in Subheading 3 of this chapter

6. Ultrapure water.
7. A magnetic stirrer.
8. A water bath, heat-block, thermal cycler or thermal incubator to be used at 16 °C, 37 °C, 65 °C and for slow cooling from 95 °C to room temperature (*see* **Note 1**).

2.2 Clip Preparation

1. T4 polynucleotide kinase (PNK) and 10× T4 PNK buffer (New England BioLabs).
2. T4 DNA ligase and 10× T4 DNA Ligase buffer (New England BioLabs).
3. 10 mM ATP.

2.3 Polyacrylamide Gel Components (for Optional Testing of Ligation Efficiency)

1. 5× TBE buffer: 1.1 M Tris, 900 mM borate, 25 mM ethylenediaminetetraacetic acid (EDTA), pH 8.3. To make 5 × TBE, weigh out 2.34 g disodium EDTA, 27 g Tris base and 13.75 g boric acid. Place these into a 600 mL glass beaker and add ultrapure water to 375 mL. While mixing the solution with a magnetic stirrer, adjust pH to 8.3 with concentrated HCl (*see* **Note 2**). Adjust the volume to 500 mL with ultrapure water, and store at room temperature in a closed bottle. If a white precipitate occurs, discard the solution.
2. 0.5× TBE running buffer: mix 50 mL of 5× TBE with 450 mL of ultrapure water, and store at room temperature.
3. Ammonium persulfate (APS) solution: 10% (w/v) solution in water. Store at −20 °C in 200 μL aliquots.

4. *N,N,N′,N′*-tetramethyl-ethylenediamine (TEMED).
5. SafeWhite™ DNA stain (NBS Biologicals) or GelGreen™ DNA Stain (Biotium) (*see* **Note 3**).
6. 50 bp DNA ladder (New England Biolabs).
7. Casting module, 1.0 mm spacers, glass plates, 1.0 mm casting combs, buffer tank and lid, running module (for example Bio-Rad Mini-PROTEAN).
8. Electrophoresis power supply.
9. UV or blue light transilluminator.

2.4 PCR and PCR Cleanup (for Optional Amplification of Parts Prior to Assembly)

1. Thermal cycler.
2. KOD Hot Start or similar DNA polymerase and buffer (Novagen) (*see* **Note 4**); 2 mM dNTPs; 25 mM $MgSO_4$.
3. 10 μM forward and reverse primers (components of the Clips can be used, *see* Subheading 3.6, **step 3**).
4. Binding buffer (Qiagen): 5 M guanidinium hydrochloride, 30% isopropanol.
5. Wash buffer: 10 mM Tris–HCl pH 7.5, 80% ethanol.
6. Nuclease free water prewarmed to 70 °C.
7. Zymo-Spin™ IC column.

2.5 Agarose Gel Electrophoresis (for Testing of Optional PCR Amplification)

1. 50×TAE buffer. To prepare, mix 121 g Tris base, 28.55 mL acetic acid, 50 mL 0.5 M EDTA (pH 8.0), add ultrapure water up to 500 mL, and stir until dissolved. Store at room temperature for up to a year. To make 1×TAE buffer, mix 10 mL of 50×TAE with 490 mL of ultrapure water. 1×TAE is used for preparing agarose gels and as a running buffer.
2. Agarose, molecular grade.
3. 6× gel loading buffer (New England BioLabs): 2.5% Ficoll®-400, 10 mM EDTA, 3.3 mM Tris–HCl, 0.08% SDS, 0.02% Dye 1, 0.001% Dye 2, pH 8.0 at 25 °C.
4. Electrophoresis power supply.
5. UV or blue light transilluminator.

2.6 Assembly by PCR

1. Thermal cycler.
2. KOD Hot Start or similar DNA polymerase and buffer (Novagen) (*see* **Note 4**), 2 mM dNTPs, 25 mM $MgSO_4$.

2.7 Bacterial Transformation

1. Chemically competent *E.coli* DH10B in 100 μL aliquots (*see* **Note 6**).
2. Water bath or heat-block at 42 °C.

3. Recovery medium SOC: 2% (w/v) tryptone, 0.5% (w/v) yeast extract, 10 mM NaCl, 2.5 mM KCl, 10 mM $MgCl_2$, 20 mM glucose.
4. Shaker or shaking incubator at 37 °C.
5. LB agar plates containing appropriate antibiotic or other selective agent.

3 Methods

All steps should be carried out at room temperature unless otherwise stated.

3.1 Planning the Assembly

1. For plasmid assembly, the final construct should ideally contain a selection marker such as antibiotic resistance and/or a component allowing visual screening, such as a fluorescent protein expression cassette, etc. Having a selection marker as a separate part from the origin of replication will allow easy distinction of correct clones from false-positive clones, which may occur due to circularization of the Part bearing the replication origin.
2. Parts smaller than 200 bp can be ordered as oligonucleotides and used in the assembly as intervening sequences (Fig. 2, Protocol 3.4) (*see* **Note** 7).
3. In our hands, up to 11 parts have been assembled and selected on a single antibiotic, giving a final plasmid over 12 kb.
4. When designing an expression vector it can be useful to include START and STOP codons within the promoter and terminator Parts respectively, omitting these codons from the open reading frame Part—this will allow creation of fusion proteins and/or addition of N- or C-terminal tags if later required.

3.2 Design of Clip Oligonucleotides

1. For each of the DNA parts four oligonucleotides must be obtained (Fig. 4). To make the Upstream Forward (UF) oligonucleotide choose the first ~40 bases; the base at the 3′ end should preferably be G or C. Add GCC bases at the front (5′ end). This is UF (*see* **Note 8**).
2. To make the Upstream Reverse (UR) oligonucleotide choose the first ~37 bases (three bases shorter than the region selected for the UF oligonucleotide). Create the reverse complement strand of it. This is UR.
3. For the Downstream Forward (DF) oligonucleotide, choose the last 40 bases, and check that the three bases at the start (5′ end) are not GCC or GGC (*see* **Note 9**). If they are not GCC or GGC, then use this sequence as DF. If they are GCC or GGC, then choose the last 39 bases instead. This sequence is DF.

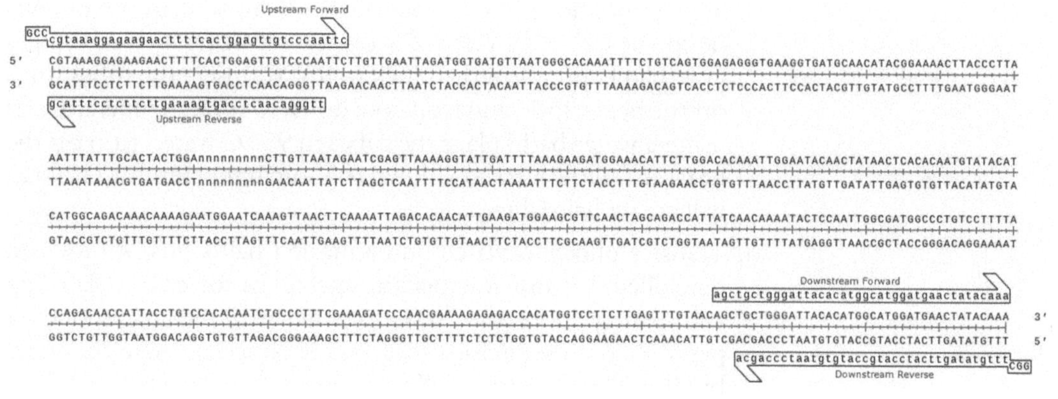

Fig. 4 An example of oligonucleotide design for PaperClip. The Upstream Forward is the first 40 bases plus GCC at the 5′ end. The Upstream Reverse is the reverse complement of the first 37 bases. The Downstream Forward is the last 40 bases of the DNA part. The Downstream Reverse is the reverse complement of the last 37 bases plus GGC at the 5′ end

4. For the Downstream Reverse (DR) oligonucleotide, choose the sequence three bases shorter than DF, and create a reverse complement strand. To this strand add GGC at the start (5′ end). This is DR.
5. Each half-Clip when annealed together creates three base sticky ends at the outer end of the Part for generation of full Clips; and three base overhangs at the inside end to prevent inappropriate ligation.

3.3 Phosphorylate and Anneal

1. Spin the tubes containing the lyophilized oligonucleotides briefly (30 s, 2000 × *g*) to ensure that the pellet is at the bottom of the tube.
2. Add nuclease free water to obtain a final concentration of 100 μM; for example, add 725 μL of water to 72.5 nmol of DNA. Mix by flicking the tube, and spin briefly (30 s, 2000 × *g*) to ensure that all liquid is at the bottom of the tube. Place the tubes on ice.
3. To phosphorylate oligonucleotides (*see* **Note 10**), combine the following in a 0.2 mL clear PCR tube: 20 μL of 100 μM forward oligonucleotide; 20 μL of 100 μM reverse oligonucleotide; 5 μL of T4 PNK buffer, 5 μL of 10 mM ATP; 0.5 μL of T4 Polynucleotide Kinase.
4. Mix by flicking the tubes, and spin briefly (30 s, 2000 × *g*) to return all liquid to the bottom of the tube. Place the tubes in a prewarmed PCR machine, water bath, or heat-block and incubate for 30 min at 37 °C.
5. To anneal the half-Clips, place the tubes with the phosphorylated oligonucleotides in a prewarmed PCR machine at 95 °C.

Slowly cool down the block from 95 °C to 4 °C using minimal ramp (or 0.5 °C/s ramp if it can be set manually). If using a heat-block, prewarm the block to 95 °C, insert the tubes, turn off the heat-block, and let it cool down to room temperature. If using a water bath, place the tubes in 95 °C water, turn off the heat, and allow it to cool down to room temperature. This step will create half-Clips.

6. Transfer phosphorylated and annealed half-Clips (40 μM) to prechilled 1.5 mL microcentrifuge tubes for ease of labeling and storing. Store at −20 °C together with the corresponding parts. Before use, defrost thoroughly on ice and spin the tubes briefly (30 s, 2000 × *g*). Keep the tubes on ice when in use.

3.4 Ligate Clips

1. For each join in the final construct one Clip must be provided. To make each Clip, to a 1.5 mL microcentrifuge tube on ice add: 4 μL of 40 μM downstream half-Clip for the upstream Part; 4 μL of 40 μM upstream half-Clip of the downstream Part; 1 μL of 10× T4 DNA ligase buffer; 0.5 μL of 10 mM ATP; 0.5 μL of T4 DNA Ligase.
2. Mix by flicking the tubes, and spin briefly (30 s, 2000 × *g*) to return all liquid to the bottom of the tube. Incubate at 16 °C for 1 h.
3. Place the tubes in a prewarmed heat-block (or water bath) and incubate at 65 °C for 10 min to inactivate T4 DNA ligase. Spin the tubes briefly (30 s, 2000 × *g*), and place on ice. Store at −20 °C. Before use, defrost thoroughly on ice and spin the tubes briefly (30 s, 2000 × *g*). Keep the tubes on ice when in use.
4. To check the quality of the Clips, analysis by polyacrylamide gel electrophoresis is recommended (*see* **Note 11**).
5. If an element is to be inserted between two parts, it should be designed and prepared in the form of four oligonucleotides: UF, UR, DF and DR, as indicated in Fig. 2 (*see* **Note 7**). These should be phosphorylated and annealed in pairs (UF with UR, DF with DR) as described in Subheading 3.3. The ligation to prepare the expanded Clip is conducted in two steps, as shown in Fig. 2. The first ligation step takes place in two separate tubes. To the first tube add: 4 μL of 40 μM downstream half-Clip 1, 4 μL of 40 μM upstream half-intervening sequence, 1 μL of 10× T4 DNA ligase buffer, 0.5 μL of 10 mM ATP, 0.5 μL of T4 DNA Ligase. To the second tube add: 4 μL of 40 μM downstream half-intervening sequence, 4 μL of 40 μM upstream half-Clip 2, 1 μL of 10× T4 DNA ligase buffer, 0.5 μL of 10 mM ATP, 0.5 μL of T4 DNA Ligase.
6. Mix by flicking the tubes, and spin briefly (30 s, 2000 × g) to return all liquid to the bottom of the tube. Incubate at 16 °C for 1 h.

7. For the second ligation step, combine 5 μL of the contents of the first tube with 5 μL of the contents of the second tube. Mix by flicking the tube, spin briefly (30 s, 2000 × *g*) and incubate for a further 1 h at 16 °C.
8. Place the tubes in a prewarmed heat-block (or water bath) and incubate at 65 °C for 10 min to inactivate T4 DNA ligase. Spin the tubes briefly (30 s, 2000 × *g*), and place on ice. Store at −20 °C. Before use, defrost thoroughly on ice and spin the tubes briefly (30 s, 2000 × *g*). Keep the tubes on ice when in use.
9. To check the quality of the Clips, analysis by polyacrylamide gel electrophoresis is recommended (*see* **Note 11**).

3.5 PAGE Analysis of Clip ligation Efficiency (Optional but Recommended)

1. The following protocol relates to preparation and use of hand-cast PAGE gels in the Bio-Rad Mini-Protean or similar system. Alternatively, precast gels may be obtained or an alternative electrophoresis system may be used.
2. Clean and dry the glass plates, set up the casting module, insert the comb at an angle.
3. To prepare two 9% w/v polyacrylamide mini gels (8.3 × 7.3 cm, 1.0 mm thickness), mix the components in a 50 mL tube in the following order: 5.9 mL ultrapure water, 3.7 mL 30% w/v acrylamide:bisacrylamide (37.5:1), 2.4 mL 5× TBE, 200 μL 10% w/v APS and 10 μL TEMED. Mix gently, avoiding introduction of bubbles.
4. Pour the gel mixture using a 1 mL pipette and insert the comb completely. Leave to set for 30–40 min at room temperature.
5. Gels can be stored for up to 1 month at 4 °C, wrapped in a tissue paper soaked with water, in a sealed bag or cling film.
6. Fit the gel in a running module according to the manufacturer's instructions. Pour the running buffer in the inner chamber to the top and in the outer chamber up to 7–10 cm. Usually 500 mL of 0.5× TBE buffer is sufficient. Carefully remove the comb and wash out the wells with running buffer using a syringe with a needle.
7. Prepare the samples by mixing 0.5 μL of the full Clip, 4.5 μL of nuclease free water, and 1 μL of SafeWhite™.
8. Prepare the DNA ladder: 1 μL of 50 bp DNA Ladder (New England BioLabs), 4 μL of nuclease free water, and 1 μL of SafeWhite™.
9. Load the samples using gel-loading tips. Apply 50 V for 10 min to allow the samples to enter the gel, then change the voltage to 100 V and run the gel for a further 60–90 min.
10. Remove the gel from the glass plates and proceed with visualization or post-staining (*see* **Note 3**).

11. Check the efficiency of ligation (*see* **Note 12**). Ideally clear bands should be present at the expected size for the full Clip (around 83 bp, or larger if additional sequences have been inserted between the half-Clips), and bands corresponding to the unligated half Clips should be faint or absent.

3.6 Amplify Parts

1. DNA Parts can be used in the form of linear PCR products, DNA fragments excised from a plasmid, or linearized or circular plasmids which contain the desired Part sequence, provided that such plasmids do not contain the same antibiotic selection marker which will be used for the final construct (*see* **Note 13**).
2. If it is desired to use linear PCR products as Parts, these may be amplified from a plasmid or genomic DNA template using the UF and DR oligonucleotides as primers. The following protocol relates to preparation of such Parts and may be omitted if Parts are to be used in other formats (such as existing plasmids). Any standard PCR protocol may be used in place of the following.
3. Dilute UF and DR oligonucleotides 1:10 to obtain a final concentration of 10 μM (use 5 μL of 100 μM stock and 45 μL of nuclease free water). Store diluted oligonucleotides at −20 °C. Before use, defrost thoroughly on ice and spin the tubes briefly (30 s at 2000 × *g*), and keep on ice when in use.
4. In a 0.2 mL clear PCR tube add: 29 μL of nuclease-free water, 5 μL of 10× KOD Hot Start Buffer, 5 μL of 2 mM dNTPs, 3 μL of 25 mM $MgSO_4$, 1 μL of ~5 ng/μL DNA template containing the required Part sequence, 3 μL of 10 μM upstream forward primer, 3 μL of 10 μM downstream reverse primer, 1 μL of 10 U/μL KOD Hot Start DNA Polymerase.
5. Flick the tube, spin briefly (30 s at 2000 × *g*), and place in a PCR machine prewarmed to 95 °C.
6. Run the following program: 2 min at 95 °C, followed by 35 cycles of 20 s at 95 °C, 10 s at the lowest Tm of the primers, and 70 °C for a period of 10 s/kb if the Part length is less than 500 bp, 15 s/kb if the Part length is 500 bp to 1 kb, 20 s/kb if the Part length is 1 kb to 3 kb, or 25 s/kb if the Part length is greater than 3 kb. The temperature should then reduce to 10 °C and hold.

3.7 Agarose Gel Electrophoresis

1. To check the success of the PCR amplification of the DNA parts, prepare a 1% (w/v) agarose gel. Place 3 g agarose (molecular biology grade) and 30 mL 1×TAE buffer in a glass beaker or heat-resistant bottle (*see* **Note 14**). Heat in the microwave until just starting to boil (e.g., ~30 s at max power 700 W). Swirl the bottle until the agarose is dissolved.

2. Cool the agarose to ~50 °C, swirling it occasionally, and add 3 μL of SafeView™ DNA stain for in-gel staining. Alternatively, the gel can be stained after electrophoresis (*see* **Note 3**).
3. Clean the gel casting tray with water, dry and wipe with 70% v/v ethanol. Cover the sides of the tray with tape to prevent gel leakage. Place the casting tray on an even surface and fit the comb. Pour agarose till the gel is 3–5 mm high (*see* **Note 15**). Leave to set for at least 30 min at room temperature. Do not move the tray while the gel is setting (*see* **Note 16**).
4. Prepare the samples by mixing 5 μL of the PCR reaction with 1 μL of 6× gel loading dye.
5. Prepare DNA ladder by mixing 1 μL of the ladder (1 kb DNA Ladder or 1 kb Plus DNA ladder) with 4 μL of nuclease free water and 1 μL of 6× gel loading dye.
6. Place the set agarose gel in the gel running chamber. Top up the chamber with 1×TAE till the gel is completely covered.
7. Carefully load the samples and the DNA ladder. Separate DNA at 60 V for 60 min.
8. Proceed to visualization or post-staining (*see* **Note 3**) if stain was not added to the gel prior to casting.

3.8 Purification of PCR-Amplified Parts

1. If the PCR resulted in a single band of the expected size, column purification can be used to remove primers, dNTPs and polymerase. If several bands are obtained either optimisation of the PCR (*see* **Note 17**) or gel-purification (*see* **Note 18**) can be performed.
2. For column purification add 250 μL of the binding buffer to the remaining PCR mix (~45 μL). Mix by pipetting.
3. Transfer the whole mixture to the Zymo-Spin™ IC column. Place the column in the collection tube.
4. Spin the column for 1 min in a microcentrifuge at 13,000 rpm (12,500 × *g*) at room temperature. Discard the flow-through.
5. Add 500 μL of the washing buffer to the column.
6. Spin the column for 1 min at 13,000 rpm at room temperature. Discard the flow-through. Spin the column again to remove residual buffer.
7. Pipette 20 μL of nuclease free water prewarmed to 70 °C directly on the resin of the column. Let stand at room temperature for 5 min.
8. Place the column in a new collection tube. Spin the column for 1 min at 13,000 rpm at room temperature. This should result in ~18 μL of >100 ng/μL DNA concentration, which is enough for ~18 assemblies if the part is around 1 kb.

Table 1
Components for Paperclip assembly

Component	Final concentration	Stock concentration	Volume
DNA parts	15 nM	2 kb part, 100 ng/μL 1 kb part, 100 ng/μL	2 μL 1 μL
DNA clips	67.2 nM	16 μM	0.21 μL
dNTPs	200 nM	2 mM each	5 μL
$MgSO_4$	1.5 mM	25 mM	3 μL
KOD Hot Start buffer (Novagen)	1×	10×	5 μL
KOD Hot Start polymerase (Novagen)	1.0 U	1.0 U/μL	1 μL
Nuclease-free water			To final volume 50 μL

9. Store DNA at −20 °C. Before use defrost thoroughly on ice and spin the tubes briefly (30 s, 2000 × *g*). Keep on ice when in use.

3.9 Assembly

1. Calculate the required volumes of all components according to Table 1. Add these components to a 0.2 mL PCR tube, adding water first and polymerase last.
2. Mix by flicking the tube, spin briefly (30 s, 2000 × *g*), and place in a PCR machine prewarmed to 95 °C.
3. Run the assembly program (a modification of CPEC [9]): 2 min denaturation at 95 °C, followed by 20 cycles of 20 s at 95 °C, slow cooling (minimal ramp) to 70 °C, and 1 min at 70 °C (for a 3 kb final product). The temperature should then decrease to 10 °C and hold.
4. To increase the number of colonies, a second "boost" PCR may be performed. In a clean 0.2 mL PCR tube add 31 μL of nuclease-free water, 5 μL of the assembly reaction from the preceding step, 5 μL of 2 mM dNTPs, 3 μL of 25 mM $MgSO_4$, 5 μL of 10× KOD Hot Start Buffer, and 1 μL of KOD Hot Start DNA Polymerase. Mix and repeat the same assembly program.
5. If the final construct is required in the form of linear DNA, add Upstream Forward (of the first Part) and Downstream Reverse (of the last part) primers to the "boost" PCR reaction and run the assembly program.

3.10 Select Correct Clones

1. The best results are obtained if the antibiotic selection marker is introduced as a separate Part from the origin of replication. If this selection marker is in a plasmid which cannot replicate in the recipient strain, all of the final clones should be correct assembly products. For example, we routinely use a kanamycin resistance cassette carried on a plasmid with a conditional origin of replication (oriR6K), and the destination strain *E.coli* DH10B, in which this plasmid is unable to propagate due to absence of the required *pir* proteins. Thus, all kanamycin resistant clones contain assembled plasmid DNA.
2. If the vector backbone contains a marker which will be lost after the assembly (RFP, *lacZ'*, etc.), then colonies containing assembled plasmids can easily be distinguished visually from those containing the vector alone.
3. Multiple antibiotic and/or color selection can be used to reduce the number of background colonies.

4 Notes

1. For Clip preparation all incubation steps can be conveniently performed in a thermal cycler. Alternatively, phosphorylation can be done at 37 °C in an incubator, a water bath, or a heat-block. Slow cooling down for annealing of half-Clips can be performed in a water bath or a heat-block by turning it off after preheating to 95 °C. Ligation can be done at 16 °C in an incubator or a cooling water bath. Ligase inactivation at 65 °C can be done in a water bath or a heat-block.
2. Adequate personal protection should be worn when handling concentrated HCl.
3. Gels can alternatively be post-stained with GelGreen™ DNA stain. Dilute GelGreen™ stain in water according to the manufacturer's instructions (15 μL for 50 mL water). Place the gel in the staining solution and incubate for 10–20 min protected from light. Shaking is not necessary. Proceed with visualization. The staining solution can be stored at room temperature, protected from light and reused during at least 1 month.
4. Add polymerase as the last component to the PCR mix, mix by flicking the tube, spin briefly, and place the tube in a thermal cycler preheated to 95 °C.
5. If deletions happen during PCR assembly, a reduced number of cycles might solve the problem (e.g., 5 cycles instead of 20). Insufficient purity of the Parts might also cause deletions during assembly; try gel purification and reamplification of the parts in which deletion occurred.

6. When handling competent cells, minimize their exposure to room temperature and keep them on ice. After the cells are thawed, add DNA straight away. Do not exceed 30 min incubation on ice and do not shake the tube during heat-shock.
7. To design oligonucleotides for intervening sequences (Fig. 2) highlight 5 bases in the middle of the sequence. Select the 5′ half of the sequence without the highlighted bases, add GCC at the 5′ end—this is UF. Select the sequence before and including the highlighted bases, and create the reverse complement—this is UR. The highlighted bases together with the 3′ half of the sequence—are DF. Select the 3′ end without the highlighted bases, create the reverse complement and add GGC at the 5′ end of it—this is DR. The annealing and ligation of the intervening sequence will be efficient if it is longer than 35 bp.
8. If the position of the start and end of the DNA part is flexible, it is recommended to choose oligonucleotides which are not likely to form hairpin structures. This can be checked at https://eu.idtdna.com/calc/analyzer.
9. It is recommended to avoid a 5′ G base in the DF design to reduce formation of Clip multimers during the ligation step.
10. Alternatively, the oligonucleotides may be ordered in phosphorylated form, usually at additional expense. If using pre-phosphorylated oligonucleotides, simply proceed with the annealing and ligation steps. It is also possible to include polynucleotide kinase in the clip ligation reactions, but this may require longer incubation periods than stated in Subheading 3.4 (e.g., overnight).
11. It is possible to use 2% (w/v) agarose gel separation to check the efficiency of ligation, but the resolution sensitivity will be less than that of PAGE.
12. If less than 50% of the DNA is ligated to create the full Clips, the ligation reaction can be extended to 16 h. If there are bands of higher molecular weight than is expected for full Clips (~80 bp), the ligation time can be reduced or oligonucleotides should be redesigned. Take extra care when choosing incompatible sticky ends on the inward facing ends of the half-Clips.
13. The best assembly results for high number of parts are obtained when using linear DNA parts after PCR amplification and DNA cleanup (see Subheading 3.8).
14. Add agarose to a dry beaker or heat-resistant bottle first, then add 1×TAE—this will result in more even dissolving of the agarose.
15. The thinner the agarose gel, the sharper DNA bands will appear on it. The best picture can be obtained if wider wells

are used with the lowest DNA sample volume (e.g., 0.75 mm × 5 mm well and 6 μL sample volume).

16. If the gel tray is moved while the agarose is setting, this will result in visible short fluorescent lines while visualizing DNA with the UV transilluminator.
17. If no bands are obtained during PCR, lower annealing temperature and longer elongation time can be used. If the DNA part has high GC content or is long and/or amplified from the genomic DNA template, addition of DMSO to a final concentration of 2–10% and increasing the concentration of $MgSO_4$ to 2 mM might be helpful.
18. Gel-purification gives the purest DNA, but usually results in low final DNA concentration. It might be useful to use a gel-purified DNA band as a template to amplify higher amount of DNA for the assembly.

Acknowledgments

This work was supported by Engineering and Physical Sciences Research Council [EP/J02175x/1].

References

1. Engler C, Kandzia R, Marillonnet S (2008) A one pot, one step, precision cloning method with high throughput capability. PLoS One 3 (11):e3647. https://doi.org/10.1371/journal.pone.0003647
2. Weber E, Engler C, Gruetzner R, Werner S, Marillonnet S (2011) A modular cloning system for standardized assembly of multigene constructs. PLoS One 6(2):e16765. https://doi.org/10.1371/journal.pone.0016765
3. Shetty R, Lizarazo M, Rettberg R, Knight TF (2011) Assembly of BioBrick standard biological parts using three antibiotic assembly. Methods Enzymol 498:311–326. https://doi.org/10.1016/B978-0-12-385120-8.00013-9
4. Casini A, MacDonald JT, De Jonghe J, Christodoulou G, Freemont PS, Baldwin GS, Ellis T (2014) One-pot DNA construction for synthetic biology: the modular overlap-directed assembly with linkers (MODAL) strategy. Nucleic Acids Res 42(1):e7. https://doi.org/10.1093/nar/gkt915
5. Gibson DG, Young L, Chuang RY, Venter JC, Hutchison CA 3rd, Smith HO (2009) Enzymatic assembly of DNA molecules up to several hundred kilobases. Nat Methods 6 (5):343–345. https://doi.org/10.1038/nmeth.1318
6. Quan J, Tian J (2009) Circular polymerase extension cloning of complex gene libraries and pathways. PLoS One 4(7):e6441. https://doi.org/10.1371/journal.pone.0006441
7. Zhang Y, Werling U, Edelmann W (2012) SLiCE: a novel bacterial cell extract-based DNA cloning method. Nucleic Acids Res 40 (8):e55. https://doi.org/10.1093/nar/gkr1288
8. Trubitsyna M, Michlewski G, Cai Y, Elfick A, French CE (2014) PaperClip: rapid multi-part DNA assembly from existing libraries. Nucleic Acids Res 42(20):e154. https://doi.org/10.1093/nar/gku829
9. Quan J, Tian J (2011) Circular polymerase extension cloning for high-throughput cloning of complex and combinatorial DNA libraries. Nat Protoc 6(2):242–251. https://doi.org/10.1038/nprot.2010.181

Chapter 11

Phytobricks: Manual and Automated Assembly of Constructs for Engineering Plants

Yao-Min Cai, Jose A. Carrasco Lopez, and Nicola J. Patron

Abstract

Phytobricks are standardized DNA parts for plants that can be assembled hierarchically into transcriptional units and, subsequently, into multigene constructs. Phytobricks each contain the sequences of one or more functional elements that comprise eukaryotic transcription units, with sequence features that enable them to be used interchangeably in one-step cloning reactions to facilitate combinatorial assembly. The simplicity and efficiency of this one-step reaction has enabled Phytobrick assembly to be miniaturized and automated on liquid handing platforms. In this method, we describe how to design and construct new Phytobricks as well as how to assemble them in both manual and nanoscale automated one-step reactions. Finally, we describe a high-throughput method for sequence verification of assembled plasmids.

Key words Plant biotechnology, Plant synthetic biology, DNA assembly, Molecular cloning, Golden Gate, Automation, Liquid handling

1 Introduction

Parallel DNA assembly methods enable multiple fragments of DNA to be joined together in a designated order in a single reaction [1, 2]. Such methods are preferable to earlier technologies that employed multiple-cloning sites or site-specific recombinases (e.g., Gateway Cloning) to insert single fragments into plasmids because they enable the rapid assembly of numerous DNA sequences into complex designs while reducing the presence of unwanted base-pairs, often known as scars, between assembled fragments. In addition, they have removed the need for laboratories to purchase proprietary vectors and reagents or retain large suites of restriction endonucleases.

Broadly, there are two approaches to parallel DNA assembly. Firstly, methods that utilize Type IIS restriction endonucleases (e.g., AarI, BpiI, BsaI, BsmBI, SapI), known as Golden Gate Assembly [3, 4]. Secondly, methods dependent on the production of linear fragments of DNA where the ends of each fragment are

Sunil Chandran and Kevin W. George (eds.), *DNA Cloning and Assembly: Methods and Protocols*, Methods in Molecular Biology, vol. 2205, https://doi.org/10.1007/978-1-0716-0908-8_11, © Springer Science+Business Media, LLC, part of Springer Nature 2020

homologous to adjacent fragments in the final assembly [5, 6]. The most widely used method to assemble these so-called overlapping fragments is isothermal assembly, more commonly known as Gibson assembly [5, 6]. Both Type IIS-dependent and overlap-dependent approaches are robust, but each has specific advantages and disadvantages. For example, overlap-dependent methods are highly flexible and have few restrictions on the sequence—DNA fragments obtained by PCR amplification from, for example, a genomic DNA template can be used directly in assembly reactions together with almost any plasmid. However, reuse of any sequence in a new assembly requires reamplification with a new pair of primers to introduce the required homology to adjacent fragments. As a result, the time required for design as well as the costs for combinatorial assemblies of even a small number of parts can be significant. In contrast, Type IIS-mediated assembly methods require the absence of any additional instances of the recognition site for the enzyme(s) used for assembly in all DNA fragments to be assembled as well as in the backbones. Depending on the specific sequence, this process, which is referred to as "domestication," can be slow. However, once a sequence has been prepared or domesticated for assembly it can be reused in any number of subsequent assemblies without any further modifications. It therefore lends itself to the production of reusable, characterized DNA "parts."

Type IIS restriction endonucleases generate single-stranded overhangs outside of their recognition site (Fig. 1). DNA parts are released from their holding plasmids by digestion and the complementary overhangs of multiple parts are then joined together using T4 ligase (Fig. 1). As the recognition sequence is not present in the cleavage site, it will also not be present in the assembled molecule. Consequently, the digestion and ligation reactions can be performed simultaneously in a one-step reaction (Fig. 1). A typical Type IIS assembly therefore only requires the user to combine multiple plasmid DNA molecules (the parts to be assembled and an plasmid acceptor) with the restriction enzyme–ligase cocktail. It is therefore relatively simple to scale up the number of reactions that can be performed at one time using laboratory automation [7].

Type IIS assembly allows DNA parts to be assembled together without introducing unwanted nucleotides or scars between them. However, this requires bespoke overhangs to be produced by the Type IIS enzyme for each pair of adjacent DNA parts. The use of standardized overhangs means that DNA parts can be reused in new assemblies with other compatible parts. To enable the exchange of interoperable and reusable DNA parts, a so-called common syntax that describes the properties of standardized DNA parts for plants (and is extensible to all eukaryotes) was agreed by the plant community [8]. This has become known as the Phytobrick standard and breaks a typical eukaryotic

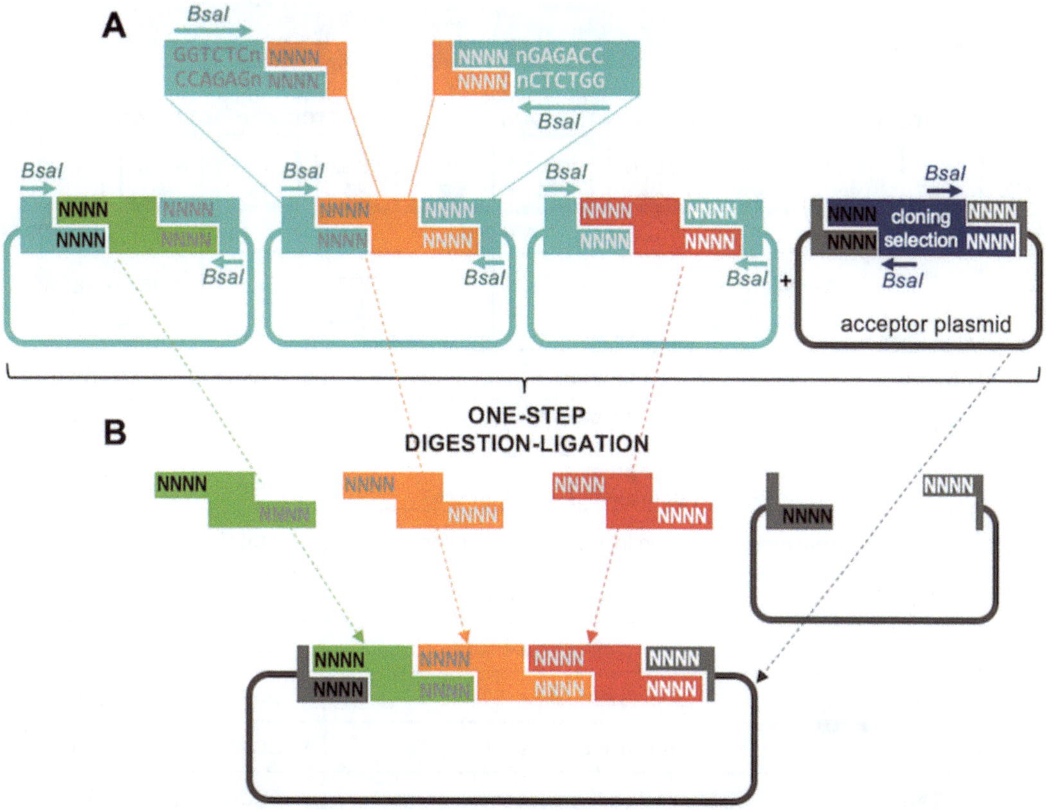

Fig. 1 (**a**) Type IIS restriction endonuclease recognition sites are nonpalindromic and cut outside of their recognition site. The resulting overhangs have no specific sequence requirements. (**b**) The recognition sequences are not present in the released fragments and are therefore absent from the final assembly. This enables simultaneous assembly of multiple DNA parts in a one-step digestion–ligation reaction, providing the overhangs are compatible

transcription unit into ten functional parts (Fig. 2). Each Phytobrick part is flanked by a pair of convergent recognition sites for *BsaI* which, when digested, result in a standard overhang. As each type of part is flanked by the same overhang, parts can be reused without modification. The inclusion of standard overhangs between parts introduces an assembly scar of up to four base pairs between parts, however, these have been selected to have minimal impact. For example, the AATG four-base overhang at the start of coding sequences includes the AUG triplet codon for methionine. Similarly the overhangs used for N and C terminal tags introduce the small amino acids, serine and glycine, which are least likely to interfere with protein structure. Although the transcriptional unit can be broken into up to ten parts, one or more of these can be contained in a larger part if desired (Fig. 2). Commonly, researchers will assemble a coding sequence part between two parts containing appropriate regulatory sequences for the species of interest.

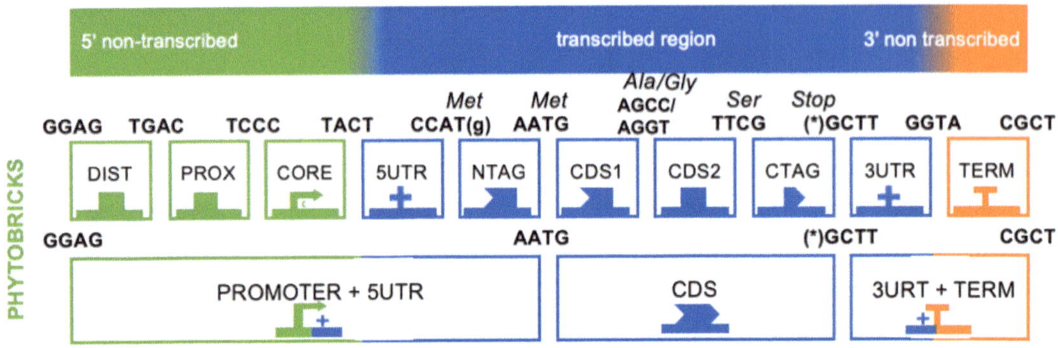

Name	Function	5' Overhang	3' Overhang
DIST	distal promoter region (*cis*-regulatory elements, enhancers)	GGAG	TGAC
PROX	proximal promoter region (*cis*-regulatory elements, enhancers)	TGAC	TCCC
CORE	minimal promoter region including transcription start site	TCCC	TACT
5UTR	5' untranslated sequence	TACT	CCAT
NTAG	N terminal coding sequence	CCAT	AATG
CDS1	coding sequence	AATG	AGCC/AGGT
CDS2	coding sequence	AGCC/AGGT	TTCG
CTAG	C terminal coding sequence	TTCG	GCTT
3UTR	3' untranslated sequence	GCTT	GGTA
TERM	terminator	GGTA	CGCT

Fig. 2 Phytobricks are standardized DNA parts for plants. The plant common syntax divides eukaryotic gene structure into ten functional basic parts. Phytobricks are comprised of one or more adjacent basic parts and are flanked by standard overhangs

Several plasmid tool kits, including MoClo [9], Golden Braid [10], Loop [11], and Mobius [12], can be used to assemble Phytobricks (Fig. 3). These kits all enable hierarchical assembly of multigene constructs into binary plasmid vectors with the necessary features for transformation of *Agrobacterium tumefaciens* in preparation for DNA-delivery to plant cells. In the first step, Phytobricks, sometimes referred to as Level 0 parts or modules, are assembled into transcriptional units using *BsaI* (Fig. 3). Subsequently, transcriptional units are assembled in one or more hierarchical assembly reactions to enable the construction of higher-order multigene assembles. The latter is done by utilising an additional Type IIS endonuclease: BpiI (MoClo), SapI (Loop), AarI (Mobius), and BsmBI (Golden Braid) (Fig. 3).

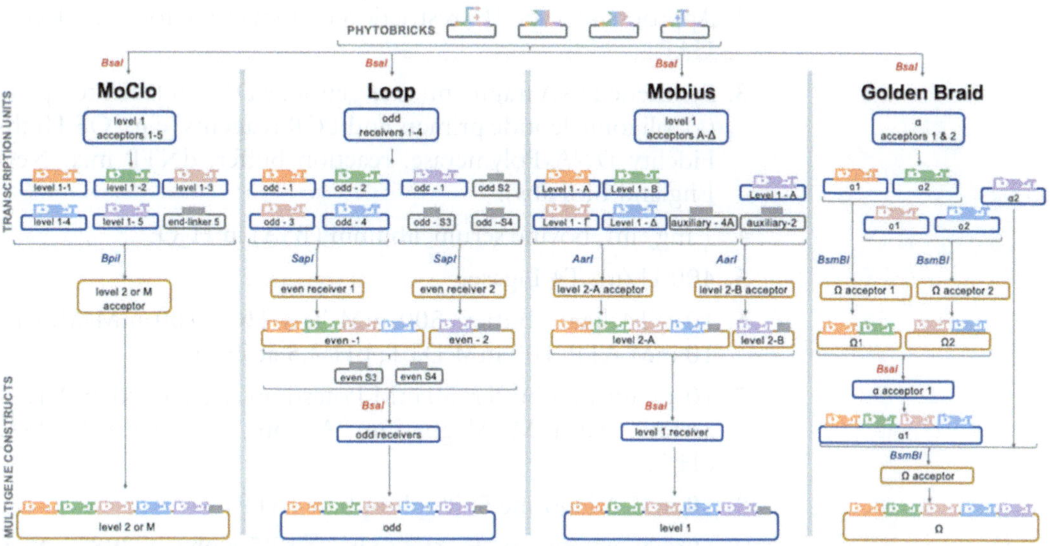

Fig. 3 Phytobricks can be assembled into transcription units and, subsequently, into multigene constructs using a number of different plasmid tool kits (MoClo, Loop, Mobius, and GoldenBraid). In this example Phytobricks are initially assembled into five acceptor plasmids in one-step digestion–ligation reactions using *BsaI*. The route to assembling these five transcription units into a single construct differs between tool kits

In this chapter, we describe methods for the design and construction of new Phytobricks and provide a generalized protocol for the manual assembly of Phytobrick parts into single and multigene constructs. We also describe protocols for high-throughput, miniaturized assembly using laboratory automation as well as sequence verification of constructs using Illumina sequencing.

2 Materials

2.1 Design of Phytobricks

1. DNA analysis and editing software. Any suitable sequence analysis and editing software package can be used. Popular examples include Benchling (www.benchling.com) and Snapgene (www.snapgene.com). Plasmid tool kit-specific software tools are also available (e.g., Loop [https://github.com/HaseloffLab/LoopDB/tree/loopdesigner] or GoldenBraid [https://gbcloning.upv.es/tools/]).

2.2 Construction of Phytobricks

1. Universal acceptor plasmid (UAP) for Phytobricks. Type IIS plasmid assembly tool kits each contain a similar UAP, each using the endonuclease specific to that system (e.g., MoClo: pUAP1 [Addgene #63674; BpiI], GoldenBraid: pUPD2 [Addgene #68161; BsmBI], Loop: pUAP4 [*see* ref. 11; SapI], Mobius: mUAV [Addgene #102680; AarI]).

2. Appropriate Type IIS restriction endonuclease for the selected UAP.
3. Synthetic DNA fragment(s) *or* genomic DNA/cDNA template (s), oligonucleotide primers and PCR reagents (e.g., Q5 High-Fidelity DNA Polymerase, reaction buffer, dNTP mix, New England BioLabs).
4. 1 mg/mL bovine serum albumin (BSA) in H_2O.
5. 400 U/μL T4 Ligase.
6. 10× T4 ligase buffer: 500 mM Tris–HCl, 100 mM $MgCl_2$, 10 mM ATP, 100 mM DTT; pH 7.5 at 25 °C.
7. 10× Cutsmart buffer: 50 mM Potassium Acetate, 20 mM Tris-acetate, 10 mM Magnesium Acetate, 100 μg/mL BSA, pH 7.9.
8. 10 mM Adenosine 5′-Triphosphate (ATP).
9. Highly-competent *E. coli* (e.g., NEB5-alpha, C2987, New England BioLabs; *see* **Note 1**).
10. LB-agar: 10 g peptone 140, 5 g yeast extract, 5 g sodium chloride, 12 g agar, 1 L H_2O. Autoclave to sterilise, cool to <50 °C and add appropriate antibiotics (50 μg/mL kanamycin or 100 μg/mL carbenicillin or 25 μg/mL chloramphenicol or 50 μg/mL spectinomycin). If blue/while selction is required, also add 500 μM isopropyl-β-D-thiogalactoside (IPTG) and 20 μg/mL 5-bromo-4-chloro-3-indolyl β-D-galactopyranoside (X-Gal).
11. Thermal cycler.

2.3 Manual Assembly of Transcription Units

1. Phytobrick parts (*see* **Note 2**).
2. Acceptor plasmid(s) from the MoClo (Addgene Kit #1000000044), Golden Braid (Addgene Kit # 1000000076), Loop [11], or Mobius (Addgene Kit #1000000134) tool kits.
3. 10 U/μL BsaI (Eco31I) (e.g., Eco31I, Thermo Fisher Scientific or Bsa*I*-HF v2, New England BioLabs; *see* **Note 3**).
4. 1 mg/mL bovine serum albumin in H_2O.
5. 400 U/μL T4 Ligase.
6. 10× T4 ligase buffer: 500 mM Tris–HCl, 100 mM $MgCl_2$, 10 mM ATP, 100 mM DTT; pH 7.5 at 25 °C.
7. Highly competent *Escherichia coli* (e.g., NEB5-alpha, New England BioLabs; *see* **Note 1**).
8. LB agar: 10 g peptone 140, 5 g yeast extract, 5 g sodium chloride, 12 g agar, 1 L H_2O. Autoclave to sterilise, cool to <50 °C and add appropriate antibiotics (50 μg/mL kanamycin or 100 μg/mL carbenicillin or 25 μg/mL chloramphenicol or 50 μg/mL spectinomycin). If blue/while selction is required,

also add 500 μM isopropyl-β-D-thiogalactoside (IPTG) and 20 μg/mL 5-bromo-4-chloro-3-indolyl β-D-galactopyranoside (X-Gal).

9. Thermal cycler.

2.4 Manual Assembly of Multigene Constructs

1. Assembled transcription units in MoClo, Golden Braid, Loop or Mobius plasmid backbones.
2. Compatible acceptor plasmids and relevant end-linkers (if required) from the same tool kit.
3. Appropriate restriction endonuclease(s) e.g., AarI, BsaI, BpiI, BsmBI or SapI for the acceptor plasmid(s) from the selected plasmid tool kit (see Fig. 3).
4. 1 mg/mL bovine serum albumin in H_2O.
5. 400 U/μL T4 Ligase.
6. 10× T4 ligase buffer: 500 mM Tris–HCl, 100 mM $MgCl_2$, 10 mM ATP, 100 mM DTT; pH 7.5 at 25 °C.
7. 10× Cutsmart buffer: 50 mM potassium acetate, 20 mM Tris-acetate, 10 mM magnesium acetate, 100 μg/mL BSA, pH 7.9.
8. 10 mM adenosine 5′-triphosphate (ATP).
9. Highly competent *E. coli* (e.g., NEB5-alpha, New England BioLabs) (*see* **Note 1**).
10. LB-agar: 10 g peptone 140, 5 g yeast extract, 5 g sodium chloride, 12 g agar, 1 L H_2O. Autoclave to sterilize, cool to <50 °C and add appropriate antibiotics (50 μg/mL kanamycin or 100 μg/mL carbenicillin or 25 μg/mL chloramphenicol or 50 μg/mL spectinomycin). If blue/while selction is required, also add 500 μM isopropyl-β-D-thiogalactoside (IPTG) and 20 μg/mL 5-bromo-4-chloro-3-indolyl β-D-galactopyranoside (X-Gal).
11. Thermal cycler.

2.5 Automated, Nanoscale Assembly of Transcription Units

1. 10 nM Phytobrick parts.
2. 10 nM acceptor plasmid(s) from the MoClo (Addgene Kit #1000000044), Golden Braid (Addgene Kit # 1000000076), Loop [11] or Mobius (Addgene Kit #1000000134) tool kits.
3. PCR grade water.
4. 1 mg/mL bovine serum albumin in H_2O.
5. 20,000 U/mL BsaI (*see* **Note 3**).
6. 400,000 U/mL T4 Ligase.
7. 10× T4 ligase buffer: 500 mM Tris–HCl, 100 mM $MgCl_2$, 10 mM ATP, 100 mM DTT; pH 7.5 at 25 °C.
8. Ultra-competent *E. coli* (e.g., XL-Gold Ultra, Agilent).

9. LB-agar: 10 g peptone 140, 5 g yeast extract, 5 g sodium chloride, 12 g agar, 1 L H_2O. Autoclave to sterilise, cool to <50 °C and add appropriate antibiotics (50 μg/mL kanamycin or 100 μg/mL carbenicillin or 25 μg/mL chloramphenicol or 50 μg/mL spectinomycin). If blue/while selction is required, also add 500 μM isopropyl-β-D-thiogalactoside (IPTG) and 20 μg/mL 5-bromo-4-chloro-3-indolyl β-D-galactopyranoside (X-Gal).
10. Super Optimal broth with Catabolite repression (SOC): 2% tryptone, 0.5% yeast extract, 10 mM NaCl, 2.5 mM KCl, 10 mM $MgCl_2$, 10 mM $MgSO_4$, 20 mM glucose.
11. High-throughput plasmid purification kit or reagents (e.g., Macherey-Nagel Nucleospin 96 kit).
12. 8-well plates (e.g., 8-well Nunc rectangular dishes).
13. A flexible, tip-based liquid automated handing platform that can transfer volumes of 2 μL to 4 mL (e.g., Hamilton STARplus).
14. 96 well low-volume storage plate compatible with automated liquid-handling platform.
15. 10 or 20 μL tips compatible with automation platform.
16. 1 mL tips compatible with tip-based automation platform.
17. 300 μL tips compatible with tip-based automation platform.
18. Nanoscale automated liquid-handling platform (e.g., Labcyte Echo 550).
19. 384-well plates compatible with nanoscale automated liquid-handling platform.
20. Adhesive gas-permeable plate seals (e.g., AB0718, ABgene).
21. Automated colony picker or colony picking module compatible with automation platform (e.g., EasyPick on the Hamilton STARplus).
22. Thermal cycler with 384-well block.

2.6 High Throughput Sequence Verification

1. 10 mM Tris pH 8.5 (*see* **Note 4**).
2. 2 N NaOH.
3. PCR-grade water.
4. Nextera XT DNA Library Preparation Kit (Illumina) containing: Amplicon Tagment Mix (ATM), Tagment DNA Buffer (TD), NexteraPCR Master Mix (NPM), and Neutralization Tagment Buffer (NT).
5. Nextera XT Index Kit v2 Sets A, B, C, and D (Illumina) (*see* **Note 5**).
6. PhiX control kit V3 (Illumina).

7. KAPA Illumina ABI Library quantification kit (Anachem).
8. Paramagnetic beads (e.g., Solid Phase Reversible Immobilization (SPRI) AMPure XP beads, Beckman Coulter).
9. Nanoscale automated liquid-handling platform (e.g., Labcyte Echo 550).
10. 384-well plates compatible with nanoscale automated liquid-handling platform (e.g., 384LDV, Labcyte; 384PP, Labcyte; 4ti-1384 PCR plate, 4titude).
11. A flexible, tip-based automated liquid handing platform that can transfer volumes of 1–100 μL (e.g., Hamilton STARplus). Alternatively, a multichannel pipette.
12. 10 or 20 μL tips compatible with automation platform.
13. Fluorometer (e.g., Qubit, Thermo Fisher Scientific).
14. Instrument for precise quantitative electrophoresis of DNA fragments (e.g., Bioanalyzer 2100, Agilent).
15. MiSeq Reagent Kit v3 (Illumina)—not required if libraries are sent to a sequencing service.
16. Illumina MiSeq platform (Illumina)—not required if libraries are sent to a sequencing service.
17. Real-time PCR instrument.
18. Sequence analysis software capable of working with FASTQ files and comparing reads to a reference template (e.g., Geneious, https://www.geneious.com; MiSeq reporter [Illumina]).

3 Methods

3.1 Design of Phytobricks

1. Each Phytobrick will be flanked by an inverted pair of recognition sequences for BsaI. The identity of the 4 bp overhangs is defined by the biological function of the part (Fig. 2). For example, Phytobricks containing a 3′ UTR + terminator sequence will be flanked by pair of inverted BsaI sites that release the fragment with GCTT and CGCT overhangs—note that the sequence of overhangs are always read 5′–3′ from the top-strand. To maintain the frame of coding sequences, the following rules are used:
   (a) If the 5′ overhang is AATG (where the last three bases code for methionine), the part will contain coding sequence and the overhang should be followed by a complete codon. This is usually the first codon after the native ATG.

(b) If the 5′ overhang is AGCC, AGGT, or TTCG (where the last three base pairs code for alanine, glycine and serine, respectively), the part will contain coding sequence and the overhang must be followed by a complete codon, starting at the first position.

(c) If the 5′ overhang is CCAT, the part will contain coding sequence and should start with a "G" to complete the codon for methionine. This should be followed by a complete codon. This is usually the first codon after the native ATG.

(d) Any GCTT 3′ overhang should be immediately preceded by a stop codon.

(e) Stop codons should not be included in parts with AATG, AGCC/AGGT or TTCG 3′ overhangs. Further, to maintain the open reading frame there are two options: First, the addition of two base pairs following the final codon of the part before the 3′ overhang to generate an additional amino acid. For example, the addition of GG prior to TTCG will result in GGU, UCG (glycine-serine). Second, the removal of the third base of the final codon enabling the codon to be completed by the first base of the overhang. This will almost certainly introduce a different amino acid, the identity of which should be assessed for undesirable influences on structure and function of the resulting protein.

2. All parts must be examined for the presence of unwanted recognition sequences for Type IIS enzymes that will be used in the assembly process. The presence of internal instances of recognitions sites for BsaI is illegal in Phytobricks. To be compatible with all commonly used assembly tool kits (GoldenBraid, Loop, Mobius, MoClo) it is desirable to mutate the sequence to remove all instances of BpiI, BsmBI, BsaI, AarI and SapI. However, most researchers will only introduce mutations to remove instances of BsaI and the additional enzyme used by the specific assembly tool kit they intend to use: BpiI (MoClo), BsmBI (Golden Braid), SapI (Loop), and AarI (Mobius). When mutating recognition sequences, care should be taken not to introduce rare codons or further undesirable recognition sequences. In regulatory sequences it is also desirable to avoid mutating functional elements such as transcription-factor binding sites if at all possible.

3. If heterologous proteins are to be expressed in plant cells, it may be desirable to recode the coding sequence to optimize codon usage to the desired plant species. Several commercial providers of synthetic DNA provide access to software for codon optimization, for example, GenSmart (GenScript) or GeneOptimizer (Thermo Fisher).

3.2 Construction of Phytobricks

1. Once the final sequence of the Phytobrick has been designed, there are two options for construction. The first option is to purchase the desired sequence from a commercial DNA synthesis company. This is essential for sequences that differ substantially from any natural sequence (e.g., codon-optimized, removal of multiple illegal sites) or when a template for amplification is unavailable. The second option is to amplify the desired sequence from a DNA or cDNA template as described in **steps 3–9**, below.
2. If the sequence is to be synthesized, a convergent pair of Type IIS recognition sequences can be included to release the fragment with the desired overhangs. It may be desirable to obtain a sequence-verified fragment, flanked by convergent *BsaI* sites cloned into plasmid backbone compatible with the assembly tool kit to be used. Alternatively, the sequence can be obtained as one or more linear fragments that can be cloned to a universal acceptor plasmid (UAP) to create the Phytobrick part (*see* **Note 6**). If the fragment is to be cloned into a UAP, the part (s) must be synthesized flanked by the appropriate restriction enzyme recognition sequences (see **step 3** below).
3. The identity of the UAP into which new parts are to be cloned will determine which restriction endonuclease recognition sites will be used (Table 1). The use of any UAP will ultimately result in Phytobrick parts flanked by an inverted pair of BsaI recognition sequences. However, the MoClo UAP (pUAP1) requires the use of BpiI to create the Phytobrick, the GoldenBraid UAP (pUPD2) requires the use of BsmBI, the Loop UAP (pUAP4) requires the use of SapI and the Mobius UAP, (mUAV) requires the use of AarI (Table 1). A convergent pair of appropriate recognition sequences must flank the part so that the part will have overhangs compatible with the chosen UAP. Figure 4 illustrates the creation of new Phytobricks using the Loop UAP (pUAP4).
4. To amplify the part from template DNA, a pair of primers must be designed to anneal to the chosen fragment of DNA. Each primer must also contain a 5′ extension to introduce the appropriate overhangs and restriction endonuclease recognition sites (Fig. 4).
5. If the sequence to be amplified contains an illegal recognition site, it can be amplified in two parts using an addition pair of primers to introduce a mutation in the unwanted site (Fig. 4).
6. The desired sequence should be amplified using a proofreading polymerase according to the manufacturer's instructions. For example, a 25 μL reaction containing: 5 μL 5× Q5 Reaction Buffer; 0.5 μL 10 mM dNTP; 1.25 μL 10 μM forward primer; 1.25 μL 10 μM reverse primer; 0.25 μL Q5 High-Fidelity DNA Polymerase; 100 ng template DNA (*see* **Note 7**).

Table 1
Features of universal acceptor plasmids

	Cloning tool kit	Restriction endonuclease	Antibiotic resistance	Cloning selection
pUAP1	MoClo	BpiI	Chloramphenicol	Pink/white (RFP)
pUPD2	GoldenBraid	BsmBI	Chloramphenicol	Blue/white (LacZ)
pUAP4	LOOP	SapI	Chloramphenicol	Blue/white (LacZ)
mUAV	Mobius	AarI	Chloramphenicol	Indigo/white (amilCP)

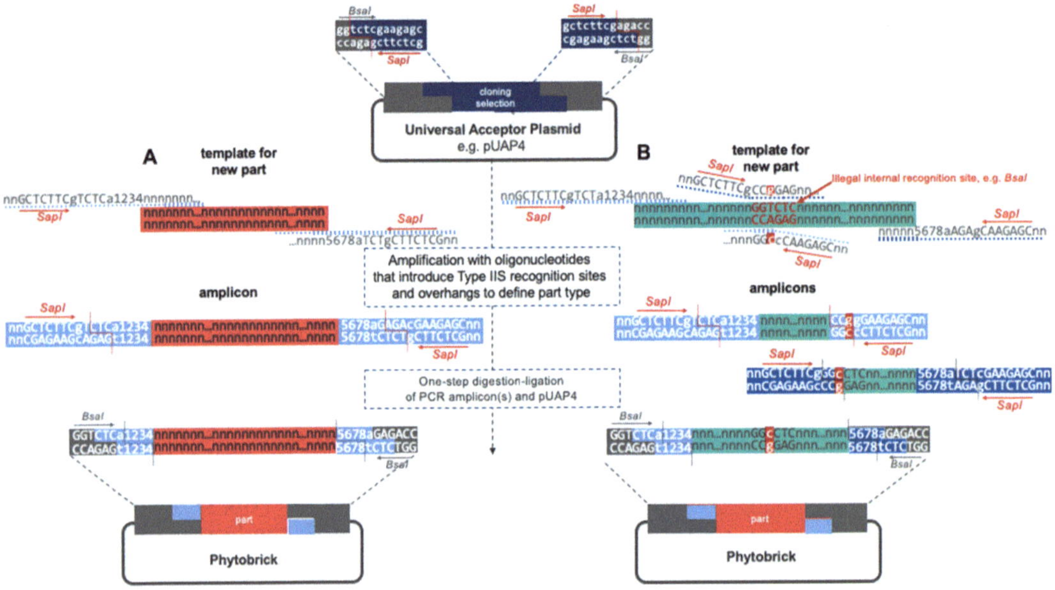

Fig. 4 New Phytobrick parts are created by cloning one or more DNA fragments into a universal acceptor plasmid (UAP). In this example, target sequences are amplified using oligonucleotide primers (dashed lines) that introduce the necessary sequences to enable a one-step digestion–ligation reaction with pUAP4 from the Loop plasmid tool kit. (**a**) A single sequence with not illegal sites is amplified (**b**). A sequence is amplified in two fragments to introduce a mutation into an illegal recognition site for a restriction endonuclease. The two amplicons are assembled together in pUAP4

7. Incubate the reaction in a thermal cycler for 98 °C for 30 s followed by 30 cycles of 98 °C for 10 s, 58–62 °C for 30 s, and 72 °C for 30 s per kilobase. The size of the amplicon should be verified by agarose gel electrophoresis and purified prior to cloning.
8. One or more linear fragments can be cloned into UAP acceptors in a one-step digestion–ligation (dig-lig) reaction. The inserted fragment(s) and acceptor plasmid must be present at a 2:1 molar ratio. Mix the following reagents into a 20 μL reaction containing 1 nM acceptor plasmid; 2 nM each DNA fragment; 2 μL 10× T4 DNA ligase buffer (or 2 μL 10×

CutSmart buffer and 2 μL 10 mM ATP, *see* **Note 8**); 4 μL 1 mg/mL BSA; 1 μL T4 Ligase (400 U/μL); 1 μL of the appropriate restriction endonuclease (10 U/μL) (*see* **Note 9**).

9. Incubate the reaction in thermocycler for 37 °C for 20 s followed by 26 cycles of 37 °C for 3 min and 16 °C for 4 min, with final incubation at 50 °C for 5 min (*see* **Note 10**) and then 80 °C for 5 min.
10. Transform the completed Level 0 assembly reaction into competent *E. coli* cells and select on LB agar containing the appropriate antibiotic for the acceptor plasmid (Table 1).
11. Select 1–3 white colonies and check for correct assembly using standard methods (restriction analysis and sequencing).

3.3 Assembly of Transcriptional Units

1. Select a plasmid assembly tool kit. Phytobricks can be assembled into acceptor plasmids from the MoClo, GoldenBraid, Loop, and Mobius tool kits. However, once transcriptional units have been assembled, these will no longer be interchangeable with acceptors from alternative tool kits.
2. Verify that the overhangs of each are compatible, such that parts will assemble into a complete transcription unit within the acceptor plasmid.
3. If assembled TUs are to be subsequently assembled into multigene constructs, select appropriate acceptors from the plasmid tool kit to define the position within the final assembly (Fig. 3).
4. All Phytobricks plasmids and the acceptor plasmid are combined in a 20 μL dig-lig reaction containing: 1 nM acceptor plasmid; 2 nM each Phytobrick plasmid; 2 μL10× T4 DNA ligase buffer; 4 μL 1 mg/mL BSA; 1 μL 400 U/μL T4 Ligase; 1 μL 10 U/μL BsaI.
5. Incubate the reaction in a thermocycler: 37 °C for 20 s followed by 26 cycles of 37 °C for 3 min and 16 °C for 4 min, with final incubation at 50 °C for 5 min and then 80 °C for 5 min.
6. Transform the completed assembly reaction into competent *E. coli* cells and select on LB agar containing the appropriate antibiotic for the acceptor plasmid.
7. Select 1–3 white colonies and check for correct assembly using standard methods (restriction analysis and sequencing).

3.4 Assembly of Multigene Constructs

1. Select the appropriate Transcription Units (TUs) and acceptor plasmids. The numbers of TUs that can be assembled at each step varies with each plasmid tool kit (Fig. 3).
2. If necessary, select the appropriate spacers or end-linkers to bridge between TUs and the overhangs of the acceptor plasmid (Fig. 3).

3. Once the correct plasmids have been selected these can be assembled using the exactly the same protocol for assembling TUs (*see* Subheading 3.3, **steps 4–7**), substituting TUs for Phytobrick parts and using the appropriate restriction endonuclease for the acceptor plasmids (Fig. 3, *see* **Note 8**).

3.5 Nanoscale Automated Assembly Reactions

1. Automated assembly has exactly the same requirements for the design and selection of parts and plasmids. However, to facilitate working with large number of plasmids, all Phytobrick parts and acceptor plasmids should be diluted to 10 nM.
2. A source plate containing 10 μL of each plasmid (Phytobricks and acceptors) to be used should be prepared in a 384-well plate compatible with the automation platform (Fig. 5). The final number of source plates will depend on the total number of Phytobricks and acceptors used. The following steps describe how to set up the source plate.
3. Each assembly reaction will be carried out in a 1 μL reaction. A master-mix of 300 nL per sample containing PCR-grade water, ligase buffer, bovine serum albumin, T4 ligase, and the appropriate Type IIS enzyme should be prepared. The total volume of the master-mix is calculated considering both the number of assemblies and the dead volume (the volume in each well that cannot be accessed) of the automation platform. For example, on the Labcyte Echo, LDV plates have a dead-volume of 3 μL and therefore at least an additional 3 μL of master-mix must be prepared for each well on the source plate. The total volume will determine how many wells of the source plate are required for master-mix. The compositions of master-mix for both 96 and 384 1 μL reactions are given in Table 2. The master-mix should be prepared manually and mixed well by vortexing followed by centrifugation before being distributed to the required number of wells on the source plate (Table 2) (*see* **Note 11**). An example set-up for a source plate is shown in Fig. 5.
4. Each reaction will require 100 nL of the appropriate acceptor plasmid (10 nM) and 200 nL of each Phytobrick plasmid (10 nM). The total volume of acceptor plasmid required considers both the number of assemblies and the dead volume (*see* Table 3 and Fig. 5).
5. Transfer 10 nM stocks of each Phytobrick to a 384 well source plate compatible with nanoscale automated platform again considering the number of assemblies each part will be used in and the dead volume of the plate (*see* **Notes 11** and **12**). If the number of Phytobricks is very large then additional source plates can be prepared as required.

	1	2	3	4	5	6	7	8	9	10	11	12	13	14	15	16	17	18	19	20	21	22	23	24
A	M	M		A		P1		T1		1	2	3	4	5	6									
B	M			A		P1		T1		7	8	9	10	11	12									
C	M			A		P1		T1		13	14	15	16	17	18									
D	M			A						19	20	21	22	23	24									
E	M			A		P2		T2		25	26	27	28	29	30									
F	M					P2		T2		31	32	33	34	35	36									
G	M					P2		T2		37	38	39	40	41	42									
H	M									43	44	45	46	47	48									
I	M					P3		T3		49	50	51	52	53	54									
J	M					P3		T3		55	56	57	58	59	60									
K	M					P3		T3		61	62	63	64	65	66									
L	M									67	68	69	70	71	72									
M	M					P4		T4		73	74	75	76	77	78									
N	M					P4		T4		79	80	81	82	83	84									
O	M					P4		T4		85	86	87	88	89	90									
P	M									91	92	93	94	95	96									

	1	2	3	4	5	6	7	8	9	10	11	12	13	14	15	16	17	18	19	20	21	22	23	24
A	1	2	3	4	5	6	1	2	3	4	5	6	1	2	3	4	5	6	1	2	3	4	5	6
B	7	8	9	10	11	12	7	8	9	10	11	12	7	8	9	10	11	12	7	8	9	10	11	12
C	13	14	15	16	17	18	13	14	15	16	17	18	13	14	15	16	17	18	13	14	15	16	17	18
D	19	20	21	22	23	24	19	20	21	22	23	24	19	20	21	22	23	24	19	20	21	22	23	24
E	25	26	27	28	29	30	25	26	27	28	29	30	25	26	27	28	29	30	25	26	27	28	29	30
F	31	32	33	34	35	36	31	32	33	34	35	36	31	32	33	34	35	36	31	32	33	34	35	36
G	37	38	39	40	41	42	37	38	39	40	41	42	37	38	39	40	41	42	37	38	39	40	41	42
H	43	44	45	46	47	48	43	44	45	46	47	48	43	44	45	46	47	48	43	44	45	46	47	48
I	49	50	51	52	53	54	49	50	51	52	53	54	49	50	51	52	53	54	49	50	51	52	53	54
J	55	56	57	58	59	60	55	56	57	58	59	60	55	56	57	58	59	60	55	56	57	58	59	60
K	61	62	63	64	65	66	61	62	63	64	65	66	61	62	63	64	65	66	61	62	63	64	65	66
L	67	68	69	70	71	72	67	68	69	70	71	72	67	68	69	70	71	72	67	68	69	70	71	72
M	73	74	75	76	77	78	73	74	75	76	77	78	73	74	75	76	77	78	73	74	75	76	77	78
N	79	80	81	82	83	84	79	80	81	82	83	84	79	80	81	82	83	84	79	80	81	82	83	84
O	85	86	87	88	89	90	85	86	87	88	89	90	85	86	87	88	89	90	85	86	87	88	89	90
P	91	92	93	94	95	96	91	92	93	94	95	96	91	92	93	94	95	96	91	92	93	94	95	96
	M A P1 T1						M A P2 T2						M A P3 T3						M A P4 T4					

Fig. 5 Example layout for source (above) and destination (below) plates for 384 assembly reactions. In this example, 96 coding sequence parts (1–96) are each being assembled into an acceptor plasmid (A) with each of four pairs of Promoter + 5′ UTR (P) and 3′ UTR + Terminator (T) parts. M: master mix (see Table 2 for components of M)

Table 2
Composition and quantities of reagents of the master-mix required for 96 and 384 nanoscale automated assembly reactions

Component	Volume of master mix for 96 assemblies (μL)	Volume of master-mix for 384 assemblies (μL)
PCR grade water	15	51
10× T4 Buffer	20	68
1 mg/mL BSA	5	17
400 U/μL T4 ligase	10	34
20 U/μL Type IIS enzyme	10	34
Total volume	60	204
Number of wells in source plate required for master-mix	5 (11 μL/well)	17 (11 μL/well)

The total volume is calculated in excess to minimize human and pipetting errors

Table 3
Volumes of acceptor and Phytobrick plasmids required in the source plate for 384 nanoscale automated assembly reactions

Component		Volume required (μL)	Number of wells in source plate
Acceptor plasmid (10 nM)		55	5 (11 μL/well)
Phytobricks	Promoter 1 (10 nM)	33	3 (11 μL/well)
	Promoter 2 (10 nM)	33	3 (11 μL/well)
	Promoter 3 (10 nM)	33	3 (11 μL/well)
	Promoter 4 (10 nM)	33	3 (11 μL/well)
	Terminator 1 (10 nM)	33	3 (11 μL/well)
	Terminator 2 (10 nM)	33	3 (11 μL/well)
	Terminator 3 (10 nM)	33	3 (11 μL/well)
	Terminator 4 (10 nM)	33	3 (11 μL/well)
	Coding sequences (96 plasmids, each at 10 nM)	11	96 (11 μL/well)
Master mix (*see* Table 2)		187	17 (11 μL/well)

In this example, 96 coding sequence parts are each being assembled with four different pairs of promoter and terminator parts into the same acceptor plasmid

6. The automationsoftware must be provided with the quantities of each well to transfer to the destination plate. On the Labcyte Echo 550, the Echo Reformat software is used to transfer 300 nL of master mix from the source plate into each well of the 384-well destination plate. Following this, the Echo CherryPick software is used to transfer the required volume of water

(if needed) followed by 100 nL of acceptor and the required volumes of each DNA part to the destination plate (*see* **Notes 11–13**).

7. Once the automation platform has completed the transfers to the destination plate (*see* **Notes 11** and **12**), the plate is transferred (automatically or manually) to a thermal cycler and incubated at the following temperatures: 35 cycles of 37 °C for 3 min and 16 °C for 4 min. This is followed by single incubations at 37 °C for 5 min, 50 °C for 5 min and 85 °C for 5 min.
8. The complete dig-lig reaction is then used to transform 2 μL of ultra-competent *E. coli* cells. To set-up the transformation, centrifuge the 384-well destination plate at max rpm for 30 s to collect the reaction and place it on ice. Using either a multichannel pipette or automated liquid-handling platform, aliquot 2 μL of cells to each well and centrifuge at 100 × *g* for 30 s. Cover the plate with a lid or seal and incubate on ice for 30 min. Transfer the plate to a 42 °C heat bath for 40 s. Transfer the plate back to ice for a further 2 min.
9. Using either a multichannel pipette or automated liquid-handling platform, prepare a sufficient number of 96-well plates for each dig-lig reaction by aliquoting 40 μL SOC to each well.
10. Transfer 10 μL of SOC to each well of the 384-well reaction plate and then transfer 10 μL from each well of the 384-well reaction plate into 96-well plates containing 40 μL/well of SOC.
11. Apply a gas-permeable plate seal to the 96-well plates with and place in a shaking incubator for 1 h at 37 °C and 265 rpm.
12. Prepare a sufficient number of 8-well plates with 3–4 mL LB agar (containing with appropriate antibiotics and X-gal/IPTG, if required). The recovered cells can be plated automatically on some automation platforms. For example, on the Hamilton STARplus, 8-well LB plates can be placed on the deck carriers and 50 μL of cells transferred to each well by "pipetting on the fly" (the pipetting head doesn't stop moving while it dispenses). The plates are then immediately covered and placed on a shaker to distribute the cells. Once plating of cells is complete, the 8-well plates can be incubated at 37 °C overnight.
13. Prepare a sufficient number of 24-well plates with 4 mL LB with appropriate antibiotics in each well.
14. Either manually, or using an automated colony-picker, inoculate each well with a single white colony. Apply a gas-permeable plate seal and incubate at 255 rpm and 37 °C for approximately 18 h.

15. Collect the cells by centrifugation at 3500 × *g* for 10 min. Remove the seal and invert the block to remove the media, retaining the cell pellets.
16. A number of different protocols and kits to extract plasmid DNA are available and can be automated using tip-based automated liquid-handing machines following the manufacturer's instructions.

3.6 Verification of Parts and Assemblies by Nanoscale Illumina Sequencing

1. Normalize the concentration of all plasmids to 0.2 ng/μL.
2. Set your liquid handing system to transfer 10 μL of each normalized plasmid into to a separate well of a 384-well plate compatible with your nanoscale liquid-handling platform.
3. Using a fresh 384-well plate compatible with your nanoscale liquid-handling platform, create source plate compatible with the liquid-handling platform (*see* **Note 11**). To this plate, add suitable quantities of tagment DNA buffer (TD), amplicon tagment mix (ATM), neutralization tagment buffer (NT) and Nextera PCR master mix (NPM) from the Nextera XT DNA library prep kit as well as the i7 and i5 index adapters. The total volume of each reagent is calculated considering both the number of assemblies and the dead volume. The volumes of each reagent and number of wells required for 384 sequencing reactions are provided in Table 4.
4. Use the nanoscale automated liquid-handling platform to transfer 200 nL of DNA, 400 nL TD and 200 nL ATM to each well of the destination plate (*see* **Notes 11** and **12**). On the Labcyte Echo 550, this is done using the CherryPick software.

Table 4
Volumes of sequencing reagents required in the source plate for 384 nanoscale sequencing reactions

Component	Volume required (μL) for 384 sequencing reactions	Number of wells in source plate
TD	520 μL	8 wells (65 μL/well)
ATM	260 μL	4 wells (65 μL/well)
NT	260 μL	4 wells (65 μL/well)
i7 index adapters	30 μL each (×24)	1 well (30 μL/well)
i5 index adapters	30 μL each (×16)	1 well (30 μL/well)
NPM	780 μL	12 wells (65 μL/well)

5. Incubate the plates in a thermal cycler at 55 °C for 5 min followed by a 10 °C hold. As soon as the plate reaches 10 °C, remove from the thermal cycler and centrifuge for 1 min at 280 × *g*, 20 °C.
6. Return the plates to the nanoscale automated liquid-handling platform and transfer 200 nL of NT to each well of the destination plate and centrifuge the plate for 1 min at 280 × *g* and 20 °C.
7. Incubate for a further 5 min.
8. Transfer 200 nL of each adapter to every reaction and centrifuge the plate for 1 min at 280 × *g* and 20 °C.
9. Finally, transfer 600 nL of NPM to every reaction.
10. Use a thermocycler to incubate the destination plate at the following temperatures: 72 °C for 3 min, 95 °C for 30 s followed by 12 cycles of 95 °C for 12 s, 55 °C for 30 s, and 72 °C for 45 s. This is followed by final a final incubation at 72 °C for 5 min.
11. Pool the samples by transferring 1 μL of each of 16 samples to a single well (i.e., 16 rows to 16 wells of a 384-well plate). Then pool four wells into one such that there are four pools in total.
12. Add 77 μL of Ampure XP beads to 96 μL of pooled libraries and transfer the final DNA samples into a pooled fresh tube.
13. Libraries should contain DNA fragments of approximately 300–800 bp. The concentration and average length should be checked using, for example, a fluorometer such as the Qubit and quantitative electrophoresis platform such as the Bioanalyzer. Alternatively, this may be performed by your sequencing service provider.
14. Quantify the libraries by qPCR by comparing the libraries to a standard. Plot the average Cq score for each DNA standard to generate a standard curve and calculate the concentrations of the libraries using absolute quantification.
15. The libraries can be provided to a sequencing service or diluted in 10 mM Tris (*see* **Note 4**) denatured with 1 μL 2 N NaOH and loaded onto an Illumina MiSeqfollowing the manufacturer's instructions.
16. Following the sequencing run, verify the quality scores and generate FASTQ files. Alternatively obtain FASTQ files from your sequencing provider.
17. Align the FASTQ files to reference sequences for the expected assemblies using sequence analysis software identifying any mis-assemblies or mutations.

4 Notes

1. Most common laboratory strains of *E. coli* will be suitable. Either electrocompetent or chemically competent may be cells may be purchased or prepared using established protocols. Note that the colour obtained with some cloning selection cassettes, for example the RFP cassette in pUAP1 may vary from pink-red to orange-red depending on the bacterial strain.
2. In place of one or more cloned Phytobrick parts, it is possible to use linear, double stranded DNA fragments (e.g., PCR amplicons or synthetic fragments) with the equivalent inverted pairs of recognition sites for BsaI that produce complementary overhangs. However, the use of linear fragments tends to negatively impact the efficiency of the reactions, particularly when many parts are being assembled.
3. Some providers may sell more than one version of BsaI (e.g., Eco31I (BsaI), BsaI-HF, and BsaI-HFv2). Some versions are less suitable for the one-step digestion–ligation reactions. Check the description of the product to ensure suitability. At the time of writing, we obtain the best results with Eco31I (Thermo Fisher Scientific) and Bsal-HF v2 (New England BioLabs).
4. This buffer can be used to elute and/or dilute DNA libraries but PCR grade water can also be used.
5. Each Illumina Index kit has unique indexes for 96 samples. The use of all four kits will provide unique indexes for 384 total samples. For 96 samples only one set is required.
6. Some DNA sequences, particularly those with regions of low-complexity or repeats may be difficult to synthesize, particularly as linear fragments. Such fragments can either be ordered as cloned, sequenced verified fragments or the fragment can be split into two or more shorter fragments flanked with inverted pairs of Type IIS enzymes that create bespoke, scarless overhangs that enable them to be assembled together during the cloning reaction with the UAP as described in Subheding 3.2, **step 8**.
7. For templates with high structural complexity where the target sequence may be in low abundance (such as genomic DNA) use 1 ng–1 μg. For simple templates where the target sequence is likely to be well-represented (such as plasmid DNA) 1 pg–1 ng will be sufficient.
8. When the restriction endonuclease SapI is used in the one-step digestion–ligation reaction, include 2 μL 10× CutSmart buffer (New England BioLabs) and 2 μL 10 mM ATP instead of 2 μL 10× T4 ligase.

9. The T4 ligase and restriction endonuclease can be reduced to 5 U per reaction (0.5 μL of 10 U/μL) with no significant impact on reaction efficiency.
10. A final incubation temperature of 37 °C was described in the instructions for the MoClo and Mobius plasmid tool kits.
11. After reagents are transferred to either source or destination plates they must always be centrifuged at ~1000 x *g* for 2 min to remove air bubbles.
12. Automation errors observed during transfer are most often related to dead volume. They may be caused by insufficient volumes or the presence of air bubbles in the source plate. These issues can be solved by adding more reagents to the source plate, consolidating wells of the same reagent or by centrifugation of the source plates.
13. Larger volumes may take longer to transfer. We recommend that the reagent with the largest volume is transferred first so that any problems related to automated transfer can be identified. Typically, this will be the master mix, however, if the assembly reaction contains only part and one acceptor, water will be a large component of the mix and should be transferred first.

References

1. Patron NJ (2014) DNA assembly for plant biology: techniques and tools. Curr Opin Plant Biol 19:14–19
2. Casini A, Storch M, Baldwin GS, Ellis T (2015) Bricks and blueprints: methods and standards for DNA assembly. Nat Rev Mol Cell Biol 16:568–576
3. Engler C, Kandzia R, Marillonnet S (2008) A one pot, one step, precision cloning method with high throughput capability. PLoS One 3: e3647
4. Engler C, Gruetzner R, Kandzia R, Marillonnet S (2009) Golden Gate shuffling: a one-pot DNA shuffling method based on type IIs restriction enzymes. PLoS One 4:e5553
5. Gibson DG, Young L, Chuang R-Y et al (2009) Enzymatic assembly of DNA molecules up to several hundred kilobases. Nat Methods 6:343–345
6. Gibson DG (2011) Enzymatic assembly of overlapping DNA fragments. Methods Enzymol 498:349–361
7. Kanigowska P, Shen Y, Zheng Y, Rosser S, Cai Y (2015) SmartDNA fabrication using sound waves: applying acoustic dispensing technologies to synthetic biology. J Lab Automat 21 (1):49–56
8. Patron NJ, Orzaez D, Marillonnet S et al (2015) Standards for plant synthetic biology: a common syntax for exchange of DNA parts. New Phytol 208:13–19
9. Weber E, Engler C, Gruetzner R et al (2011) A modular cloning system for standardized assembly of multigene constructs. PLoS One 6:e16765
10. Sarrion-Perdigones A, Falconi EE, Zandalinas SI et al (2011) GoldenBraid: an iterative cloning system for standardized assembly of reusable genetic modules. PLoS One 6:e21622
11. Pollak B, Cerda A, Delmans M et al (2019) Loop assembly: a simple and open system for recursive fabrication of DNA circuits. New Phytol 222:628–640
12. Andreou AI, Nakayama N (2018) Mobius Assembly: a versatile Golden-Gate framework towards universal DNA assembly. PLoS One 13:e0189892

Chapter 12

Mobius Assembly

Andreas I. Andreou and Naomi Nakayama

Abstract

Mobius Assembly is a versatile and user-friendly DNA Assembly method, which facilitates rapid and simple generation of DNA constructs. Mobius Assembly combines high cloning capacity and vector toolkit simplicity to streamline combinatorial assemblies. It is a two-level hierarchical modular cloning system that enables quadruple assembly augmentation. It adopts the 4 bp standard overhangs defined by Phytobricks to promote standard part sharing, and it can be made compatible with different chassis. Furthermore, Mobius Assembly reduces domestication requirements and uses chromogenic proteins to facilitate the identification of positive assemblies.

Key words Mobius Assembly, Phytobricks, Golden Gate Assembly, DNA assembly, Molecular cloning, Synthetic biology, Chromoproteins, AarI

1 Introduction

DNA assembly is a fundamental technology in molecular biology to physically link DNA fragments. Golden Gate Assembly exploits Type IIS restriction endonucleases to cut double-stranded DNA outside their recognition sites [1, 2]. Consequently, the recognition sites are removed during cloning, while the generated overhangs do not depend on the restriction enzyme. Ligation and digestion are carried out simultaneously, and the unique overhangs are set by users allowing unidirectional, scarless assembly of multiple DNA fragments in a one-tube reaction.

Several variations of Golden Gate Assembly have been developed over the last decade, each addressing specific limitations of the method [3–6]. However, most of them compromise at least at one of the following: (1) cloning capacity, (2) vector toolkit simplicity, (3) elimination of internal restriction sites (domestication), (4) standardization, (5) dependence on specific strains for propagation, and/or (6) negative screening requires additives in the selection medium.

Sunil Chandran and Kevin W. George (eds.), *DNA Cloning and Assembly: Methods and Protocols*, Methods in Molecular Biology, vol. 2205, https://doi.org/10.1007/978-1-0716-0908-8_12, © Springer Science+Business Media, LLC, part of Springer Nature 2020

We developed Mobius Assembly to tackle the tradeoffs of the preexisting systems, ensuring user-friendliness [7]. Mobius Assembly has a high cloning capacity (exponential assembly) even though the vector toolkit is minimalistic. There is a part storage level (Level 0) and two cloning levels (Level 1 and Level 2). Level 0 has one plasmid (mUAV), which converts a functional DNA fragment (e.g., promoter, coding sequence, tag) into a standard part. In Level 1, there are four Acceptor Vectors (A-Δ), with which up to four transcriptional units (TUs) can be formed in the first cloning round. In Level 2, there are four Acceptor Vectors (A-Δ), as well as seven Auxiliary Plasmids that contain a unique design that switches the overhangs and allows continuation of the cloning back to Level 1. In Level 2, up to four TUs can be assembled. Cloning can further continue by switching back and forth between the Level 1 and Level 2, leading to expansion of multi-TUs according to the geometric sequence: 1, 4, 16, 64, ... Mobius Assembly uses two Type IIS restriction endonucleases: BsaI (from Level 0 to Level 1 and from Level 2 to Level 1) and AarI (in Level 0 and from Level 1 to Level 2). We used AarI since it is a rare cutter (7-bp recognition sequence instead of popularly used 6-bp cutters to reduce domestication requirements).

Mobius Assembly further simplifies cloning protocols via negative selection screening carried out by constitutively expressed chromogenic proteins. Each level is demarcated by a specific color (purple: amilCP-Level 0, pink: spisPink-Level 1, and yellow: sfGFP-Level 2). This feature eliminates the need for IPTG and X-Gal and expensive blue/white screening.

In order to promote standard part sharing, Mobius Assembly adopts the Phytobrick common syntax, a predefined set of 4 bp overhangs flanking the different types of standard parts (Fig. 1, [8]). Recently, to simplify the generation of combinatorial assembly libraries, we introduced a new feature based on DNA methylation (Fig. 4). Mobius Assembly toolkits are available for *E. coli* and plant systems so far; however, it can be easily adapted to different chassis by transferring the Mobius Assembly cassettes to different backbones.

2 Materials

2.1 Primer Design/DNA Synthesis

1. DNA construct design tool (e.g., SnapGene; Benchling, which is free for academic use).
2. DNA synthesis services (e.g., Twist Bioscience; Integrated DNA Technologies) (Optional).
3. Oligo synthesis service (e.g., Integrated DNA Technologies; Sigma-Aldrich).

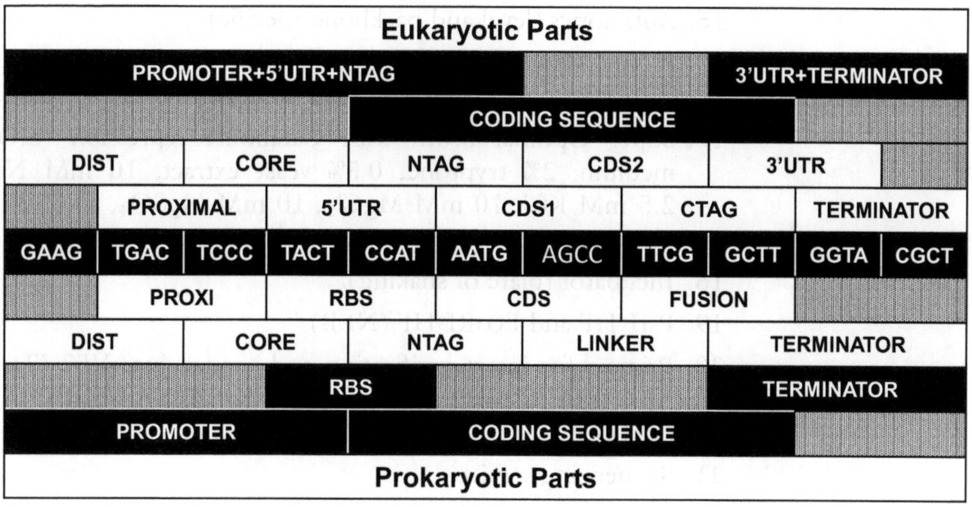

Fig. 1 Phytobrick overhangs. Phytobrick are standard DNA parts which contain a common genetic syntax for their flanking overhangs (NNNN). The standardized syntax facilitates part sharing within the community. Top half depicts the major and minor functional parts of Phytobricks for eukaryotic genes, while the bottom half denotes the ones for prokaryotic genes

2.2 Molecular Biology Techniques

1. Milli-Q sterilized water.
2. DNA Template (genomic DNA; complementary DNA; plasmid DNA).
3. High-Fidelity DNA Polymerase (e.g., Q5® High-Fidelity DNA Polymerase, NEB).
4. 200 μl PCR tubes.
5. 1.5 ml microcentrifuge tubes.
6. Agarose gel: 1% agarose in 1×TAE (40 mM Tris-Base, 20 mM acetic acid and 1 mM EDTA).
7. Gel Imaging System.
8. PCR Purification Kit (e.g., Monarch® Nucleic Acid Purification Kits, NEB).
9. (Optional) Gel Purification Kit (e.g., Monarch® DNA Gel Extraction Kit, NEB).
10. Plasmid Miniprep Kit (e.g., Monarch® Miniprep Kit, NEB; PureYield™ Plasmid Miniprep System, Promega).
11. Plasmid Midiprep Kit (e.g., PureYield™ Plasmid Midiprep System, Promega).
12. Microvolume spectrophotometer.
13. DNA ladder (e.g., HyperLadder™ 1 kb, Bioline).
14. Lysogeny Broth (LB) Medium: 1% tryptone, 0.5% yeast extract, and 1% NaCl adjusted to pH 7.0. For plates, 1.5% agar is added.

15. Antibiotics (level and backbone-specific).
16. Home-Made competent cells (e.g., DH10B, DH5α, JM109, TOP10) prepared with the TSS method [9].
17. Super Optimal broth with Catabolite repression (S.O.C) medium: 2% tryptone, 0.5% yeast extract, 10 mM NaCl, 2.5 mM KCl, 10 mM $MgCl_2$, 10 mM $MgSO_4$, and 20 mM glucose.
18. Incubator (plate or shaking).
19. PstI-HF and EcoRI-HF (NEB).
20. Primers for insert verification and sequencing: VF2 (TGCC ACCTGACGTCTAAGAA) and VR (ATTACCGCCTTTG AGTGAGC).
21. Sequencing service.

2.3 Mobius Assembly Cloning

1. Milli-Q sterilized water.
2. Thermocycler.
3. BSA, Molecular Biology Grade (NEB).
4. T4 DNA Ligase and 10× T4 DNA Ligase buffer (e.g., Thermo Fisher Scientific, NEB).
5. For Level 0 and Level 2 cloning: OPTIZYME™ AarI, Fisher BioReagents™; Thermo Scientific™ AarI (both supplied with 50× Oligos).
6. For Level 1 cloning: Eco31I, Thermo Fisher Scientific; BsaI-HF®v2, NEB.

2.4 Plasmid Construction and Methylation

1. Isothermal Assembly kit (e.g., Gibson Assembly®, NEB; NEBuilder® HiFi DNA Assembly, NEB).
2. DpnI (e.g., NEB, Life Technologies).
3. Thermocycler.
4. CpG Methyltransferase (*M.SssI*) (NEB).

3 Methods

3.1 Primer Design/DNA Synthesis

DNA fragments should be designed to carry the selected standard overhangs and AarI restriction sites for cloning into the Mobius Universal Acceptor Vector (mUAV) (Fig. 2). They can either be synthesized from a DNA synthesis company, or PCR amplified from genomic (*see* **Note 1**) or cDNA. For primer design or DNA synthesis, follow the guidelines below:

1. Select your sequence and open it in a DNA visualizer/editor software.

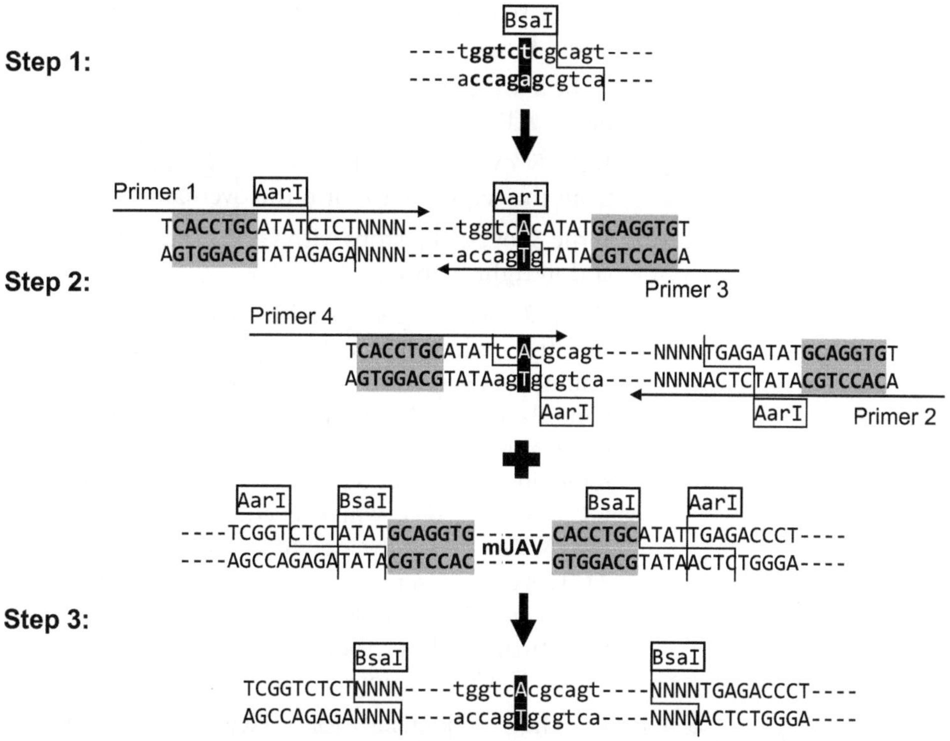

Fig. 2 Sequence domestication and standard part (Phytobrick) cloning into mUAV. The coding sequence shown here bears an internal *BsaI* restriction site, which needs to be removed (i.e. domesticated). In Step 1, remove in silico the recognition site by changing A to T, resulting in a silence mutation of serine (TCT to TCA). In Step 2, split the sequence into two parts, both of which share an overhang that brings together the mutation (tcAc). At the split point add the *AarI* recognition sequence (GCAGGTG). Also, at the 5′ of the first fragment and the 3′ of the second fragment, add the *AarI* recognition sequence, the overhangs for cloning into mUAV (CTCT-TGAG), and the Phytobrick overhangss(NNNN). We design primers and PCR amplify the sequence in two parts (Primer 1 and 3 for the first fragrment and Primer 4 and 2 for the second fragment). In the last step (Step 3), the two fragments are fused together and cloned into the mUAV in one tube-reaction, generating a Phytobrick

2. Add the sequence 5′-T**CACCTGC**ATATCTCT *NNNN*-3′ at the beginning (5′ end) of your DNA fragment and 3′-*NNNN* TGAGATAT**GCAGGTG**T-5′ at the end (3′ end). **CACCTGC** and **GCAGGTG** are the AarI recognition sites, CTCT and TGAG (the 5′ and 3′ overhangs) which pair with the overhangs of mUAV and *NNNN*, the Phytobrick Overhangs, respectively (Table 1).
3. For primer design, **step 2** should be modified to Forward Primer: 5′-T**CACCTGC**ATATCTCT *NNNN* + (17–25 bp)-3′ and Reverse Primer: 3′-(17–25 bp) + *NNNN*TGAG A-TAT**GCAGGTG**T-5′.
4. For noncoding sequences, check the Phytobrick specification (Fig. 1) to select the overhangs (*NNNN*) according to the type of the part (e.g., GGAG-AATG for a promoter).

5. For coding sequences, in addition to the overhangs, further rules should be considered:
   (a) If the 5′ overhang starts with AATG, add one A next to the native ATG start codon to form the overhang.
   (b) If the 5′ overhang starts with CCAT remove the native AT from the sequence to form the overhang.
   (c) If the 3′ overhang stops with GCTT, be sure there is a stop codon right before.
   (d) If the 3′ overhang stops with AGCC or TTCG, remove the stop codon right before. GCC and TCG encode for alanine and serine, respectively, so A or T from the overhang should be the last amino acid of a triplet to preserve the reading frame. This can be done by (a) completely removing the stop codon and adding two extra bases (GG, GC, TC, or AG), which encode for small amino acids (GGA, GGT == glycine; GCA, GCT == alanine; TCA, TCT, AGT == serine), (b) completely removing the stop codon and the last base pair from the codon before, if the two remaining base pairs combined with A (for AGCC) or T (for TTCG) do not change the amino acid.
6. Check the sequence for illegal recognition sites (**CACCTGC** == **AarI, GGTCTC** == **BsaI**). If the part is a coding sequence, remove the illegal recognition site with a silent point mutation (domesticate) (*see* **Note 2**).
7. If the part is a noncoding sequence (e.g., promoter) try to avoid changing known functional elements (e.g., transcription factor binding sites). If it is not possible, ensure that the mutation does not affect the function.
8. If you PCR amplify the part, you should introduce the point mutation by splitting the sequence, forming overhangs (*see* **Note 3**) and adding AarI recognition sites. The point mutation should be introduced in the overhangs (Fig. 2).
9. If you PCR amplify the sequence, run 2 μl of the reaction in gel electrophoresis to verify the presence and the correct size of the fragment.
10. Purify the PCR products using a column-based kit or carry out gel purification if there are multiple bands.

3.2 Level 0 Cloning

In Level 0 cloning, DNA fragments generated as above are cloned into mUAV to form Phytobricks (Fig. 2). Negative cloning screening is performed by dropout of amilCP (*see* **Note 4**); the successfully cloned constructs will result in white colonies, the unsuccessful in purple. Set up the Level 0 reaction as follows (*see* **Note 5**):

1. Total reaction volume is 10 μl (*see* **Note 6**). Calculate the volumes of the reagents and add the sterile Milli-Q water first.

2. Add 50 ng of mUAV. We usually add 0.5 μl of 100 ng/μl stock.
3. Add the insert(s). Follow the 2:1 insert–vector ratio.
4. Sequentially add: 1 μl BSA, 1 μl T4 DNA ligase Buffer (*see* **Note** 7), 0.5 μl T4 DNA ligase, 0.5 μl AarI, and 0.2 μl 50× AarI oligonucleotides (0.025 mM).
5. Gently mix well and spin down in a microcentrifuge.
6. Incubate in a thermal cycler with the program: 5× (5 min at 37 °C + 10 min at 16 °C), 5 min at 37 °C, 5 min at 80 °C and hold at 10 °C.
7. Transform 5 μl of the reaction to 100 μl of competent cells. Incubate for 30 min on ice, heat-shock in a 42 °C heat block or water bath for 90 s, cool for 3 min on ice, add 400 μl S.O.C media, and incubate in a 37 °C shaker for 1 h.
8. Plate 100 μl of the transformation culture on each LB agar plate containing 25 μg/ml chloramphenicol and incubate overnight at 37 °C.
9. Run colony PCR for three white colonies to confirm the size of the part (Optional).
10. Pick one colony for LB culture containing 25 μg/ml chloramphenicol and incubate overnight in a 37 °C shaker.
11. Miniprep the cultures and verify the construct by restriction digestion analysis with *Pst*I-HF and *Eco*RI-HF.
12. Send the part for sequencing using the primers VF2 and VR.

3.3 Level 1 Cloning

In the first round of Level 1 cloning, Phytobricks from Level 0 are combined in a Level 1 Acceptor Vector to form a Transcriptional Unit (TU) (Fig. 3). Second round of Level 1 cloning can be done by assembling multi-TUs back from Level 2. There are four Level 1 Acceptor Vectors, A, B, Γ and Δ. We always start with the first acceptor vector (A) and sequentially use the other ones when we need to assemble multiple TUs. Successful assembly will result in white colonies because of the replacement of the pink chromoprotein spisPink, which is the negative screening marker. The backbones for Level 1 cloning cannot have chloramphenicol resistance and they should have different antibiotic resistance from Level 2 Vectors. Set the Level 1 reaction as follows:

1. Total reaction volume is 10 μl. Calculate the volumes of the reagents and add the sterile Milli-Q water first.
2. Add 20 fmol of Level 1 Acceptor Vector.
3. Add 40 fmol of the Phytobricks.
4. Add sequentially: 1 μl BSA, 1 μl T4 DNA ligase Buffer, 0.5 μl T4 DNA ligase, and 0.5 μl BsaI.
5. Gently mix well and spin down in a microcentrifuge.

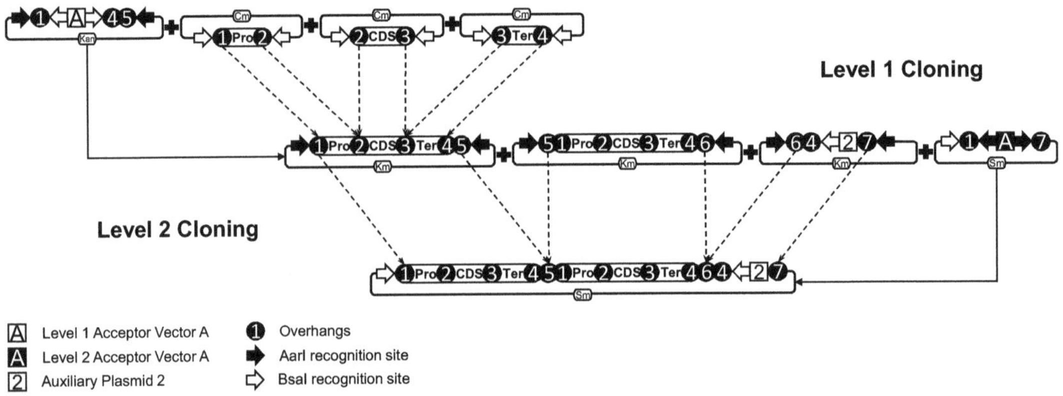

Fig. 3 Level 1 and Level 2 cloning. Phytobricks are released by *BsaI* digestion and assembled in the Level 1 reaction to form a TU. The resulting TUs are then released by *AarI* and assembled in the Level 2 reaction to form multi-TU constructs. Level 2 cloning is assisted by Auxiliary Plasmids, which provide the overhangs for cloning back to Level 1. In this example two transcriptional units are assembled. When necessary, the assembly can further continue by switching back and forth between the two levels, quadrupling the number of TUs every time

6. Incubate in a thermal cycler with the program: 5–10× (5 min at 37 °C + 10 min at 16 °C) (*see* **Note 8**), 5 min at 37 °C, 5 min at 80 °C and hold at 10 °C.
7. Transform 5 μl of the reaction to 100 μl of competent cells. Incubate for 30 min on ice, heat shock in a 42 °C heat block or water bath for 90 s, cool for 3 min on ice, add 400 μl S.O.C media, and incubate in a 37 °C shaker for 1 h.
8. Plate 100 μl of the transformation on LB agar plates containing the appropriate antibiotic and incubate overnight at 37 °C.
9. Run colony PCR for three white colonies to confirm the size of the part (optional).
10. Pick one colony for LB culture containing the appropriate and incubate overnight at 37 °C.
11. Miniprep the cultures and verify the construct by restriction digestion analysis with *Pst*I-HF and *Eco*RI-HF.
12. Submit samples to sequencing only if:
    (a) There are tandem repeats on the construct, and you need to check possible deletions.
    (b) The cloning in Level 2 is failing due to the rare scenario in which there are mutations in the overhangs or the AarI recognition sites.

3.4 Level 2 Cloning

In Level 2 cloning, multiple TUs from Level 1 are fused in a Level 2 Acceptor Vector with the help of the Auxiliary Plasmids (Fig. 3). There are four Level 2 Acceptor Vectors, A, B, Γ, and Δ. Every Acceptor Vector in Level 2 can take up to four TUs from Level

1. The backbones for Level 2 cloning should have different antibiotic resistance from the one for Level 1; and it should not be Kanamycin resistance because Auxiliary plasmids confer Kanamycin resistance. We always start with the Acceptor Vector A and sequentially use the other ones. Successful assembly will result in white colonies because of the inserts replace the yellow pigmentation by *sfGFP* (*see* **Note 9**). The Auxiliary Plasmids provide missing overhangs and help the Level 2 cloning (see examples below). Set the Level 2 reaction as follows:

   1. Total volume reaction is 10 μl (*see* **Note 10**). Calculate the volumes of the reagents and add the sterile Milli-Q water first.
   2. Add 20 fmol of Level 2 Acceptor Vector.
   3. Add 40 fmol of Level 1 TUs.
   4. Add 40 fmol of the appropriate Auxiliary plasmid.
   5. Add sequentially: 1 μl BSA, 1 μl T4 DNA ligase Buffer, 0.5 μl T4 DNA ligase, 0.5 μl AarI, and 0.2 μl 50× AarI oligonucleotides (0.025 mM).
   6. Gently mix well and spin down in a microcentrifuge.
   7. Incubate in a thermal cycler with the program: 5–10× (5 min at 37 °C + 10 min at 16 °C), 5 min at 37 °C, 5 min at 80 °C and hold at 10 °C.
   8. Transform 5 μl of the reaction to 100 μl of competent cells. Incubate for 30 min on ice, heat shock in a 42 °C heat block or water bath for 90 s, cool for 3 min on ice, add 400 μl S.O.C media, and incubate in a 37 °C shaker for 1 h.
   9. Plate 100 μl of the transformation on LB agar plates containing the appropriate antibiotic and incubate overnight at 37 °C.
   10. Run colony PCR for three white colonies to confirm the size of the part (Optional).
   11. Pick one colony (*see* **Note 11**) for LB culture containing the appropriate and incubate overnight at 37 °C.
   12. Miniprep the cultures and verify the construct by restriction digestion analysis with *Pst*I-HF and *Eco*RI-HF.
   13. Sequencing is necessary only if:
       (a) There are tandem repeats on the construct, and you need to check possible deletions.
       (b) The cloning in Level 2 is failing due to the rare scenario in which there are mutations in the overhangs or the BsaI recognition sites.

3.5 Examples for How to Use the Auxiliary Plasmids

The Auxiliary Plasmids 4A, 4B, 4Γ, and 4Δ are used when four TUs are assembled in Level 2, and they correspond to the four Level 2 Acceptor Vectors. For example, if you want to make a 4-TU construct comprising from the TUs ***a***, ***b***, ***c***, and ***d***, first you assemble your TUs in Level 1 Acceptor Vectors as follows:

TU*a*	Level 1 Vector A
TU*b*	Level 1 Vector B
TU*c*	Level 1 Vector Γ
TU*d*	Level 1 Vector Δ

Then, TUs ***a***, ***b***, ***c***, and ***d*** will be cloned in the Level 2 Vector A to form the 4-TU ***abcd***, and in the reaction, you will use **Auxiliary Plasmid 4A** (as you assemble 4-TUs in Level 2 Acceptor Vector A).

TU a + TU b + TU c + TU d + Aux 4A	Level 2 Vector A

When you clone fewer than four TUs in your Level 2 Acceptor Vector, you use the **Auxiliary Plasmids 1**, **2**, and **3** corresponding to the number of TUs you clone: one, two or three.

Let's say you just want to assemble three TUs ***a***, ***b***, and ***c***. Again, you will first clone them in Level 1 Acceptor Vectors:

TU*a*	Level 1 Vector A
TU*b*	Level 1 Vector B
TU*c*	Level 1 Vector Γ

As before TUs ***a***, ***b***, and ***c*** will be cloned in the Level 2 Vector A to create the 3-TU construct ***abc*** but this time you will add the Auxiliary plasmid 3 in the reaction (as you are assembling three TUs).

TU a + TU b + TU c + Aux 3	Level 2 Vector A

A more complex example: Let's say you want to make a 10-TU construct comprising of the TUs ***a***, ***b***, ***c***, ***d***, ***e***, ***f***, ***g***, ***h***, ***i***, ***j***. In the first cloning round you will assemble each of your TUs in a Level 1 Acceptor Vector:

TU*a*	Level 1 Vector A
TU*b*	Level 1 Vector B
TU*c*	Level 1 Vector Γ
TU*d*	Level 1 Vector Δ
TU*e*	Level 1 Vector A
TU*f*	Level 1 Vector B
TU*g*	Level 1 Vector Γ

(continued)

TU*h*	Level 1 Vector Δ
TU*i*	Level 1 Vector A
TU*j*	Level 1 Vector B

In the second cloning round, TUs ***a***, ***b***, ***c***, ***d*** will be cloned in Level 2 Acceptor Vector A (+ Auxiliary 4A, as you assemble 4-TUs in Level 2 Acceptor Vector A), TUs ***e***, ***f***, ***g***, ***h*** in Level 2 Acceptor Vector B (+ Auxiliary 4B, as you assemble 4-TUs in Level 2 Acceptor Vector B) and TUs ***i***, ***j*** in Level 2 Acceptor Vector Γ (+ Auxiliary 2, as you assemble two TUs):

TUa + TUb + TUc + TUd + Aux 4A	Level 2 Vector A
TUe + TUf + TUg + TUh + Aux 4B	Level 2 Vector B
TUi + TUj + Aux 2	Level 2 Vector Γ

Lastly, in the third cloning round the multi-TUs ***abcd***, ***efgh***, and ***ij*** will be assembled in a Level 1 Vector A to form the 10-TU construct.

4-TU abcd + 4-TU efgh + 2-TU ij	Level 1 Vector A

Exemption to this rule applies when we make constructs of **8** or **12** TUs, where in the last Level 2 Acceptor Vector, we **always** add **Auxiliary Plasmid 4Δ**, which provides the necessary overhangs for the cloning back to Level 1. Cloning in Level 1 is skipped below and directly showing the assembly in Level 2.

Eight-TU construct:

TUa + TUb + TUc + TUd + Aux 4A	Level 2 Vector A
TUe + TUf + TUg + TUh + Aux 4Δ	Level 2 Vector B

Which back in Level 1 will form the 8-TU ***abcdefgh***:

4-TU abcd + 4-TU efgh	Level 1 Vector A

And 12-TU construct:

TUa + TUb + TUc + TUd + Aux 4A	Level 2 Vector A
TUe + TUf + TUg + TUh + Aux 4B	Level 2 Vector B
TUi + TUj + TUk + TUl Aux 4Δ	Level 2 Vector Γ

Which back in Level 1 will form the 12-TU ***abcdefghijkl***

4-TU abcd + 4-TU efgh + 4-TU ijkl	Level 1 Vector A

3.6 MethylAble Feature

The assembly of combinatorial libraries is a tedious task when it is not fully automated. Numerous constructs need to be built for each level. We developed an additional feature based on the methylation of the BsaI recognition site, to bypass the Level 1 cloning and directly feed into Level 2 with standard parts (Fig. 4). According to Rebase CGGTCTC^{m5}G/GC$_{m5}$CAGAGC methylation provides strong protection against BsaI digestion. We exploited this property to design an *amilCP* gene flanked by standard overhangs, which are bordered by inward and outward facing BsaI restriction sites. The outward restriction sites are designed to be susceptible to CpG methylation (CGGTCTC^{m5}G/GC$_{m5}$CAGAGC) and thus protected from BsaI digestion while the inward facing sites are not (TGGTCTC^{m5}T/AC$_{m5}$CAGAGA). Consequently, the *amilCP* gene is propagated to Level 2 with the intact outward facing BsaI sites and overhangs. The Level 2 construct can then be fused with the Level 0 library of Phytobricks in a Level 1 reaction. The MethylAble feature can be designed for any standard part or combination of parts (*see* **Note 12**).

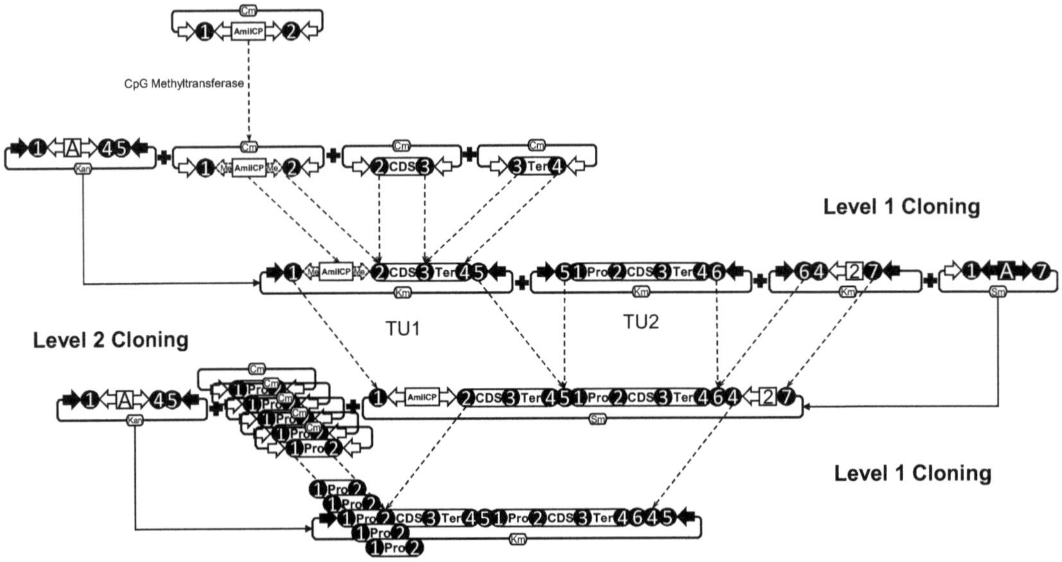

Fig. 4 MethylAble feature. It is a feature that ease the generation of combinatorial libraries and enables direct assembly of standard parts with Level 2 constructs. The MethylAble plasmid has outward facing *BsaI* sites that are designed to be blocked with CpG methylation and are propagated to Level 2 constructs. It has the cognate overhangs of the part that replaces and amiCP as screening marker. In this way, a library of standard parts can be fused witha multi-TU in Level 2 to create a library of TUs, bypassing the generation of library of Level 1 constructs. In this example the MethylAble plasmid takes the place of the promoter, which is assembled with a coding

MethylAble plasmids are built via isothermal assembly:

1. PCR amplify the *amilCP* cassette to insert the appropriate standard overhangs and the BsaI recognition sites. Use mUAV as the template and the primers: Forward: 5′-TCAGG TCTCT**NNNN** CGAGACC GTTTACGGCTAGCTCAGTC-3′ and Reverse: 3′-GGGTGGGCCTTTCTGCGCGGTCTCG **NNNN**TGAGACCCTG-5′.
2. PCR amplify the backbone using mUAV as the template and the primers: Forward: 5′-GCGCGGTCTC G**NNNN** TGA-GACCCTGCAGTCCGGCAAAAA-3′ and Reverse: 3′-AAGG ATGATTTCTGGAATTCAGGTCTCT**NNNN**CGAGACCG TT-5′.
3. Analyze 2 μl of the reaction by gel electrophoresis, to verify the correct sizes of the fragments.
4. Mix 8 μl of PCR reaction with 1 μl of 10× Cutsmart Buffer and 1 μl of DpnI.
5. Incubate at 37 °C for 30 min and deactivate at 80 °C for 20 min.
6. PCR purify the reaction.
7. Proceed with the isothermal assembly according to the manufacturer's instructions.
8. Transform 5 μl of the reaction to 100 μl of competent cells. Incubate for 30 min on ice, heat shock in a 42 °C heat block or water bath for 90 s, cool for 3 min on ice, add 400 μl S.O.C media, and incubate in a 37 °C shaker for 1 h.
9. Plate 200 μl of the transformation on LB agar plates containing 25 μg/ml chloramphenicol and incubate overnight at 37 °C.
10. Pick a purple colony for LB culture containing 25 μg/ml chloramphenicol and incubate overnight at 37 °C.
11. Miniprep the cultures.
12. Send the plasmid for sequencing using VF2 and VF primers to verify the restriction sites and overhangs.

Before applying the MethylAble feature, you need to in vitro-methylate the plasmid.

1. Isolate plasmid DNA using midi-prep kit and following the manufacturer's instructions.
2. In a PCR tube add nuclease-free water to make the final volume 20 μl.
3. Add 2 μl methyltransferase Reaction Buffer (10×).
4. Add 2 μl of 6.4 mM SAM (diluted from the stock).
5. Add 1–2 μg of plasmid DNA (no more than 5 μl).

6. Add 2 μl of CpG methyltransferase.
7. Incubate for 4 h at 37 °C and stop the reaction by heating at 65 °C for 20 min.
8. PCR-purify the reaction.

The methylated plasmid is ready to be used in Mobius Assembly cloning. Let us say you are building a promoter library to test the strengths of 20 previously uncharacterized promoters in plant systems. Promoters will be driving the expression of GFP, which is normalized by the expression of the reference protein RFP. The final constructs will be PromoterX:GFP:HSPter:UBQ10pro:RFP: MASter, which is comprised of two TUs. Without MethylAble, you will need to make 20 constructs of PromoterX:GFP:HSPter in the Level 1 Acceptor Vector A and one construct, UBQ10pro:RFP: MASter, in the Level 1 Acceptor Vector B. Then, in 20 more Level 2 reactions, the two TUs will be combined together. Instead, the methylated plasmid will replace the promoter in the Level 1 reaction, resulting in just one construct, amilCP:nluc:HSPter. You will then select for purple colonies (due to the expression of amilCP) and isolate the plasmid DNA, which will be used for the Level 2 reaction to form one multi-TU construct, amilCP:nluc:HSPter: UBQ10pro:fluc:UBQ5ter. Lastly, the assembly will finish in a Level 1 reaction. You can make a mastermix of all the reagents for 20 reactions, along with a Level 1 Acceptor Vector A and amilCP: nluc:HSPter:UBQ10pro:fluc:UBQ5ter. Aliquot the mastermix and add one of the 20 Level 0 promoters in each PCR tube.

3.7 Plasmid Construction

For different chassis or experimental requirements Mobius Assembly cassettes can be adapted to any backbone. They can be either PCR amplified or synthesized. This can be done for the whole toolkit or just for the final destination vectors. The easiest way to swap backbones is by isothermal assembly which joins overlapping DNA fragments.

1. Retrieve the sequences of Mobius Assembly cassettes from: https://www.addgene.org/kits/nakayama-mobius-assembly-toolkit/.
2. Select your destination vector(s) and open it in your DNA editing software.
3. Copy the Mobius Assembly cassette sequence and paste it into the destination vector. A Level 1 cassette is the sequence between 5′-GGAATTCCACCTGCATAT-3′ and 3′-ATATGC AGGTGCTGCAG-5′; A Level 2 cassette is the sequence between 5′-TTCTGGAATTCGGTCTCA-3′ and 3′-CTGC ATATACCCCTGCAG-5′.
4. Design primers to PCR amplify the Mobius Assembly cassettes. For Level 1 vectors: Forward: 5′-(~10 bp) + GGAATTCCACC

TGCATAT and Reverse: 3′-ATATGCAGGTGCTGCAG + (~10 bp)-5′. For Level 2 vectors: Forward: 5′--(~10 bp) + TTCTGGAATTCGGTCTCA-3′ and Reverse: 3′-CTGCATATACCCCTGCAG + (~10 bp)-5′. The 10 bp sequences come from the destination vector backbone, and they will form the 20 bp overlapping DNA fragments.

5. Design primers to PCR amplify the selected backbone. For Level 1 vectors: Forward: 5′-GGTGCTGCAG + (17–25 bp)-3′ and Reverse 3′-(17–25 bp) + GGAATTCCAC-5′. For Level 2 vectors: Forward: 5′-ACCCCTGCAG + (~17–25 bp)-3′ and Reverse (17–25 bp) + TTCTGGAATT.
6. Set the PCR reaction following the manufacturer's instructions. Usually an annealing temperature of 55 °C and 30 cycles work well for the amplification.
7. Analyze 2 μl of the reaction by gel electrophoresis to verify correct size of the products.
8. If the destination vector shares the same antibiotic resistance with the donor plasmids, perform DpnI digestion:
   (a) Mix 8 μl of PCR reaction with 1 μl of 10× CutSmart Buffer and 1 μl of DpnI.
   (b) Incubate at 37 °C for 30 min and deactivate at 80 °C for 20 min.
9. PCR purify the reaction (optional).
10. Proceed with the isothermal assembly following the manufacturer's instructions.
11. Transform 5 μl of the reaction to 100 μl of competent cells. Incubate for 30 min on ice, heat shock in a 42 °C heat block or water bath for 90 s, cool for 3 min on ice, add 400 μl S.O.C media, and incubate in a 37 °C shaker for 1 h.
12. Plate 100–400 μl of the transformation on LB agar plates containing the appropriate antibiotic and incubate overnight at 37 °C (*see* **Note 13**). If the reaction is not efficient you will need to plate the whole content of the tube.
13. Pick a pink (Level 1 Vectors) or yellow (Level 2 Vectors) colony for LB culture containing the appropriate antibiotics and incubate overnight at 37 °C.
14. Miniprep the cultures and verify the plasmid by restriction digestion analysis with *Pst*I-HF and *Eco*RI-HF.
15. Send the plasmid for sequencing to verify the Mobius Assembly restriction sites and overhangs.

3.8 Mobius Assembly Troubleshooting

Successful assembly will result in mostly white colonies. Bear in mind that generation of complex and large constructs will reduce the number of white colonies. When you face problems with your cloning, follow the steps bellow to troubleshoot.

1. If there are no or very few colonies on the plates:
   (a) Check that you used the correct antibiotic for selection.
   (b) Check the competency of your cells.
   (c) Change the *E. coli* strain (*see* **Note 14**).
   (d) Use fresh T4 DNA ligase and/or T4 DNA buffer.
   (e) Check if there is a mutation in the overhangs of the vector or the enzyme recognition site (a rare scenario).
2. If there are many negative colonies and few positives, use fresh restriction enzymes.
3. For **steps 1** and **2** and especially for the assembly of large constructs:
   (a) Check that you added the right plasmids, enzymes, and buffers for each reaction.
   (b) Use more digestion/ligation cycles, which can be combined with halving digestion/ligation times.
   (c) Be sure that you use the recommended molar ratios.
   (d) Be sure about the DNA quantification (*see* **Note 15**).
   (e) Double the amount of the enzymes.
   (f) Increase the volume of the reaction to dilute possible reaction inhibitors.
   (g) Avoid old and possibly degraded plasmid DNA.
4. Toxic or unstable inserts might cause mutations in the assembled constructs.
   (a) Screen more colonies.
   (b) Try different *E. coli* strains.
   (c) Use a backbone with a low copy number origin of replication.
   (d) Incubate at lower temperatures.
   (e) Place toxic coding sequences under an inducible promoter.

4 Notes

1. Sometimes amplification from genomic DNA might be tricky, and you will need to first amplify the fragment with primers without the Mobius Assembly extension sequence.
2. Avoid using rare codons. Refer to the codon usage table of the organism you are working with.

3. Avoid using the same overhangs as Phytobricks or palindromic sequences.
4. To isolate plasmid DNA from any Acceptor Vector, use a single colony from a streaked agar plate to inoculate a liquid culture. If glycerol stock is used directly as inoculum, the color of the negative selection marker sometimes does not properly develop.
5. Before proceeding to an in vitro experiment in every level, always perform an in silico simulation using your vector mapping software to check that you use the correct parts, vectors and overhangs.
6. If the concentration of the DNA is low or the parts are several you might need to increase the volume of the reaction to 15 or 20 μl.
7. Always aliquot the T4 DNA ligase buffer to avoid repetitive thaw-freeze cycles which degrade ATP.
8. For the first round of Level 1 reaction, 5 cycles are adequate unless several parts with considerable size differences are being assembled, where the number of the cycles should be increased (up to 10). For the further rounds of Level 1 reaction, the number of the cycles should be increased as the constructs become larger.
9. For the isolation of Level 2 Acceptor Vectors (especially with high copy number backbones), the use of Monarch®Plasmid Miniprep Kit or PureYield™ Plasmid Miniprep (Promega) is recommended. We tested GeneJET Plasmid Miniprep Kit (Thermofisher) and QIAprep® Spin Miniprep Kit (Qiagen), but the quantity and purity of the DNA was low, since sfGFP binds on the membrane of the columns and coelute with the plasmid DNA. For midi prep we use PureYield™ Plasmid Midiprep System (Promega) without any issues.
10. If the concentration of the DNA is low or you are cloning several parts, you might need to increase the volume of the reaction to 15 or 20 μl.
11. For large and complex constructs, it is recommended to screen at least two colonies.
12. The only limitation of this feature is that MethylAble parts starting with GGAG (promoter) should be placed at the beginning of the construct (in the first TU) and parts ending with CGCT (terminator) should be placed at the end (in the fourth TU). This rule prevents incorporating extra GGAG or CGCT overhangs in the constructs, as all the Mobius Assembly Acceptor vectors already have those overhangs at the beginning (GGAG) and at the end (CGCT) of the cloning cassettes.

13. Sometimes, high copy number vectors expressing spisPink do not grow well at 37 °C right after the isothermal assembly. Incubating at 30 °C or using different *E. coli* strains may help.
14. We noticed that some backbones give no or very few colonies when having large inserts and transformed into specific strains. (e.g., pGreen with TOP10 cells).
15. RNA and genomic DNA contaminations lead to overestimation of the plasmid concentration. They can be spotted on agarose gel electrophoresis as low molecular weight bands for the former and smear for the latter. RNA contamination occurs when RNase in the buffer is not working, and genomic DNA contamination may result from shearing of host chromosomal DNA.

References

1. Engler C, Gruetzner R, Kandzia R, Marillonnet S (2009) Golden gate shuffling: a one-pot DNA shuffling method based on type IIs restriction enzymes. PLoS One 4:e5553. https://doi.org/10.1371/journal.pone.0005553
2. Engler C, Kandzia R, Marillonnet S (2008) A one pot, one step, precision cloning method with high throughput capability. PLoS One 3: e3647. https://doi.org/10.1371/journal.pone.0003647
3. Sarrion-Perdigones A, Vazquez-Vilar M, Palaci J, Castelijns B, Forment J, Ziarsolo P et al (2013) GoldenBraid 2.0: a comprehensive DNA assembly framework for plant synthetic biology. Plant Physiol 162:1618–1631. https://doi.org/10.1104/pp.113.217661
4. Weber E, Engler C, Gruetzner R, Werner S, Marillonnet S (2011) A modular cloning system for standardized assembly of multigene constructs. PLoS One 6:e16765. https://doi.org/10.1371/journal.pone.0016765
5. Moore SJ, Lai HE, Kelwick RJR, Chee SM, Bell DJ, Polizzi KM et al (2016) EcoFlex: a multifunctional MoClo kit for E. coli synthetic biology. ACS Synth Biol 5:1059–1069. https://doi.org/10.1021/acssynbio.6b00031
6. Iverson SV, Haddock TL, Beal J, Densmore DM (2016) CIDAR MoClo: improved MoClo assembly standard and new E. coli part library enable rapid combinatorial design for synthetic and traditional biology. ACS Synth Biol 5:99–103. https://doi.org/10.1021/acssynbio.5b00124
7. Andreou AI, Nakayama N (2018) Mobius Assembly: a versatile Golden-Gate framework towards universal DNA assembly. PLoS One 13:e0189892. https://doi.org/10.1371/journal.pone.0189892
8. Patron N et al (2015) Standards for plant synthetic biology: a common syntax for exchange of DNA parts. New Phytol 208:13–19
9. Chung CT, Miller RH (1993) Preparation and storage of competent Escherichia coli cells. Methods Enzymol 218:621–627

Chapter 13

Design and Implementation of Multi-protein Expression Constructs and Combinatorial Libraries using Start-Stop Assembly

George M. Taylor and John T. Heap

Abstract

Start-Stop Assembly is a multi-part, modular, Golden Gate-based DNA assembly system with two key features which distinguish it from previous DNA assembly methods. Firstly, coding sequences are assembled with upstream and downstream sequences via overhangs corresponding to start and stop codons, avoiding unwanted 'scars' in assembled constructs at coding sequence boundaries. Scars at these crucial, sensitive locations can affect mRNA structure, activity of the ribosome binding site, and potentially other functional RNA features. Start-Stop Assembly is therefore both functionally scarless (an advantage usually only achieved using bespoke, overlap-based assembly methods) and suitable for efficient, unbiased and combinatorial assembly (a general advantage of Golden Gate-based methods). Secondly, Start-Stop Assembly has a new, streamlined assembly hierarchy, meaning that typically only one new vector is required in order to assemble constructs for any new destination context, such as a new organism or genomic location. This should facilitate more rapid and convenient development of engineered metabolic pathways for diverse nonmodel organisms in order to exploit their applied potential. This chapter explains both design considerations and practical procedures to implement multi-part, hierarchical assembly of multi-protein expression constructs, either individually or as combinatorial libraries, using Start-Stop Assembly.

Key words DNA assembly, Synthetic biology, Metabolic engineering, Gene expression, Cloning

1 Introduction

Start-Stop Assembly is a multi-part, modular DNA assembly system that builds upon the principles of Golden Gate assembly [1, 2], in which type IIS restriction endonucleases are used to generate unique non-palindromic cohesive DNA ends (referred to as fusion sites) allowing the ordered and irreversible assembly of multiple DNA fragments into a single construct using a single type IIS restriction endonuclease and DNA ligase in an efficient 'one-pot' assembly reaction.

Start-Stop Assembly uses hierarchical assembly of up to four levels (Level 0, Level 1, Level 2, and Level 3) which allows up to

Sunil Chandran and Kevin W. George (eds.), *DNA Cloning and Assembly: Methods and Protocols*, Methods in Molecular Biology, vol. 2205, https://doi.org/10.1007/978-1-0716-0908-8_13, © Springer Science+Business Media, LLC, part of Springer Nature 2020

15 'expression units' (each comprising one coding sequence (CDS) and other elements defined below) to be assembled into a single construct from up to 60 DNA parts or mixtures of DNA parts. In this hierarchy, individual parts are usually stored at Level 0 in the storage vector pStA0. Individual Level 0 parts can be assembled into an expression unit at Level 1, up to five Level 1 expression units can be assembled at Level 2, and finally up to 15 expression units can be assembled at Level 3 (Fig. 1).

DNA parts are stored at Level 0 with the appropriate fusion sites and flanking sequences for multi-part Level 1 assembly of expression units from four modular parts with a pre-defined architecture: the promoter, 5′-untranslated region/ribosome-binding site (UTR/RBS), CDS, and transcriptional terminator (Fig. 2). These four parts are assembled using the Level 1 fusion sites α (alpha), β (beta), γ (gamma), δ (delta), and ε (epsilon) as shown in Fig. 2 (sequences shown in Table 1). To allow functionally scarless assembly of expression units, the start codon (sequence ATG) and the stop codon (sequence TAA) of each CDS are used as fusion sites (γ and δ, respectively). As every CDS requires a start and stop codon anyway, this design avoids introduction of any additional scar sequence. Scarless assembly at the CDS boundaries is a key advantage, as these are highly sensitive sites where scars affect mRNA structure and the activity of the RBS and potentially other functional RNA features [3].

Start-Stop Assembly uses a new, 'streamlined' assembly hierarchy in which the Level 2 vector does not need to be changed to accommodate different numbers of expression units being assembled at that level, unlike other Golden Gate-based methods. Level 2 supports assembly of up to five expression units, which in many cases is sufficient, so Level 2 is often the destination (final) level. Consequently, it is usually only necessary to construct one additional Level 2 vector in order to apply Start-Stop Assembly to a new destination context, such as a new organism or genomic location, and the rest of the assembly is facilitated by the existing core vectors. This improvement over other methods means that Start-Stop Assembly can be thought of as essentially 'organism-agnostic', as it is equally suitable for assembly of constructs for any organism. In contrast, many assembly methods use systems of vectors tailored to only one or a narrow range of organisms and require additional cloning effort to construct sets of vectors for any new organism or other destination context. The streamlined hierarchy is achieved by always using the same pair of acceptor fusion sites, A and Z (sequences shown in Table 1), for the Level 2 vectors (Fig. 3). Therefore the first fusion site of the first Level 1 part must always be A and the last fusion site of the last Level 1 part must always be Z, irrespective of the number of expression units being assembled (up to five expression units at Level 2). When more than one expression unit is being assembled the B, C, D, and E fusion sites

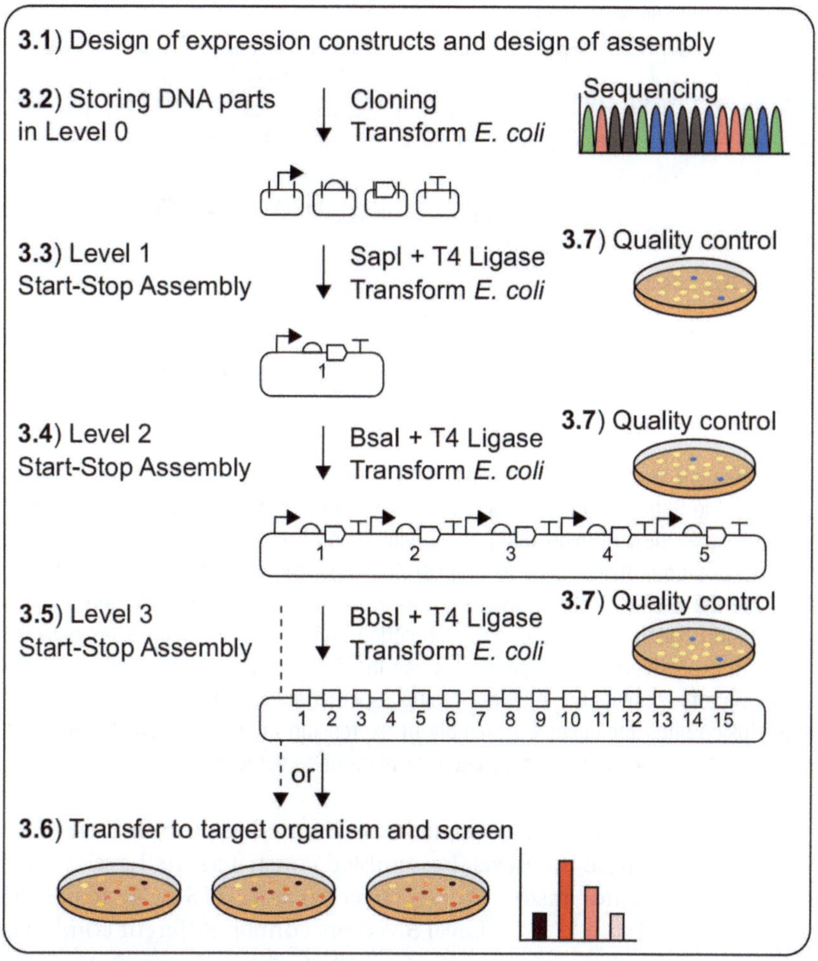

Fig. 1 Schematic overview and workflow of Start-Stop Assembly. Schematic illustration of the overall Start-Stop Assembly process, in which constructs containing up to 15 expression units can be hierarchically assembled from basic DNA parts. The workflow has been divided into key sections, numbered to correspond with the main text: (Subheading 3.1) design of expression constructs and design of assembly; (Subheading 3.2) storing DNA parts in Level 0 using the storage vector pStA0; (Subheading 3.3) Level 1 assembly of individual expression units from Level 0 parts; (Subheading 3.4) Level 2 assembly of constructs containing up to five expression units from Level 1 parts; (Subheading 3.5) Level 3 assembly of constructs containing up to 15 expression units from Level 2 parts; (Subheading 3.6) transfer of assembled constructs to the target organism and screening; and finally the (Subheading 3.7) quality control at each level

(sequences shown in Table 1) are used for part-part junctions between expression units. To allow the last fusion site of the last expression unit in each series to be assembled to always be Z, Start-Stop Assembly uses alternative Level 1 'Z vectors' for the last expression unit in each series. In Z vectors, the second of the two fusion sites is replaced by a Z fusion site (shown in Fig. 3, and more explicitly in Fig. 3 of Taylor et al. [4]).

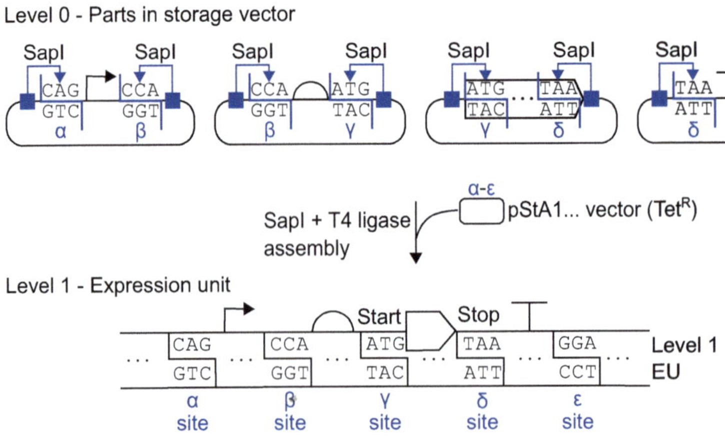

Fig. 2 Level 1 architecture and assembly of expression units from Level 0 parts. Parts stored in Level 0 are flanked by inward-facing SapI restriction sites (blue boxes) with corresponding donor fusion sites (staggered blue lines) for Level 1 assembly of expression units. Each type of part (promoter, UTR/RBS, CDS, terminator) uses a different pair of unique fusion sites (α-ε sites) that allow the correct ordered assembly of expression units. Promoters use the α site (CAG) and β site (CCA). RBSs use the β site (CCA) and γ site (ATG). Coding sequences use the start codon ATG (γ site) and stop codon TAA (δ site) as fusion sites allowing functionally scarless assembly of expression units. Terminators use the δ site (TAA) and the ε site (GGA). The assembly cassette in Level 1 vectors contains two outward-facing SapI recognition sites with corresponding α (CAG) and ε (GGA) acceptor fusion sites for Level 1 assembly of expression units. Between the SapI sites in a Level 1 empty vector is a *lacZα* gene which can be used for blue-white screening

Level 2 constructs can be used as parts in Level 3 assembly. To excise the Level 2 assembled constructs the Level 2 vectors contain donor fusion sites (referred to as 1, 2, 3 or 4; sequences shown in Table 1). The Level 3 vectors contain different combinations of the corresponding acceptor fusion sites for assembling the Level 2 constructs in a Level 3 vector (Fig. 3).

This chapter provides a protocol for the multi-part assembly of expression constructs using Start-Stop Assembly. The protocol has been divided into key sections (Fig. 1): (Subheading 3.1) design of individual expression constructs or libraries of expression constructs and design of assemblies; (Subheading 3.2) storage of individual DNA parts in the storage vector, pStA0; (Subheading 3.3) assembly of expression units at Level 1; (Subheading 3.4) assembly of up to five expression units at Level 2; (Subheading 3.5) assembly of up to 15 expression units at Level 3; (Subheading 3.6) transfer to the target organism and screening; and finally (Subheading 3.7) quality control at each level.

Table 1
Acceptor and donor fusion sites

Level	Acceptor/donor	Fusion site	Sequence
0	Acceptor	F	TGTG
0	Acceptor	R	GACC
0	Donor	α	CAG
0	Donor	β	CCA
0	Donor	γ	ATG
0	Donor	δ	TAA
0	Donor	ε	GGA
1	Acceptor	α	CAG
1	Acceptor	ε	GGA
1	Donor	A	GGAG
1	Donor	B	AATG
1	Donor	C	AGGT
1	Donor	D	GCTT
1	Donor	E	CGCT
1	Donor	Z	TACT
2	Acceptor	A	GGAG
2	Acceptor	Z	TACT
2	Donor	1	TGCC
2	Donor	2	ACTA
2	Donor	3	TTAC
2	Donor	4	CGAG
3	Acceptor	1	TGCC
3	Acceptor	3	TTAC
3	Acceptor	4	CGAG

2 Materials

1. Start-Stop Assembly Toolkit of vectors and parts (Addgene; *see* **Note 1**).
2. Nuclease-free water (Qiagen).
3. 10 mM deoxynucleotide (dNTP) Solution Mix (NEB).
4. Phusion High-Fidelity DNA polymerase (NEB), supplied with 5× HF buffer (NEB).

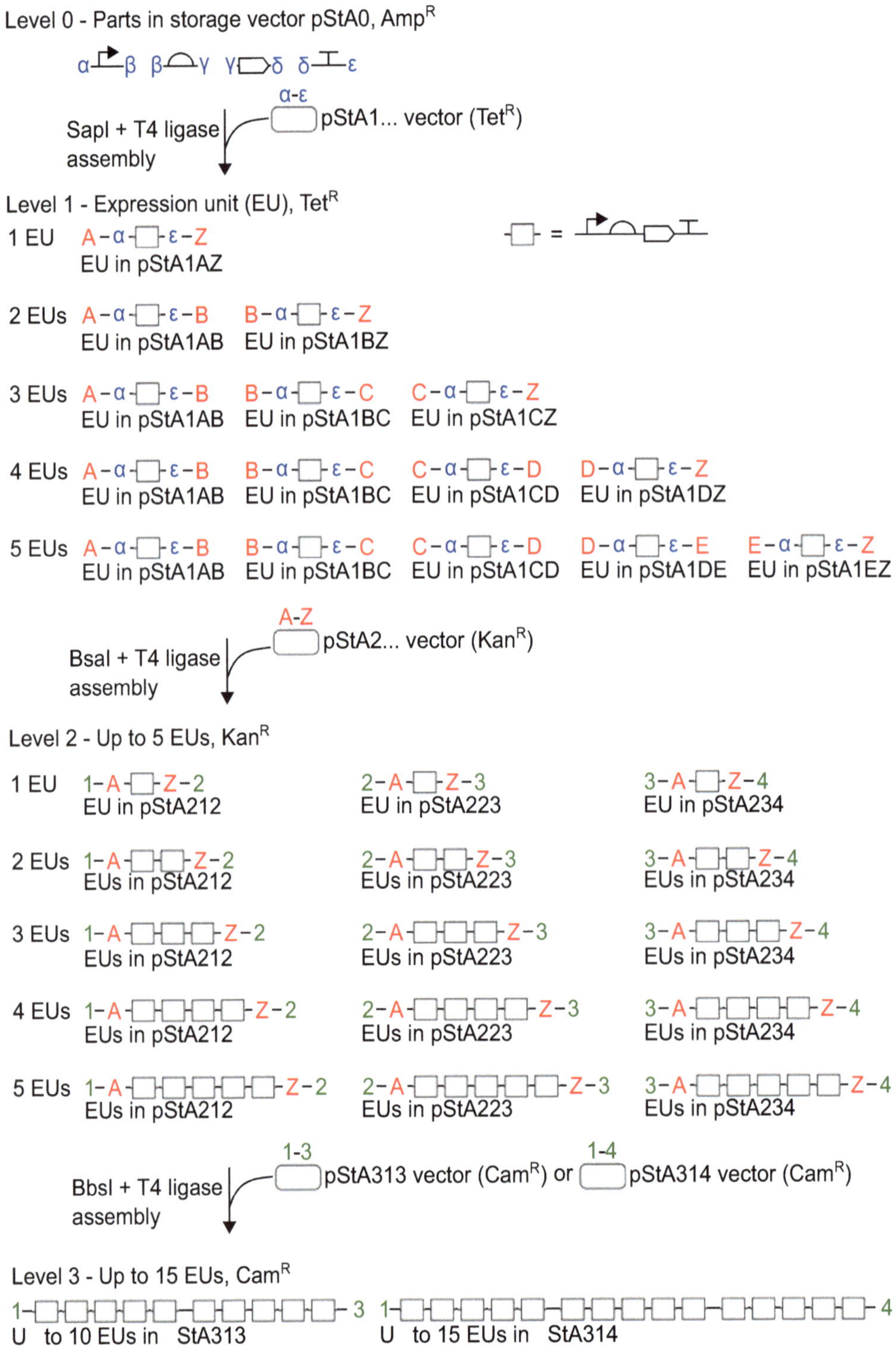

Fig. 3 Detail of multipart hierarchical assembly and fusion sites used in Start-Stop Assembly. Up to 15 expression units can be hierarchically assembled from basic DNA parts (promoters, UTR/RBSs, CDSs, and terminators) stored at Level 0. Level 1 expression units are assembled from four Level 0 parts, Level

5. QIAquick Gel Extraction Kit (Qiagen).
6. QIAprep Spin Miniprep kit (Qiagen).
7. SapI (NEB), 10,000 U/ml (*see* **Note 2**).
8. BsaI (NEB), 10,000 U/ml.
9. BbsI (NEB), 10,000 U/ml.
10. T4 DNA Ligase (NEB), 400,000 U/ml, supplied with 10× T4 DNA ligase buffer: 500 mM Tris-HCl, pH 7.5, 100 mM $MgCl_2$, 10 mM ATP, 100 mM DTT.
11. Commercial or homemade competent cells of a strain of *Escherichia coli* with an alpha-complementable (alpha-acceptor) *lacZ* mutation, such as *lacZ*ΔM15 (*see* **Notes 3** and **5**).
12. If using electroporation: MF-Millipore™ Membrane Filter, 0.22 μm pore size (Merck-Millipore) for dialysis of the reaction product (*see* **Note 10**).
13. If using electroporation: 1 mm-gap electroporation cuvettes.
14. LB medium: tryptone 10 g/l, yeast extract 5 g/l, and NaCl 5 g/l, ddH_2O.
15. LB agar: tryptone 10 g/l, yeast extract 5 g/l, NaCl 5 g/l, and agar 15 g/l, ddH_2O.
16. Antibiotics: where appropriate LB agar should be supplemented after autoclaving with either 100 μg/ml ampicillin, 10 μg/ml tetracycline, 50 μg/ml kanamycin, or 25 μg/ml chloramphenicol.
17. Blue-white screening: where appropriate LB agar should be supplemented with 40 μg/ml 5-bromo-4-chloro-3-indolyl-β-galactopyranoside (X-Gal) and 0.1 mM isopropyl β-D-1-thiogalactopyranoside (IPTG) after autoclaving.
18. Oligonucleotides for use as sequencing primers and spacers are given in Tables 2 and 3, respectively.

Fig. 3 (continued) 2 constructs are assembled from up to five Level 1 expression unit parts, and Level 3 constructs are assembled from up to three Level 2 constructs. Different type IIS restriction endonucleases (SapI, BsaI, or BbsI) and different antibiotic-resistance markers (Amp^R, Tet^R, Kan^R, or Cam^R) are used for each assembly level as shown. Acceptor fusion sites are shown inside donor fusion sites at Level 1 and Level 2. Only donor fusion sites are shown at Level 0, and only acceptor fusion sites are shown at Level 3. Alternative sets of vectors are used at Level 1 depending upon the number of expression units being assembled, such that the first and last Level 1 donor fusion sites are always A and Z, respectively, which are always the Level 2 acceptor fusion sites. Multiple Level 2 vectors (pStA212, pStA223, pStA234) with different donor fusion sites are only needed if the assembly will be continued to Level 3, in order to assemble more than five expression units. Alternative Level 3 vectors (pStA313, pStA314) with different acceptor fusion sites allow for assembly of two or three Level 2 constructs

Table 2
List of sequencing primers used in Start-Stop Assembly

Primer	Sequence (5′–3′)	Comments
OligoGT234	GGGGAAACGCCTGGTATCT	Level 0 Forward sequencing primer
OligoGT235	AGCAAAAACAGGAAGGCAAA	Level 0 Reverse sequencing primer
OligoGT339	GTTGAGGACCCGGCTAGG	Level 1 Forward sequencing primer
OligoGT340	TGTGACGGAAGATCACTTCG	Level 1 Reverse sequencing primer
OligoGT573	CCTCGGTGAGTTTTCTCCTTC	Level 2 Forward sequencing primer
OligoGT486	GATTACGCGCAGACCAAAAC	Level 2 Reverse and Level 3 Forward sequencing primer
OligoGT487	AAACGGTTAGCGCTTCGTTA	Level 3 Reverse sequencing primer

3 Methods

3.1 Design of Expression Constructs and Design of Assembly

The design of expression constructs and design of assembly (Fig. 4) includes: (1) specifying an individual or combinatorial design, and individual parts or mixtures of parts for each position in the construct; (2) specifying the configuration (monocistronic, operon, or a hybrid of both) of the expression units in the construct; (3) assigning expression units to positions in the hierarchy, and therefore potentially into groups at Level 2; and (4) identifying the Start-Stop Assembly vectors required for the assembly.

3.1.1 Individual or Combinatorial Design

- Start-Stop Assembly can be used to assemble individual expression constructs or combinatorial libraries of expression constructs (Fig. 4).
- Individual expression constructs can be assembled by using a single promoter, UTR/RBS, CDS, and terminator for each expression unit, whereas combinatorial libraries of expression constructs can be assembled by using mixtures of these parts instead of individual parts at one or more positions [4–7].
- Combinatorial assembly of expression constructs offers an ideal approach for exploring the 'expression space' of a pathway or other system. The resulting combinatorial library will typically contain many poorly performing and/or burdensome constructs, but within the library there may be constructs with more optimal expression profiles, giving improved performance with lower metabolic burden. Such optimal constructs are very difficult to design a priori purely rationally but can often be identified using combinatorial, empirical approaches.

Table 3
Spacers implemented as double-stranded linkers

<table>
<tr><th>Spacer</th><th>Spacer sequence in double-stranded linker</th><th>Forward oligonucleotide</th><th>Reverse oligonucleotide</th></tr>
<tr><td>Spacer 1
α-β format</td><td>5′-CAGTGGTCAGCGACT-3′
||||||||||||
3′-ACCAGTCGCTGAGGT-5′</td><td>oligoGT538:
CAGTGGTCAGCGACT</td><td>oligoGT539:
TGGAGTCGCTGACCA</td></tr>
<tr><td>Spacer 1
δ-ε format</td><td>5′-TAATGGTCAGCGACT-3′
||||||||||||
3′-ACCAGTCGCTGACCT-5′</td><td>oligoGT540:
TAATGGTCAGCGACT</td><td>oligoGT541:
TCCAGTCGCTGACCA</td></tr>
<tr><td>Spacer 2
α-β format</td><td>5′-CAGGCTGCCGTGAAT-3′
||||||||||||
3′-CGACGGCACTTAGGT-5′</td><td>oligoGT542:
CAGGCTGCCGTGAAT</td><td>oligoGT543:
TGGATTCACGGCAGC</td></tr>
<tr><td>Spacer 2
δ-ε format</td><td>5′-TAAGCTGCCGTGAAT-3′
||||||||||||
3′-CGACGGCACTTACCT-5′</td><td>oligoGT544:
TAAGCTGCCGTGAAT</td><td>oligoGT545:
TCCATTCACGGCAGC</td></tr>
<tr><td>Spacer 3
α-β format</td><td>5′-CAGGGCACGCTCAAT-3′
||||||||||||
3′-CCGTGCGAGTTAGGT-5′</td><td>oligoGT546:
CAGGGCACGCTCAAT</td><td>oligoGT547:
TGGATTGAGCGTGCC</td></tr>
<tr><td>Spacer 3
δ-ε format</td><td>5′-TAAGGCACGCTCAAT-3′
||||||||||||
3′-CCGTGCGAGTTACCT-5′</td><td>oligoGT548:
TAAGGCACGCTCAAT</td><td>oligoGT549:
TCCATTGAGCGTGCC</td></tr>
<tr><td>Spacer 4
α-β format</td><td>5′-CAGAGTCCGTGCTCA-3′
||||||||||||
3′-TCAGGCACGAGTGGT-5′</td><td>oligoGT550:
CAGAGTCCGTGCTCA</td><td>oligoGT551:
TGGTGAGCACGGACT</td></tr>
<tr><td>Spacer 4
δ-ε format</td><td>5′-TAAAGTCCGTGCTCA-3′
||||||||||||
3′-TCAGGCACGAGTCCT-5′</td><td>oligoGT552:
TAAAGTCCGTGCTCA</td><td>oligoGT553:
TCCTGAGCACGGACT</td></tr>
<tr><td>Spacer 5
α-β format</td><td>5′-CAGATTCTGTGCCGC-3′
||||||||||||
3′-TAAGACACGGCGGGT-5′</td><td>oligoGT554:
CAGATTCTGTGCCGC</td><td>oligoGT555:
TGGGCGGCACAGAAT</td></tr>
<tr><td>Spacer 5
δ-ε format</td><td>5′-TAAATTCTGTGCCGC-3′
||||||||||||
3′-TAAGACACGGCGCCT-5′</td><td>oligoGT556:
TAAATTCTGTGCCGC</td><td>oligoGT557:
TCCGCGGCACAGAAT</td></tr>
<tr><td>Spacer 6
α-β format</td><td>5′-CAGATCAACGCCTGC-3′
||||||||||||
3′-TAGTTGCGGACGGGT-5′</td><td>oligoGT558:
CAGATCAACGCCTGC</td><td>oligoGT559:
TGGGCAGGCGTTGAT</td></tr>
<tr><td>Spacer 6
δ-ε format</td><td>5′-TAAATCAACGCCTGC-3′
||||||||||||
3′-TAGTTGCGGACGCCT-5′</td><td>oligoGT560:
TAAATCAACGCCTGC</td><td>oligoGT561:
TCCGCAGGCGTTGAT</td></tr>
<tr><td>Spacer 7
α-β format</td><td>5′-CAGATCTGCGGCAAC-3′
||||||||||||
3′-TAGACGCCGTTGGGT-5′</td><td>oligoGT562:
CAGATCTGCGGCAAC</td><td>oligoGT563:
TGGGTTGCCGCAGAT</td></tr>
<tr><td>Spacer 7
δ-ε format</td><td>5′-TAAATCTGCGGCAAC-3′
||||||||||||
3′-TAGACGCCGTTGCCT-5′</td><td>oligoGT564:
TAAATCTGCGGCAAC</td><td>oligoGT565:
TCCGTTGCCGCAGAT</td></tr>
<tr><td>Spacer 8
α-β format</td><td>5′-CAGTGCGACCTGACT-3′
||||||||||||
3′-ACGCTGGACTGAGGT-5′</td><td>oligoGT566:
CAGTGCGACCTGACT</td><td>oligoGT567:
TGGAGTCAGGTCGCA</td></tr>
<tr><td>Spacer 8
δ-ε format</td><td>5′-TAATGCGACCTGACT-3′
||||||||||||
3′-ACGCTGGACTGACCT-5′</td><td>oligoGT616:
TAATGCGACCTGACT</td><td>oligoGT617:
TCCAGTCAGGTCGCA</td></tr>
</table>

(continued)

Table 3
(continued)

<table>
<tr><td>Spacer 9
α-β format</td><td>5′-CAGAGGTGTCTCGCA-3′
||||||||||||
3′-TCCACAGAGCGTGGT-5′</td><td>oligoGT618:
CAGAGGTGTCTCGCA</td><td>oligoGT619:
TGGTGCGAGACACCT</td></tr>
<tr><td>Spacer 9
δ-ε format</td><td>5′-TAAAGGTGTCTCGCA-3′
||||||||||||
3′-TCCACAGAGCGTCCT-5′</td><td>oligoGT620:
TAAAGGTGTCTCGCA</td><td>oligoGT621:
TCCTGCGAGACACCT</td></tr>
<tr><td>Spacer 10
α-β format</td><td>5′-CAGGCTACAGGCTGC-3′
||||||||||||
3′-CGATGTCCGACGGGT-5′</td><td>oligoGT622:
CAGGCTACAGGCTGC</td><td>oligoGT623:
TGGGCAGCCTGTAGC</td></tr>
<tr><td>Spacer 10
δ-ε format</td><td>5′-TAAGCTACAGGCTGC-3′
||||||||||||
3′-CGATGTCCGACGCCT-5′</td><td>oligoGT624:
TAAGCTACAGGCTGC</td><td>oligoGT625:
TCCGCAGCCTGTAGC</td></tr>
<tr><td>Spacer 11
α-β format</td><td>5′-CAGTCAGACGGCACT-3′
||||||||||||
3′-AGTCTGCCGTGAGGT-5′</td><td>oligoGT626:
CAGTCAGACGGCACT</td><td>oligoGT627:
TGGAGTGCCGTCTGA</td></tr>
<tr><td>Spacer 11
δ-ε format</td><td>5′-TAATCAGACGGCACT-3′
||||||||||||
3′-AGTCTGCCGTGACCT-5′</td><td>oligoGT628:
TAATCAGACGGCACT</td><td>oligoGT629:
TCCAGTGCCGTCTGA</td></tr>
<tr><td>Spacer 12
α-β format</td><td>5′-CAGATCGCAACTGGC-3′
||||||||||||
3′-TAGCGTTGACCGGGT-5′</td><td>oligoGT630:
CAGATCGCAACTGGC</td><td>oligoGT631:
TGGGCCAGTTGCGAT</td></tr>
<tr><td>Spacer 12
δ-ε format</td><td>5′-TAAATCGCAACTGGC-3′
||||||||||||
3′-TAGCGTTGACCGCCT-5′</td><td>oligoGT632:
TAAATCGCAACTGGC</td><td>oligoGT633:
TCCGCCAGTTGCGAT</td></tr>
<tr><td>Spacer 13
α-β format</td><td>5′-CAGGGCAATCGTGCT-3′
||||||||||||
3′-CCGTTAGCACGAGGT-5′</td><td>oligoGT634:
CAGGGCAATCGTGCT</td><td>oligoGT635:
TGGAGCACGATTGCC</td></tr>
<tr><td>Spacer 13
δ-ε format</td><td>5′-TAAGGCAATCGTGCT-3′
||||||||||||
3′-CCGTTAGCACGACCT-5′</td><td>oligoGT636:
TAAGGCAATCGTGCT</td><td>oligoGT637:
TCCAGCACGATTGCC</td></tr>
<tr><td>Spacer 14
α-β format</td><td>5′-CAGATTGCCTGCGTC-3′
||||||||||||
3′-TAACGGACGCAGGGT-5′</td><td>oligoGT638:
CAGATTGCCTGCGTC</td><td>oligoGT639:
TGGGACGCAGGCAAT</td></tr>
<tr><td>Spacer 14
δ-ε format</td><td>5′-TAAATTGCCTGCGTC-3′
||||||||||||
3′-TAACGGACGCAGCCT-5′</td><td>oligoGT640:
TAAATTGCCTGCGTC</td><td>oligoGT641:
TCCGACGCAGGCAAT</td></tr>
<tr><td>Spacer 15
α-β format</td><td>5′-CAGGCACCAATCGCT-3′
||||||||||||
3′-CGTGGTTAGCGAGGT-5′</td><td>oligoGT642:
CAGGCACCAATCGCT</td><td>oligoGT643:
TGGAGCGATTGGTGC</td></tr>
<tr><td>Spacer 15
δ-ε format</td><td>5′-TAAGCACCAATCGCT-3′
||||||||||||
3′-CGTGGTTAGCGACCT-5′</td><td>oligoGT644:
TAAGCACCAATCGCT</td><td>oligoGT645:
TCCAGCGATTGGTGC</td></tr>
<tr><td>Spacer 16
α-β format</td><td>5′-CAGAGCAATCCACGC-3′
||||||||||||
3′-TCGTTAGGTGCGGGT-5′</td><td>oligoGT646:
CAGAGCAATCCACGC</td><td>oligoGT647:
TGGGCGTGGATTGCT</td></tr>
<tr><td>Spacer 16
δ-ε format</td><td>5′-TAAAGCAATCCACGC-3′
||||||||||||
3′-TCGTTAGGTGCGCCT-5′</td><td>oligoGT648:
TAAAGCAATCCACGC</td><td>oligoGT649:
TCCGCGTGGATTGCT</td></tr>
</table>

Spacers can be used in either α-β format in place of a promoter, or δ-ε format in place of a terminator. Here we show each of the 16 spacers as double-stranded linkers in both the α-β and δ-ε configurations. Fusion site cohesive ends are shown in bold. Spacers are obtained as two single-stranded oligonucleotides and then mixed and annealed together (described in Supplementary Materials and Methods of Taylor et al.) to generate the spacer as a linker part that can be used directly in Level 1 assembly reactions. This table and legend is reproduced from Table S1 of Taylor et al. [4]

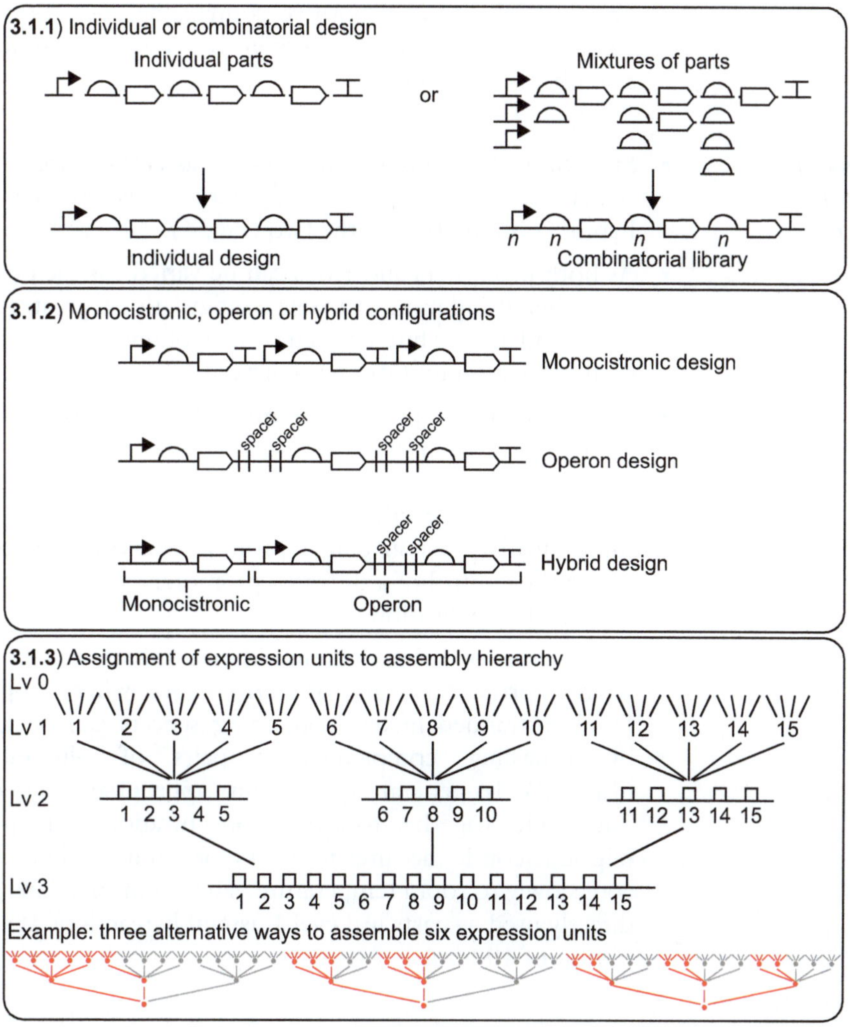

Fig. 4 Designing expression constructs and design of assembly. Top panel, which corresponds to Subheading 3.1.1: Individual or combinatorial design: Constructs can be assembled individually from individual parts, or combinatorial libraries of constructs can be assembled from mixtures of parts. Middle panel, which corresponds to Subheading 3.1.2: Monocistronic, operon, or hybrid configurations: Expression constructs can be assembled in these different configurations by using spacers in place of promoters and/or terminators where appropriate. Bottom panel, which corresponds to Subheading 3.1.3: Assignment of expression units to assembly hierarchy: Expression units can be assigned to different positions in the assembly hierarchy by specifying which parts are assembled with which other parts at each level, and by using the appropriate vectors from Level 1 onward. Expression units can therefore can be placed in different groups at Level 2, which may be useful to allow groups encoding different functional components to be optimized and/or re-used independently

- Mixtures of promoters and RBSs with a proven wide range of expression levels in *E. coli* and stored at Level 0 are available from Addgene (*see* **Note 1**) [4]. It may be helpful for first time users to assemble a carotenoid pathway library originally described in

Taylor et al. [4] as a positive control, because the variety of colored *E. coli* colonies conveniently shows that combinatorial assembly is occurring.

3.1.2 Monocistronic, Operon, and Hybrid Configurations

- Start-Stop Assembly can be used to assemble expression constructs in monocistronic configurations, as operons, or as a hybrid configuration of the two (Fig. 4).
- As both promoters and RBSs can be varied for each expression unit when using monocistronic configurations, greater independence and a more fine-grained control of the expression levels of each expression unit can be achieved.
- Operon configurations allow multiple expression units to be co-regulated by a single promoter, which may be useful to partially couple expression levels in combinatorial designs, and/or to enable efficient optimisation of a construct under the control of a convenient promoter before later placing it under the control of a different promoter more appropriate for a particular application.
- For operon or hybrid configurations, spacers can be used to replace promoters and/or terminators where appropriate. Double-stranded linkers containing spacers can be generated by annealing complementary pairs of oligonucleotides (Table 3). To generate spacers, mix appropriate pairs of oligonucleotides together to a final concentration of 10 μM. Heat oligonucleotide mixtures to 95 °C for 5 min, and then allow to cool slowly to room temperature. These annealed linkers can be directly used as parts in Level 1 assembly reactions at a concentration of 40 fmol.

3.1.3 Assignment of Expression Units to Assembly Hierarchy

- Start-Stop Assembly supports assembly of up to 15 expression units, so the assembly hierarchy effectively has 15 different positions for expression units at Level 1 (Fig. 4).
- Assigning expression units to these positions determines the route they will take through the assembly hierarchy, and which expression units will be grouped together at Level 2.
- Expression units are assigned to positions by using the appropriate vectors, and by specifying which parts are assembled with which other parts at each level.
- Grouping expression units together in this way may be useful to allow groups encoding different functional components (such as a transporter and a metabolic pathway) to be optimized separately from other groups, and re-used independently.

3.1.4 Vector Choice

- Once the expression construct is designed and expression units are assigned to the hierarchy, then the necessary vectors for the assembly can be determined using Figs. 3 and 4.
- Detailed examples of specific assemblies including all vectors are provided in several supplementary figures of Taylor et al. [4].
- The 15 core Start-Stop Assembly vectors (shown in Fig. 3 and listed in Table 4) are available from Addgene (*see* **Note 1**).
- To construct an alternative destination vector the appropriate Start-Stop Assembly cassette needs to be inserted into the alternative destination vector backbone. *See* Fig. S6 of Taylor et al. [4] for an example.

Table 4
List of core Start–Stop Assembly vectors

Plasmid name	Accession number	Level	Acceptor fusion sites	Donor fusion sites	Comments
pStA0	MG649420	0	F-R (BsaI)	α-ε (SapI)	Amp^R, pMB1, *lacZα*, ID = pGT400
pStA1AZ	MG649422	1	α-ε (SapI)	A-Z (BsaI)	Tet^R, p15A, *lacZα*, ID = pGT401
pStA1AB	MG649421	1	α-ε (SapI)	A-B (BsaI)	Tet^R, p15A, *lacZα*, ID = pGT402
pStA1BZ	MG649424	1	α-ε (SapI)	B-Z (BsaI)	Tet^R, p15A, *lacZα*, ID = pGT403
pStA1BC	MG649423	1	α-ε (SapI)	B-C (BsaI)	Tet^R, p15A, *lacZα*, ID = pGT404
pStA1CZ	MG649426	1	α-ε (SapI)	C-Z (BsaI)	Tet^R, p15A, *lacZα*, ID = pGT405
pStA1CD	MG649425	1	α-ε (SapI)	C-D (BsaI)	Tet^R, p15A, *lacZα*, ID = pGT406
pStA1DZ	MG649428	1	α-ε (SapI)	D-Z (BsaI)	Tet^R, p15A, *lacZα*, ID = pGT407
pStA1DE	MG649427	1	α-ε (SapI)	D-E (BsaI)	Tet^R, p15A, *lacZα*, ID = pGT408
pStA1EZ	MG649429	1	α-ε (SapI)	E-Z (BsaI)	Tet^R, p15A, *lacZα*, ID = pGT409
pStA212	MG649430	2	A-Z (BsaI)	1-2 (BbsI)	Kan^R, p15A, *lacZα*, ID = pGT417
pStA223	MG649431	2	A-Z (BsaI)	2-3 (BbsI)	Kan^R, p15A, *lacZα*, ID = pGT418
pStA234	MG649432	2	A-Z (BsaI)	3-4 (BbsI)	Kan^R, p15A, *lacZα*, ID = pGT419
pStA313	MG649433	3	1-3 (BbsI)	N/A	Cam^R, p15A, *lacZα*, ID = pGT415
pStA314	MG649434	3	1-4 (BbsI)	N/A	Cam^R, p15A, *lacZα*, ID = pGT416

The 15 vector names reflect their level in the hierarchy (number in blue) and their donor fusion sites (letters or numbers in red) or in the case of Level 3 vectors their acceptor fusion sites (numbers in green). Each vector contains an assembly cassette (including a *lacZα* gene for blue-white screening), a resistance marker (Amp^R, Tet^R, Kan^R, or Cam^R) and a replicon (high-copy pMB1 or low-copy p15A). Each also has a unique ID number (pGT. . .). Genbank accession numbers are shown. This table is reproduced from Table 1 of Taylor et al. [4]

3.2 Storing DNA Parts in Level 0

DNA parts can be stored in the Level 0 storage vector, pStA0. DNA parts must be cloned into pStA0 in the appropriate format for Start-Stop Assembly to ensure that excision of each DNA part by SapI will result in cohesive DNA ends with 3 bp 5′ overhangs corresponding to the 3 bp fusion sites. This ensures compatibility between all the DNA parts in the assembly of an expression unit at Level 1. For this purpose, standard prefix and suffix sequences have previously been defined (Table 5). Appropriately formatted DNA parts could be generated by any applicable method such as PCR or DNA synthesis, and could be used for assembly as either linear DNA fragments or PCR products, or circular plasmid DNA. Note S2 of Taylor et al., [4] provides details on three approaches often used: (a) inverse PCR, adding part sequences using primer tails; (b) PCR amplification followed by one-pot assembly with pStA0 using Level 0 acceptor fusion sites (Table 1); or (c) DNA synthesis followed by one-pot assembly with pStA0 using Level 0 acceptor fusion sites.

Level 0 Start-Stop Assembly reaction:

1. Prepare Level 0 Start-Stop Assembly reaction mixtures by combining 20 fmol of empty vector pStA0 plasmid DNA, 40 fmol of insert (PCR product or synthetic DNA), T4 DNA Ligase buffer, 400 U of T4 DNA Ligase (typically use 1 μl), and 10 U of BsaI (typically 1 μl); make up to a total reaction volume of 20 μl.

Table 5
Prefix and suffix sequences

Part	Prefix		Suffix
Promoter	GGTCTCA**TGTG**GCTCTTCG**CAG** BsaI F SapI α	-Promoter-	**CCA**TGAAGAGC**GACC**AGAGACC β SapI R BsaI
RBS	GGTCTCA**TGTG**GCTCTTCG**CCA** BsaI F SapI β	-RBS-	**ATG**TGAAGAGC**GACC**AGAGACC γ SapI R BsaI
CDS	GGTCTCA**TGTG**GCTCTTCG**ATG** BsaI F SapI γ	-CDS-	**TAA**TGAAGAGC**GACC**AGAGACC δ SapI R BsaI
Terminator	GGTCTCA**TGTG**GCTCTTCG**TAA** BsaI F SapI δ	-Terminator-	**GGA**TGAAGAGC**GACC**AGAGACC ε SapI R BsaI

Prefix and suffix sequences should be added to genetic parts to allow cloning into Level 0 storage vector pStA0 and subsequent use in Start–Stop Assembly. They can be added by PCR using primer tails or included in the design of synthetic DNA sequences. The prefix and suffix sequences include inward-facing BsaI recognition sites (grey box) and corresponding storage fusion sites (F and R; bold) for cloning parts into Level 0 storage vector pStA0 using a Level 0 assembly reaction, as well as inward-facing SapI recognition sites (grey box) with corresponding donor fusion sites (α, β, γ, δ, or ε depending on the type of part; bold) for subsequent multipart assembly of expression units in Level 1 assembly. This table is reproduced from Table S1 of Taylor et al. [4]

2. Incubate reactions using a thermocycler for 30 two-step cycles each of 37 °C for 5 min then 16 °C for 5 min, before a single final denaturation step at 65 °C for 20 min.
3. Transform *E. coli* with the assembly reaction product. Ideally, to maximize transformation efficiency, use electroporation: dialyze reaction product using a membrane filter over de-ionized water for 1 h, thaw one aliquot of electro-competent *E. coli* on ice, then add 1–10 μl of dialyzed reaction product (*see* **Note 11**). Transfer cells to a cold 0.1 cm gap electroporation cuvette and then electroporate (*see* **Note 12**). Following electroporation add 250 μl of LB broth and allow the cells to recover for 1 h at 37 °C, shaking. After recovery plate onto LB agar plates containing 100 μg/ml ampicillin, 0.1 mM IPTG, and 40 μg ml X-Gal.
4. Pick single white (*see* Subheading 3.7) colonies, purify plasmids, and sequence insert.

3.3 Level 1 Start-Stop Assembly

Level 1 reactions can be carried out to assemble expression unit(s) from four modules, the promoter, UTR/RBS, CDS, and transcriptional terminator, using the five fusion sites α, β, γ, δ, and ε. As described in Subheading 3.1 and shown in Fig. 4, mixtures of DNA parts (typically promoters and/or UTR/RBSs) can be used at appropriate positions for combinatorial assemblies; and/or promoters and/or terminators can be replaced with spacer sequences at appropriate positions when using operon or hybrid configurations.

Level 1 Start-Stop Assembly reaction:

1. Prepare Level 1 Start-Stop Assembly reaction mixtures by combining 20 fmol of Level 1 destination vector plasmid DNA, 40 fmol of each insert (Level 0 plasmid DNA constructs and/or annealed oligonucleotides), T4 DNA ligase buffer, 400 U of T4 DNA ligase (typically use 1 μl), and 10 U of SapI (typically use 1 μl); make up to a total reaction volume of 20 μl. For combinatorial assemblies, use an equimolar mix of the DNA parts at a working concentration of 40 fmol.
2. Incubate reactions using a thermocycler for 30 two-step cycles each of 37 °C for 5 min then 16 °C for 5 min, before a single final denaturation step at 65 °C for 20 min.
3. Transform *E. coli* with assembly reaction product (as described in Subheading 3.2), allow cells to recover, then plate onto LB agar plates containing 10 μg/ml tetracycline, 0.1 mM IPTG, and 40 μg/ml X-Gal.
4. Pick single white colonies (*see* Subheading 3.7). To validate constructs by sequencing and/or to use in subsequent assembly levels, prepare overnight cultures and miniprep plasmid DNA. For combinatorial assemblies harvest all white colonies

from transformation plates and immediately resuspend the mixture in P1 buffer, miniprep.

3.4 Level 2 Start-Stop Assembly

Level 2 Start-Stop Assembly reactions can be carried out using purified plasmid DNA of individual or mixtures of Level 1 constructs, and the appropriate Level 2 destination vector.

Level 2 Start-Stop Assembly reaction:

1. Prepare Level 2 Start-Stop Assembly reaction mixtures by combining 20 fmol of Level 2 destination vector plasmid DNA, 40 fmol of each insert (Level 1 plasmid DNA constructs), T4 DNA Ligase buffer, 400 U of T4 DNA Ligase (typically use 1 μl), and 10 U of BsaI (typically use 1 μl); make up to a total reaction volume of 20 μl.
2. Incubate reactions using a thermocycler for 30 two-step cycles each of 37 °C for 5 min then 16 °C for 5 min, before a single final denaturation step at 65 °C for 20 min.
3. Transform *E. coli* with assembly reaction product (as described in Subheading 3.2), allow cells to recover, then plate onto LB agar plates containing 50 μg/ml kanamycin, 0.1 mM IPTG, and 40 μg/ml X-Gal.
4. Pick single white colonies (*see* Subheading 3.7). To validate constructs by sequencing and/or to use in subsequent assembly levels, prepare overnight cultures and miniprep plasmid DNA. For combinatorial assemblies, harvest all white colonies from transformation plates and immediately resuspend in P1 buffer, miniprep.

3.5 Level 3 Start-Stop Assembly

Level 3 Start-Stop Assembly reactions can be carried out using purified plasmid DNA of individual or mixtures of Level 2 constructs and the appropriate Level 3 destination vector.

Level 3 Start-Stop Assembly reaction:

1. Prepare Level 3 Start-Stop Assembly reaction mixtures by combining 20 fmol of Level 3 destination vector plasmid DNA, 40 fmol of each insert (Level 2 plasmid DNA constructs), T4 DNA Ligase buffer, 400 U of T4 DNA Ligase (typically use 1 μl), and 10 U of BbsI (typically use 1 μl); make up to a total reaction volume of 20 μl.
2. Incubate reactions using a thermocycler for 30 two-step cycles each of 37 °C for 5 min then 16 °C for 5 min, before a single final denaturation step at 65 °C for 20 min.
3. Transform *E. coli* with assembly reaction product (as described in Subheading 3.2), allow cells to recover, then plate onto LB agar plates containing 25 μg/ml chloramphenicol, 0.1 mM IPTG, and 40 μg/ml X-Gal.

4. Pick single white colonies (*see* Subheading 3.7). To validate constructs by sequencing, prepare overnight cultures and miniprep plasmid DNA.

3.6 Transfer to Target Organism and Screen

1. Transfer the final constructs or libraries of constructs into the target organism (if the target organism is not *E. coli*), using an appropriate DNA transfer method.
2. Screen colonies of the target organism for the presence or activity of assembled constructs (*see* **Notes 8** and **9**).

3.7 Quality Control at Each Level

1. Use blue-white screening at all Start-Stop Assembly levels, if possible. Unmodified 'empty' vectors give blue colonies. Correctly assembled constructs and misassembled constructs both give white colonies (*see* **Notes 3** and **4**).
2. If transforming a strain which is not alpha-complementable (and therefore not compatible with blue-white screening) use a small fraction of the assembly reaction product in parallel to transform an alpha-complementable strain of *E. coli*, allowing for blue-white screening to be carried out (*see* **Note 7**).
3. Count the number of white colonies for combinatorial assemblies, as this provides a measure of the coverage of the combinatorial library at each level.
4. Verify constructs by sequencing (*see* **Note 6**).

4 Notes

1. The Start-Stop Assembly Toolkit of core vectors and *E. coli* parts can be obtained from Addgene (https://www.addgene.org/kits/heap-start-stop-assembly/).
2. LguI (Invitrogen) can be used in place of SapI (NEB).
3. All 15 core vectors contain an assembly cassette including an alpha-complementing (alpha-donor) *lacZα* gene. In strains of *E. coli* like DH10B or DH5α, with an alpha-complementable (alpha-acceptor) *lacZ* mutation such as *lacZ*ΔM15, the presence of the *lacZα* gene leads to a functional beta-galactosidase activity and a blue colony color on agar plates containing X-Gal. During assembly the *lacZα* gene is replaced, so assembled constructs give white colonies.
4. Blue-white screening is recommended to provide an indication of the success of the assembly reaction but also because unexpectedly high blue-white ratios can indicate that the intended construct is toxic or burdensome to cells, or otherwise problematic. This is especially relevant to guide iterative modification of combinatorial designs.

5. Transformation by electroporation is recommended to benefit from the high transformation efficiencies, especially for combinatorial assemblies.
6. Sequencing is valuable to validate any assembly but is also typically necessary to determine the variation and bias in combinatorial libraries.
7. In some cases it may be desirable to use strains which are not alpha-complementable, such as an auxotrophic or other strain of *E. coli* which allows direct selection of the phenotype or activity of interest, or host organisms other than *E. coli*.
8. Any assembly method can generate misassembled constructs as well as the intended constructs. Such misassembled constructs are likely to be over-represented among populations of transformant colonies as they may impose much lower metabolic burden on the cell than the intended complete constructs. It is therefore always advisable to pick multiple clones to screen for this reason.
9. For combinatorial libraries, it is generally preferable to screen large numbers of colonies in order to achieve a good sample of the design space.
10. If using electroporation, dialysis of the reaction product is recommended prior to the transformation.
11. When assembling an individual construct, transformation of *E. coli* using 1 μl assembly reaction product is sufficient. However, when assembling a combinatorial library, larger volumes of assembly reaction product are better, in order to maximize coverage of the combinatorial library. Dialysis of assembly reaction product allows larger volumes to be used in electroporation without undesirable arcing occurring during electroporation.
12. *E. coli* can be electroporated using a Bio-Rad MicroPulser electroporator with parameter preset Ec1.

Acknowledgments

The authors thank Lara Sellés Vidal and Rhiannon Leyden-Preece for critical reading of the manuscript. This work was supported by the Biotechnology and Biological Sciences Research Council [BB/M002454/1], an Imperial College London Schrödinger Scholarship, and the Imperial College Excellence Fund for Frontier Research.

References

1. Engler C, Gruetzner R, Kandzia R, Marillonnet S (2009) Golden gate shuffling: a one-pot DNA shuffling method based on type IIs restriction enzymes. PLoS One 4:e5553
2. Engler C, Kandzia R, Marillonnet S (2008) A one pot, one step, precision cloning method with high throughput capability. PLoS One 3: e3647
3. Mutalik VK, Guimaraes JC, Cambray G et al (2013) Quantitative estimation of activity and quality for collections of functional genetic elements. Nat Methods 10:347–353
4. Taylor GM, Mordaka PM, Heap JT (2018) Start-Stop Assembly: a functionally scarless DNA assembly system optimized for metabolic engineering. Nucleic Acids Res 47:e17
5. Coussement P, Bauwens D, Maertens J, De Mey M (2017) Direct combinatorial pathway optimization. ACS Synth Biol 6:224–232
6. Smanski MJ, Bhatia S, Zhao D et al (2014) Functional optimization of gene clusters by combinatorial design and assembly. Nat Biotechnol 32:1241–1249
7. Zelcbuch L, Antonovsky N, Bar-Even A et al (2013) Spanning high-dimensional expression space using ribosome-binding site combinatorics. Nucleic Acids Res 41:e98

Chapter 14

BASIC: A Simple and Accurate Modular DNA Assembly Method

Marko Storch, Ari Dwijayanti, Haris Mallick, Matthew C. Haines, and Geoff S. Baldwin

Abstract

*B*iopart *A*ssembly *S*tandard for *I*dempotent *C*loning (*BASIC*) is a simple, robust, and highly accurate DNA assembly method, which provides 99% correct assemblies for a typical four-part assembly, enabling high efficiency cloning workflows (Storch et al., ACS Synth Biol, https://doi.org/10.1021/sb500356, 2015). BASIC employs standardised DNA linkers to combine bioparts, stored in the universal BASIC format. Once a new biopart is formatted into BASIC standard, defined by flanking 18 bp prefix and suffix sequences, it can be placed at any position and in any context within a designed BASIC assembly. This modularity of the BASIC approach is further enhanced by a range of functional linkers, including genetic elements like ribosomal binding sites (RBS) and peptide linkers. The method has a single tier format, whereby any BASIC assembly can create a new composite BASIC part in the same format used for the original parts; it can thus enter a subsequent BASIC assembly without the need for reformatting or changes to the workflow. This unique idempotent cloning mechanism allows for the assembly of constructs in multiple, conceptionally simple hierarchical rounds. Combined with its high accuracy and robustness, this makes BASIC a versatile assembly method for combinatorial and complex assemblies both at bench and biofoundry scale. The single universal storage format of BASIC parts enables compressed universal biopart libraries that promote sharing of parts and reproducible assembly strategies across labs, supporting efforts to improve reproducibility. In comparison with other DNA assembly standards and methods, BASIC offers a simple robust protocol, relies on a single tier format, provides for easy hierarchical assembly, and is highly accurate for up to seven parts per assembly round (Casini et al., Nat Rev Mol Cell Biol. https://doi.org/10.1038/nrm4014, 2015).

Key words DNA assembly, Hierarchical, Idempotent, Combinatorial, Pathway engineering

1 Introduction

BASIC DNA assembly is based on linker guided DNA assembly and defines a universal standard that describes the design framework for linkers and a universal format for part storage. At the conceptual DNA design level, the assembly always follows a part–linker–part–linker format. At the construction level, each linker is physically split into two sections encoded by oligonucleotides; each half linker

Sunil Chandran and Kevin W. George (eds.), *DNA Cloning and Assembly: Methods and Protocols*, Methods in Molecular Biology, vol. 2205, https://doi.org/10.1007/978-1-0716-0908-8_14, © Springer Science+Business Media, LLC, part of Springer Nature 2020

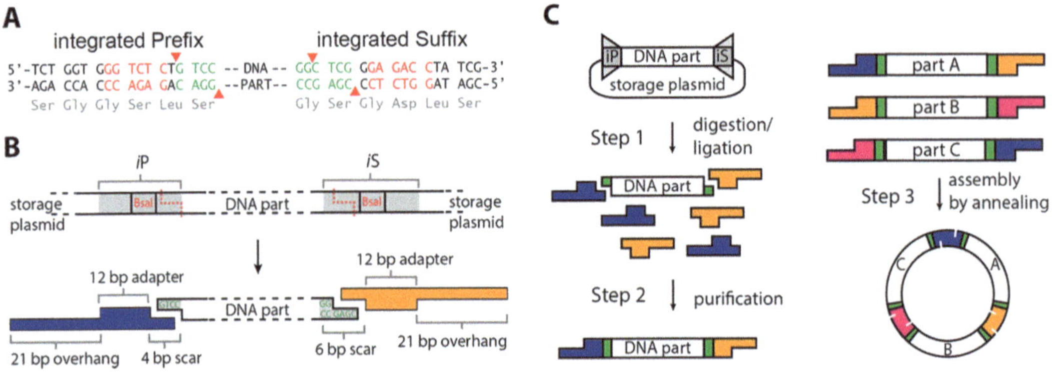

Fig. 1 BASIC standard and workflow. (**a**) Each BASIC part is flanked by prefix (*i*P) and suffix (*i*S) sequences including BsaI recognition (red) and cleavage sites (red arrows). (**b**) Upon BsaI digestion BASIC parts are released with specific four base overhangs at each end, enabling efficient ligation of partially double-stranded linkers sections. The only true scar sequences in BASIC are the 4 bp at the prefix junction and 6 bp at the suffix junction, designed to facilitate friendly codon usage. (**c**) Complete BASIC workflow in three steps. (Reprinted (adapted) with permission from [1]. Copyright 2015 American Chemical Society)

is then ligated to their respective parts in separate "clip" reactions. Following purification, these clips (half linker adducted parts) are annealed, thus combining the respective half linkers to create complete linkers between the parts, resulting in assembly of the final construct.

The BASIC standard requires the integrated prefix (*i*P; 5′-TCTGGTGGGTCTCTGTCC-3′) and suffix (*i*S; 5′-GGCTC GGGAGACCTATCG-3′) sequences to be flanking each part (Fig. 1a). Both sequences contain a BsaI recognition site, which is used to release parts from the storage vector, resulting in different four base overhangs flanking the DNA part at each end. These overhangs are used to ligate prefix- and suffix-specific half linkers, containing a double-stranded section to facilitate ligation and 21 base single-stranded overhangs: The specific sequences of these overhangs then direct the accurate annealing of parts in a defined order within the assembly. Importantly, the linker sequences are derived from an online computational algorithm that ensures they are orthogonal to the host genome sequence, do not contain secondary structures, are devoid of proscribed restriction enzyme sequences, and avoid complementarity between different linkers [3].

The BASIC assembly workflow contains three major steps (Fig. 1):

1. BASIC clip reaction (Subheading 3.3).
2. BASIC clip purification (Subheading 3.4).
3. BASIC clip assembly (Subheading 3.5).

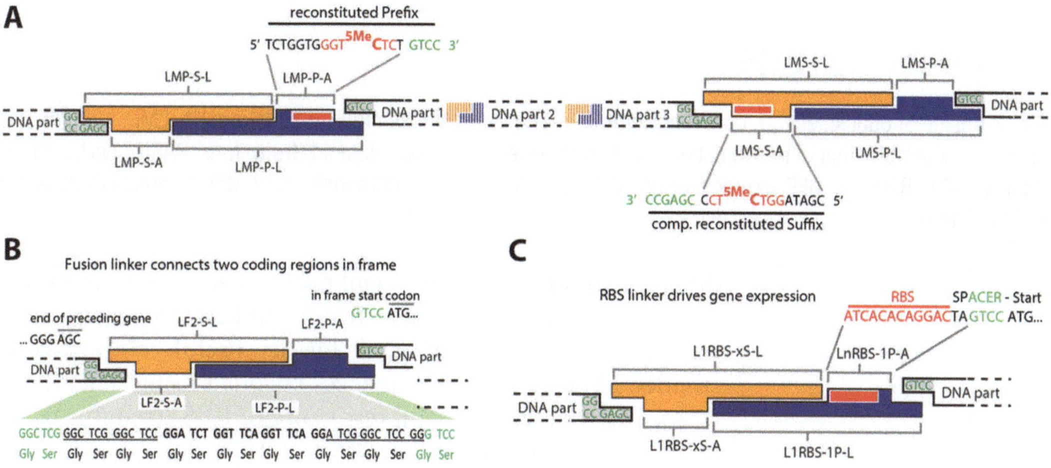

Fig. 2 BASIC functionalized linkers. (**a**) Methylated linkers LMP and LMS reconstitute BASIC prefix and suffix sequences around a set of assembled BASIC parts. (**b**) Fusion linkers can be used to join protein domains in frame, stop codon needs to be removed from the upstream ORF. (**c**) RBS encoded on specific UTR-RBS linkers enable simple RBS library creation via combinatorial assembly. (Reprinted (adapted) with permission from [1]. Copyright 2015 American Chemical Society)

An important aspect of BASIC is that the linkers required for the physical organization of the DNA assembly may also be used to encode functional elements of choice [1]. Three examples are described in Fig. 2:

1. Neutral Linkers are used to connect two interchangeable bioparts in a BASIC format.
2. Methylated linkers are used to regenerate prefix and suffix sequences flanking a composition of parts during BASIC assembly, thus providing an idempotent single-tier format: the resulting composition is returned as a single part in BASIC format flanked by *i*P and *i*S. Specific methylation of the BsaI recognition sites in these linkers prevents their cleavage during the clip reaction but is not maintained during replication following transformation into *E. coli*. The prefix and suffix sites are thus fully functional for a subsequent round of DNA assembly (Fig. 2a).
3. Fusion linkers can be designed to encode a peptide linker between two protein domains in frame to create a fusion protein between two BASIC parts. Fusion linker libraries encoding peptide linkers of varying length and flexibility allow for rapid protein engineering (Fig. 2b).
4. UTR-RBS linkers encode functional RBS sequences directly on the double-stranded portion of the linkers, driving translation of downstream open reading frame (ORF) parts. The upstream part of the linker including the overlap provides a consistent 5′UTR context for the RBS sequence. RBS sequences of

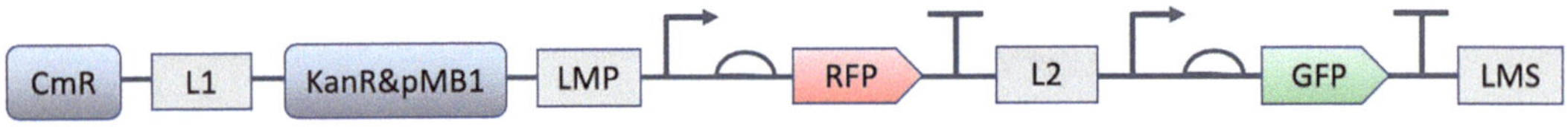

Fig. 3 Design of dual selection, dual reporter four-part assembly plasmid. BASIC linkers L1, LMP, L2, LMS are used to assemble this dual selection, dual fluorescent reporter plasmid from four BASIC parts: CmR, KanR&pMB1, RFP and GFP expression cassettes, creating a new composite RFP–GFP expression cassette in BASIC format

different strength are a powerful tool to tune gene expression in constructs assembled using BASIC and are particularly useful for operon engineering. A further useful tool are constrained libraries of RBS sequences encoded using degenerate oligos [4], enabling a useful repertoire of RBS strengths to be coded with a single linker clip reaction (Fig. 2c).

Along with the standard protocol, in **Note 1** we provide detailed sections for the construction of an example four-part assembly combining the following parts in BASIC format:

Chloramphenicol resistance cassette (68138): **CmR**.

Kanamycin resistance cassette & pMB1 origin of replication (68136): (**KanR&pMB1**).

RFP expression cassette (68141): **RFP**; GFP expression cassette (68144): **GFP**.

As shown in Fig. 3, we use four BASIC linkers to join the four BASIC parts: neutral linkers L1, L2 and methylated linkers LMP and LMS to flank the RFP and GFP expression cassettes with BASIC prefix and suffix creating a novel composite biopart in BASIC format (*see* **Note 1**).

2 Materials

1. BASIC linkers: available from www.biolegio.com and details can be found in [1, 5].
2. Neutral and Methylated linker set (Biolegio).
3. RBS linker set (Biolegio).
4. Mixed RBS linker set (Biolegio).
5. Annealing buffer: 10 mM Tris–HCl pH 7.9, 100 mM NaCl, 10 mM $MgCl_2$, dH_2O.
6. Oligonucleotides: Oligonucleotides required for the example assembly are summarized in Table 1 and can also be ordered from preferred DNA suppliers (*see* **Note 1**).
7. BASIC parts: available from Addgene (plasmids 68135–68147) with details provided in [1, 5].

Table 1
Oligonucleotides for BASIC linkers

Name			Sequence 5′->3′
L1	L1-S	L1-S-L	PO_4-CTCGTTACTTACGACACTCCGAGACAGTCAGAGGGTA
		L1-S-A	PO_4-TGTCGTAAGTAA
	L1-P	L1-P-L	PO_4-GGACTAGTTCAATAAATACCCTCTGACTGTCTCGGAG
		L1-P-A	PO_4-TTTATTGAACTA
L2	L2-S	L2-S-L	PO_4-CTCGATCGGTGTGAAAAGTCAGTATCCAGTCGTGTAG
		L2-S-A	PO_4-TTTCACACCGAT
	L2-P	L2-P-L	PO_4-GGACAGGTAATAAGAACTACACGACTGGATACTGACT
		L2-P-A	PO_4-TTCTTATTACCT
LMP	LMP-S	LMP-S-L	PO_4-CTCGGGTAAGAACTCGCACTTCGTGGAAACACTATTA
		LMP-S-A	PO_4-CGAGTTCTTACC
	LMP-P	LMP-P-L	PO_4-GGACAGAGACCCACCAGATAATAGTGTTTCCACGAAGTG
		LMP-P-A	PO_4-TCTGGTGGGT/iMe-dC/TCT
LMS	LMS-S	LMS-S-L	PO_4-CTCGGGAGACCTATCGGTAATAACAGTCCAATCTGGTGT
		LMS-S-A	PO_4-CGATAGGT/iMe-dC/TCC
	LMS-P	LMS-P-L	PO_4-GGACGATTCCGAAGTTACACCAGATTGGACTGTTATTAC
		LMS-P-A	PO_4-AACTTCGGAATC

Oligonucleotides constituting neutral BASIC linkers (L1, L2) and methylated linkers (LMP, LMS). BASIC part ligation annealing regions are shown in green; intralinker annealing regions are shown in blue. Oligonucleotides are ordered 5′ phosphorylated, HPLC purified and with an internal 5′-C methylation for LMP-P-A and LMS-S-A. Premixed linker and adapter oligos are available ready to use from Biolegio

8. BsaI restriction enzyme: BsaI-HFv2, 20 U/μl (NEB).
9. 10× NEB CutSmart buffer (NEB).
10. T4 ligase: T4 DNA Ligase, 1–3 U/μl (Promega).
11. 10× T4 buffer (Promega). Included with T4 DNA ligase.
12. dH_2O.
13. 70% ethanol (freshly made with dH_2O).
14. Magnetic beads: Agencourt AMPure XP (Beckman Coulter) or AmpliClean (NimaGen).
15. Magnetic plate, 96-well: AM10050 Ambion (Applied Biosystems) or Alpaqua (NimaGen).
16. 96-well Falcon plate, U-shaped bottom (Corning).
17. Phusion Polymerase (optional): for creating new BASIC parts or colony PCR (NEB).
18. LB medium: 10 g/l tryptone, 5 g/l yeast extract, 10 g/l NaCl.
19. LB agar: 10 g/l tryptone, 5 g/l yeast extract, 10 g/l NaCl, 15 g/l agar.
20. SOC medium.

21. Antibiotics working concentrations: 50 μg/ml kanamycin, 25 μg/ml chloramphenicol.
22. Chemically competent DH5α strain (NEB/we find the Inoue method [6] to give consistently high competency).

3 Methods

The BASIC DNA assembly method is presented in sections addressing BASIC part preparation (Subheading 3.1), BASIC linker preparation (Subheading 3.2), BASIC clip reaction (Subheading 3.3), BASIC clip purification (Subheading 3.4), BASIC clip assembly (Subheading 3.5), heat-shock transformation into *E. coli* (Subheading 3.6) and concluded with remarks on assembly confirmation (Subheading 3.7).

3.1 BASIC Biopart Preparation

Each BASIC part must be at least 120 bp long, cannot contain internal BsaI restriction sites, and is defined by the flanking BASIC prefix and suffix sequence (Fig. 1a).

Some standard BASIC parts are available from Addgene (www.addgene.com) and users can easily create their own project specific BASIC parts such as promoters, open reading frames, terminators, and backbones with a range of selection markers via PCR. We recommend new BASIC parts to be ordered directly from gene synthesis companies.

New BASIC DNA parts should be ordered already flanked by prefix and suffix sequences as double-stranded DNA fragments. While DNA vendors can provide BASIC parts sequence verified in BsaI restriction site free storage vectors of choice, it is more cost effective to assemble the supplied linear DNA into a high copy storage vector and to sequence verify the new BASIC part. We recommend a two-part BASIC assembly with methylated BASIC linkers (this will recreate the BASIC prefix and suffix around the new BASIC part) in a high-copy AmpR-pUC BASIC format backbone (any backbone can be formatted into a BASIC part following similar strategies). For workflow consistency it is useful to have a consistent antibiotic resistance type in storage vectors.

For high-fidelity workflows, it is important to sequence verify new DNA parts. However, it is worth noting that we find gene synthesis to provide consistently high accuracy in new DNA parts. In high-throughput screening workflows, such synthetic linear DNA sources may therefore be used directly in BASIC assembly without prior sequence verification, saving cost and time. Users may also mitigate the experimental cost associated with sequence fidelity by picking multiple colonies to check the consistency of plasmid function across constructs.

Since BASIC assembly typically uses sequenced plasmid DNA as the input and avoids PCR, it is also less imperative to sequence validate the final constructs. Again, this offers significant cost and time savings. In high throughput, validation of data from multiple constructs is usually easier than prior sequence validation of all constructs (*see* **Note 2**).

3.2 BASIC Linker Preparation

BASIC linkers can be ordered from Biolegio (*www.biolegio.com*) and will be delivered in a lyophilized format along with linker annealing buffer. Once restored to 200 μl with linker annealing buffer, new linkers are ready to use and can be stored at −20 °C for up to 3 months. A 200 μl, 1 μM BASIC linker solution is sufficient for 200 BASIC clip reactions, and each clip reaction has sufficient material for the assembly step of up to 30 different plasmids [1, 5] (*see* **Note 3**).

1. Spin down tubes with lyophilized linkers to ensure oligos are at bottom of the tube.
2. Set heating block to 95 °C.
3. Add 200 μl of linker annealing buffer to each linker tube and leave on bench for 1 h.
4. Vortex tubes and collect liquid at tube bottom via a quick centrifuge spin.
5. After heating block reached 95 °C, place tubes in block and switch off.
6. Allow tubes to cool down to less than 30 °C over at least 2 h in heat block.
7. Collect solution at tube bottom via a quick centrifuge spin.
8. Linkers are ready to use and can be stored at −20 °C for up to 2 months.

3.3 BASIC Clip Reaction

For each BASIC clip reaction, set up one 200 μl PCR tube with 30 μl total volume:

17 μl dH_2O, 3 μl 10× Promega T4 buffer, 1 μl of prefix linker, 1 μl of suffix linker, 200 nmol of BASIC biopart DNA (i.e., 1 μl if a 4-kb plasmid is supplied at concentration of 200 ng/μl), top up with dH_2O to 28.5 μl. Add 1 μl of BsaI-HFv2 (20 U), add 0.5 μl of T4 ligase (0.5–1.5 U), mix by pipetting up and down and incubate in PCR thermal cycler running the BASIC linker-ligation program: cycle 20× [37 °C for 2 min, 20 °C for 1 min], 60 °C for 20 min, store at 4 °C (Table 2) (*see* **Notes 4** and **5**).

3.4 BASIC Clip Purification

1. Prepare fresh 70% EtOH (0.5 ml per BASIC reaction) and bring magnetic beads stored at 4 °C back into homogeneous mixture by vortexing for 1 min and allow to warm to room temperature before use (54 μl per BASIC clip reaction; AmpureXP or AmpliClean).

Table 2
BASIC clip reaction setup and PCR thermal cycler program

30 μl BASIC clip reaction setup per BASIC part		
Reagent	*Volume*	
dH_2O	17.5 μl	
Promega T4 buffer (10×)	3 μl	
Prefix Linker	1 μl	
Suffix Linker	1 μl	
BASIC biopart	0.5–6 μl (50 ng per 1 kb total plasmid size)	
dH_2O	Add dH_2O to total 28.5 μl volume	
NEB BsaI-HF v2 enzyme (20 U/μl)	1 μl	
Promega T4 ligase (1–3 U/μl)	0.5 μl	
Mix by pipetting up and down 3×		
PCR thermal cycler program		
Temperature	*Time*	
37 °C 20 °C	2 min 1 min	*×20 cycles*
60 °C	20 min	
Store at 4 °C overnight or at −20 °C for up to 1 month		

We recommend using a 96-well Falcon plate in combination with a magnetic plate for quick magnetic bead immobilization and easy pipetting access (*see* **Note 6**).

2. Add 54 μl of magnetic beads per reaction into a 96-well Falcon plate (one well per BASIC clip reaction) and add the 30 μl BASIC linkerligation from the PCR machine step, mix by pipetting ten times, avoid creating bubbles.
3. Wait 5 min to allow DNA to bind the magnetic beads.
4. Place Falcon plate on magnetic stand and wait for magnetic beads to form immobilized rings and the solution to clear—takes roughly 2 min. The plate should remain on the magnet stand until **step 11**; removing it will disturb the immobilized beads.
5. Remove the solution with a 200 μl pipette tip from the center of each well.
6. Wash step 1: Add 190 μl 70% EtOH to each well and wait 30 s.
7. Wash step 1: Remove solution from each well (pipette set to 200 μl volume).

8. Wash step 2: Add 190 μl 70% EtOH to each well and wait 30 s.
9. Wash step 2: Remove solution from each well (pipette set to 200 μl volume).
10. Leave the plate open without lid for magnetic beads to dry for 5 min.
11. Remove Falcon plate from magnet and resuspend magnetic beads in 32 μl dH_2O.
12. Wait 1 min for DNA to elute.
13. Place Falcon plate back on magnetic stand and allow the magnetic beads to form an immobilized ring and for the solution to clear—takes roughly 1 min.
14. Pipette 30 μl of H_2O with eluted DNA into fresh 1.5 ml Eppendorf tube for direct use in assembly or storage at −20 °C for up to 1 month.

3.5 BASIC Clip Assembly

For each final BASIC assembly reaction, set up one 200 μl PCR tube with 10 μl total volume:

2 μl H_2O, 1 μl 10× NEB CutSmart buffer, 1 μl of each magnetic bead purified linker ligated BASIC biopart (clip) going into this specific assembly, top up with H_2O to 10 μl. Mix by pipetting up and down and incubate in PCR machine running the BASIC assembly program: 50 °C for 45 min (Table 3). If not used for transformation immediately, store at 4 °C overnight or at −20 °C for up to 1 month (*see* **Notes 7** and **8**).

Table 3
BASIC clip assembly setup and PCR thermal cycler program

10 μl BASIC clip assembly reaction	
Reagent	*Volume*
dH_2O	2 μl
NEB CutSmart buffer 10×	1 μl
Each purified linker ligated BASIC part (2–7)	1 μl (up to 7 μl for 7 parts)
dH_2O	Add dH_2O to total 10 μl volume
Mix by pipetting up and down 3×	
PCR thermal cyclerprogram	
Temperature	*Time*
50 °C	45 min
Store at 4 °C overnight or at −20 °C for up to 1 month	

3.6 Heat-Shock Transformation of BASIC Assemblies into High Efficiency Chemically Competent E. coli DH5α Strain

Use 50 μl of chemically competent cells DH5α with high transformation efficiency (10^9 CFU/μg pUC19) to transform 5 μl of each BASIC assembly (*see* **Note 9**):

1. Chemically competent cells are stored at −80 °C.
2. Thaw competent cells on ice (takes 5–10 min); 50 μl are required per BASIC assembly.
3. Cool 5 μl of BASIC assembly in 1.5 ml Eppendorf tube on ice.
4. Add 50 μl of competent cells to each precooled 5 μl BASIC reaction.
5. Incubate on ice for 20 min.
6. Apply heat shock in 42 °C water bath for 45 s and place back on ice for 2 min.
7. Add 200 μl SOC medium to each tube and incubate shaking at 37 °C for 1 h recovery.
8. Prewarm agar plates for plating for 1 h at 37 °C with lids partially open.
9. Spot or plate cells on agar plates with appropriate antibiotics. Depending on number of parts assembled and transformation efficiency 2–250 μl might be spotted or plated.
10. Incubate agar plates at 37 °C overnight, next day pick colony for assay, colony PCR, or miniprep (*see* **Note 10**).

3.7 Assembly Confirmation

Given the high accuracy of BASIC assembly, for most applications clones do not have to be screened. One to two clones are picked usually and can be confirmed via colony PCR (details in [1]). If the DNA is to be purified, analytical enzymatic digest or sequencing can verify the construct. In many cases, colonies can be directly used in application/functional study workflows (growth assays coupled with plate reader/flow cytometer/expression studies) without prior construct verification (*see* **Note 11**).

4 Notes

1. Specific protocol steps for the example four-part assembly described in Fig. 3:
   (a) BASIC parts can be sourced from Addgene (plasmids 68138, 68136, 68141, 68144) or ordered from synthetic DNA vendors.
   (b) BASIC linkers L1, L2, LMP, and LMS are required and the respective suffix and prefix sections L1-S, L1-P, L2-S, L2-P, LMP-S, LMP-P, LMS-S, and LMS-P are provided in the Neutral and Methylated linker set (Biolegio).

(c) For the example four-part assembly (linkers are underlined):
CmR-L1-**KanR&pMB1**-LMP-**RFP**-L2-**GFP**-LMS.

The individual BASIC parts are combined with the following prefix and suffix linker sections in their respective BASIC clip reactions according to Table 2.

LMS-P + **CmR** + L1-S; L1-P + **KanR&pMB1** + LMP-S.

LMP-P + **RFP** + L2-S; L2-P + **GFP** + LMS-S.

(d) The four reactions are purified as described under Subheading 3.4 in four individual wells, providing 30 μl ready to assemble linker ligated BASIC parts for each of the four parts.

(e) 5 μl of H_2O are combined with 1 μl of 10× NEB CutSmart buffer and 1 μl of each purified linker ligated BASIC part (clip) CmR, KanR&pMB1, RFP, GFP and annealed as described in Table 3.

(f) 5 μl of the annealed assembly mix is used to transform highly competent cells as described under Subheading 3.6. 50 μl is then plated on LB agar plates with either 50 μg/ml kanamycin or 50 μg/ml kanamycin and 25 μg/ml chloramphenicol for single or double selection respectively.

(g) Correct clones will confer resistance to both kanamycin and chloramphenicol and will show GFP and RFP fluorescence that can be assessed the next day by fluorescence scanner (for GFP and RFP fluorescence), on a blue box (for GFP fluorescence), visual inspection after 2 days storage in the fridge (for pink RFP response) or growing in liquid culture in a plate reader measuring GFP and RFP fluorescence. As a negative control, original KanR&pMB1 plasmid (Addgene 68136) can be transformed and plated on LB-agar plates with 50 μg/ml Kanamycin.

2. The BASIC standard has only one forbidden restriction site—BsaI, which is used to release bioparts from their storage vectors and create specific 4 bp overhangs to enable BASIC linker ligation. Hence, BASIC parts must not contain internal BsaI restriction sites. If new BASIC parts are created via PCR with primer extensions encoding for prefix and suffix, template sequences might contain such forbidden BsaI sites. These forbidden sites can be removed via mutagenesis PCR, usually resulting in silent mutations applied to the template or stored BASIC part. To improve the efficiency of the linker ligation, the storage backbone sequence should be free of additional BsaI restriction sites, since they will reduce the rate at which the BASIC reaction reaches completion. The ampicillin gene

contains a BsaI site, but backbones are available where this site has been removed. Using a storage backbone with a BsaI-containing AmpR gene will still work, but it has a notable impact on efficiency. With reducing synthesis costs, especially for smaller bioparts, BASIC format parts can be ordered already designed to remove internal BsaI sites and add the flanking BASIC prefix and suffix sites, often sequence verified in BsaI-free storage backbones. This provides a convenient and affordable path to project-specific BASIC libraries and is our recommended route where possible.

3. The design and quality of BASIC linkers is fundamental to BASIC's accuracy and functional modularity. Foundation for their design principles was laid through our work on the MODAL and BASIC methods and our freely available online linker design tool R2o DNA Designer [3, 5, 7 http://www.r2odna.com/]. In brief, linker sequences are optimized to contain no unwanted motifs or complementarity and perform well during the linker annealing and the final assembly step. The key aspect of the linkers is to use standardized sets that are designed relative to each other to ensure that there is no cross-hybridization. Standard linkers reduce cost and facilitate workflows and users should avoid creating customized linkers where possible.

   A complete list of linker sequences along with the oligonucleotides needed to build them is provided in the supplementary material of our BASIC paper [1]. New orthogonal linker sets can be easily created with the R2o DNA Design software tool [3]. Special linkers containing functional sequences like RBS and peptide fusion linkers are designed manually and checked for being orthogonal to other linkers being used within the same assemblies, via the R2o DNA Design reverse mode.

   Biolegio provides ready to use BASIC linkers based on high quality oligos in curated sets in quantities that make the supply convenient and affordable. Linkers can also be ordered by conventional oligonucleotide synthesis, with the 5′phosphate and 5meC. For custom synthesis we also recommend HPLC purification, which provides high fidelity and long-term storage stability (we have stored HPLC purified linkers for up to 2 years with no apparent loss in efficiency). This will provide a high yield of linkers but is a costly initial outlay.

4. Ideally, newly created BASIC parts are sequence verified and plasmid concentrations adjusted so that 1 μl added to the BASIC reaction would provide a final 2.5 nM concentration in the BASIC clip linkerligation reaction. That means for a given plasmid approximately 50 ng per 1 kb total size are required as reaction input. For instance, the concentration of

a BASIC part storage plasmid of 4 kb total size should be adjusted to ~200 ng/μl to standardize workflows and ease automation.

5. A complete digest of storage plasmids is central to a highly efficient and accurate BASIC assembly, since undigested selection cassettes in storage plasmids can give rise to background, when selected on single antibiotics. We found that a cycle number of 20 in Subheading 3.3 usually ensures a complete reaction, but users can experiment with higher cycle numbers to improve assembly efficiencies (especially if additional BsaI restriction sites are present in the storage backbone). We also found that this new cycling protocol, where relative BsaI restriction activity is enhanced at 37 °C and T4 ligase activity is dominant at 20 °C, makes the reaction more robust against enzyme batch variation. The final extended step at 60 °C is intended to inactivate the T4 DNA ligase, while BsaI digestion with the NEB BsaI-HF v2 continues, driving the reaction to completion.

   If a high background or just a few colonies are observed on agar plates the BASIC clip reaction may not be working efficiently. To diagnose, the magnetic bead purified BASIC parts (after Subheading 3.4) can be run on a diagnostic 1% agarose gel to confirm the storage vectors are completely digested into two DNA species corresponding in size to linker-ligated BASIC part and empty storage vector. If digestion looks complete on the gel, but specific BASIC reactions fail repeatedly, fresh linkers and buffers should be considered to address potential issues with cross-contamination.

   We use PCR machines for all temperature critical steps, even when isothermal. The heated lid also reduces evaporation and ensures a consistent ionic strength.

6. We found that spin columns did not remove un-ligated oligonucleotides (which are present in excess in the BASIC reactions) sufficiently, while magnetic bead purification systems worked very well. We recommend Falcon U-shaped 96-well plates in combination with a magnetic stand, since the wide rings of immobilized magnets allow for manual removal of supernatant and reactions can be performed in parallel or on an automated platform. More affordable tube magnetic stands work well when only a few BASIC clip reactions need to be purified in Subheading 3.4.

7. The buffer designed for the linker annealing uses a simple ionic solution of Na^+ and Mg^{2+} to offset the negative charge on the phosphates. The dual valency and coordination geometry of magnesium means that it is a much more effective counterion than sodium for offsetting phosphate charge on the DNA

backbone. We have observed higher efficiency in the overall process when the annealing buffer contains both Na^+ and Mg^{2+} compared to just Na^+.

We found the Promega T4 buffer works best for the BASIC clip reaction (Subheading 3.3), while a wider range of typical restriction enzyme buffers will work for the final assembly step (Subheading 3.5), which requires counterions for effective annealing; linker annealing buffer may also be used. If using electroporation you may need to control ionic strength, but be mindful of effective counterion charge compensation for accurate annealing.

8. Depending on the number of parts, one can increase the amount of DNA parts going into the final assembly to maximize the number of colonies returned after transformation. For instance, in a four-part assembly 2 μl of each purified part can be used instead of 1 μl and will increase the number of returned colonies. Thanks to the high accuracy of BASIC, maximizing the number of colonies is not so important for simple two- to seven-part assemblies, because most colonies will represent the correct construct. In contrast, for combinatorial one pot assemblies where individual clones will differ in promoter, RBS or gene variants within the same assembly framework, maximizing the number of returned colonies becomes crucial to cover diversity.
9. Using BASIC successfully for up to seven parts requires highly competent cells. There are several commercial cloning strains available offering at least 10^9 CFU/μg pUC19. As an alternative the Inoue protocol [6] proved consistently successful for preparing high efficiency chemically competent *E. coli DH5α* cells (up to 10^8 CFU/μg pUC19) in house. We updated the original protocol by using 300 ml SOC medium for the final outgrowth to OD600 0.3–0.55 at 18 °C, shaking 220 rpm.
10. We found undigested resistance cassettes for selection still in their storage vector to be the main source of background colonies after transformation (check by gel electrophoresis as in **Note 5**), while incorrect assemblies are extremely rare events. Since even very small amounts of undigested resistance cassettes will contribute a background level of around 40 colonies for any assembly, we devised double selection as a standard strategy in our lab. While the ~40 background colonies do not reduce accuracy within thousands of correct colonies for less than five-part assemblies, this source of background forces one to screen more colonies for seven-part assemblies when overall colonies returned from transformation are below 100. When two independent selection marker parts (e.g., KanR and CmR) are included in the overall assembly strategy, then double selection (on kanamycin and chloramphenicol) lift overall

accuracy up to 90% for seven-part assemblies—providing BASIC with a very high accuracy compared to other techniques.

11. Usually the functional cassettes will be flanked by methylated linkers encoding BASIC prefix and suffix within the backbone. Thus, a simple BsaI digest of miniprepped DNA will show two species corresponding to the size of the backbone and the assembled cargo cassette. Alternatively, colony PCR can be used to verify the size of the cargo cassette in between LMP and LMS using the standard primers (LMP-for: CTATTATCTGGTGGGTCTCT and LMS-rev: TTACCGATAGGTCTCCCG) or by using linker-specific colony PCR primers explained in [1].

    Since all parts are usually sequence verified in their storage plasmids and no PCR amplification step is used throughout the BASIC workflow, it is highly unlikely that the part sequences would carry mutations in the assembled construct. This minimizes the necessity for complete sequence validation of all constructs. The complete standardization of the workflow including standard test colony PCR primers leverages the advantages of BASIC assembly especially for high-throughput library construction.

References

1. Storch M, Casini A, Mackrow B et al (2015) BASIC: a new Biopart Assembly Standard for Idempotent Cloning provides accurate, single-tier DNA assembly for synthetic biology. ACS Synth Biol. https://doi.org/10.1021/sb500356
2. Casini A, Storch M, Baldwin G, Ellis T (2015) Bricks and blueprints: methods and standards for DNA assembly. Nat Rev Mol Cell Biol. https://doi.org/10.1038/nrm4014
3. Casini A, Christodoulou G, Freemont P et al (2014) R2oDNA Designer: computational design of biologically neutral synthetic DNA sequences. ACS Synth Biol. https://doi.org/10.1021/sb4001323
4. Jeschek M, Gerngross D, Panke S (2016) Rationally reduced libraries for combinatorial pathway optimization minimizing experimental effort. Nat Commun 7:11163. https://doi.org/10.1038/ncomms11163
5. Storch M, Casini A, Mackrow B et al (2017) BASIC: a simple and accurate modular DNA assembly method. Methods Mol Biol (Clifton, NJ) 1472:79–91. https://doi.org/10.1007/978-1-4939-6343-0_6
6. Inoue H, Nojima H, Okayama H (1990) High efficiency transformation of Escherichia coli with plasmids. Gene 96:23–28. https://doi.org/10.1016/0378-1119(90)90336-P
7. Casini A, Macdonald JT, Jonghe JD et al (2013) One-pot DNA construction for synthetic biology: the Modular Overlap-Directed Assembly with Linkers (MODAL) strategy. Nucleic Acids Res. https://doi.org/10.1093/nar/gkt915

Chapter 15

Joint Universal Modular Plasmids: A Flexible Platform for Golden Gate Assembly in Any Microbial Host

Marcos Valenzuela-Ortega and Christopher E. French

Abstract

Modular cloning standards based on Golden Gate DNA assembly allow for construction of complex DNA constructs over several rounds of assembly. Despite being reliable and automation-friendly, each standard uses a specific set of vectors, requiring researchers to generate new tool kits for novel hosts and cloning applications. JUMP vectors (Valenzuela-Ortega and French, bioRxiv 799585, 2019) combine the robustness of modular cloning standards with the Standard European Vector Architecture and a flexible design that allows researchers to easily modify the vector backbone via secondary cloning sites. This flexibility allows for JUMP vectors to be used in a wide variety of applications and hosts.

Key words Synthetic biology, Molecular cloning, DNA assembly, Single-pot assembly, Golden Gate cloning, PhytoBricks, JUMP

1 Introduction

Synthetic biology aims to solve problems by approaching biology as an engineering discipline. Simple DNA elements (or parts) are characterized and combined to generate biological systems with new features [1]. However, the degree to which functionality and behavior of new sequences built from basic parts can be predicted depends on their molecular context, a phenomenon widely known as context dependency [2]. Considering this, generation of new biological systems requires an iterative design–build–test cycle and robust methodologies for building complex DNA constructs [3]. DNA assembly standards based on Golden Gate methods are ideal for process automation and basic part reusability and sharing. In Golden Gate cloning [4], DNA parts are flanked by sites recognized by type IIS restriction enzymes, which cut outside the recognition site, leaving a user-defined fusion site. Parts are ligated in an ordered manner and restriction sites are removed during assembly.

Various hierarchical assembly standards based on Golden Gate have been published [5–10], collectively known as modular cloning

Sunil Chandran and Kevin W. George (eds.), *DNA Cloning and Assembly: Methods and Protocols*, Methods in Molecular Biology, vol. 2205, https://doi.org/10.1007/978-1-0716-0908-8_15,

(or MoClo) standards or tool kits. In MoClo standards, the product of one assembly level can be used as a part in the next level assembly, which is possible due to the use of alternate restriction enzymes and selection markers in the destination vectors of different assembly levels (Fig. 1a). Basic (or level 0) parts must be "domesticated," a process that removes internal restriction sites and adds flanking restriction sites with the appropriate overhang, normally also introducing the part into a plasmid for amplification and distribution. After a part has been domesticated, MoClo standards are PCR-independent, allow for the sharing and reuse of parts in different assemblies, and allow for generation of complex constructs after multiple assembly rounds without PCR.

Different MoClo standards differ in restriction enzymes, fusion sites, and vectors, but they share the same design paradigm (Fig. 1a). Each tool kit offers a set of vectors that are arranged in a rigid format with a backbone containing an origin of replication, selection markers, and a modular cloning site where multiple parts from lower level vectors are combined and can then be transferred to higher level backbones. The first limitation of this paradigm is the nature of the vector: each host and application has its own specific vector, which has led to the proliferation of MoClo tool kits. The second limitation comes from the fact that more assembly steps are needed for complex constructs, where there are common assembled modules for multiple genes or sequences (Fig. 1b). Some tool kits include common elements or auxiliary factors in the vector backbone, simplifying the use of the tool kit for a specific application. For example, plant tool kits include left and right border sequences flanking the cloning site to allow for *Agrobacterium*-mediated plant transformation [5, 6, 8, 9]. Another interesting additional feature is the "secondary module" in the EcoFlex tool kit [7]. Multiple transcriptional units (TUs) can be introduced in the secondary module, using an alternative type IIS nuclease, followed by the assembly of two or three TUs in the main site. Moore et al. showed that secondary sites increase assembly efficiency by decreasing the number and size of parts in an assembly, which is highly desirable when generating libraries. While the potential of secondary sites was demonstrated, in EcoFlex this feature is restricted to special level 2 vectors that can only receive two or three TU in the main site. Furthermore, the TUs must be preassembled into a level 2 vector before subcloning them into the secondary site, limiting the flexibility and usefulness of this feature.

The Joint Universal Modular Plasmids (JUMP) offer a flexible platform to overcome the limitations of current modular cloning systems. JUMP vectors have been designed to be easily modified: orthogonal assemblies allow for introduction of any sequence or assembly into the vector backbone (Fig. 1c) without affecting the main modular cloning site. Additionally, JUMP vectors are based on the Standard European Vector Architecture (SEVA) [11] and

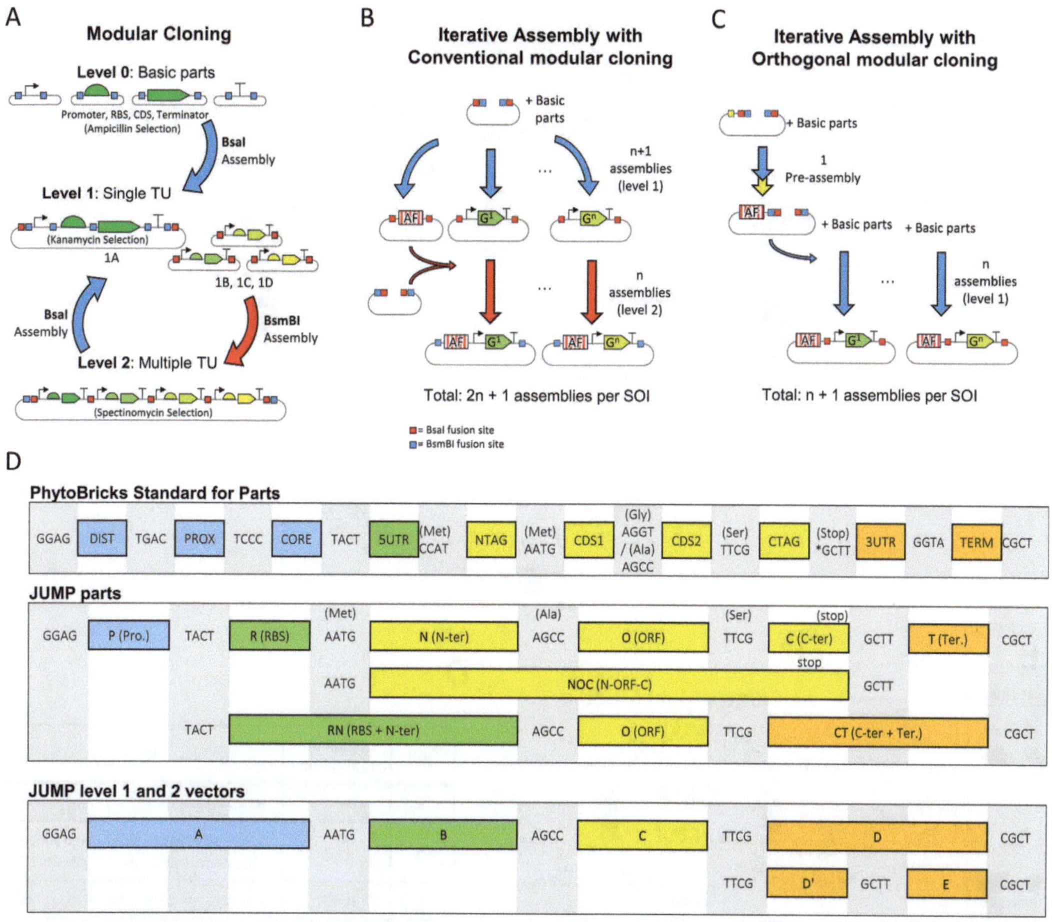

Fig. 1 Modular cloning and PhytoBrick standardization of parts. (**a**) Basic parts are contained in level 0 vectors, which are assembled to form a single transcription unit (TU) in a level 1 vector. To simplify cloning and screening, the level 1 destination vector contains a different selection marker than level 0 vectors and a reporter gene replaced by the assembled TU. Multiple level 1 assembly products can be combined in a level 2 vector, due to the use of an alternative type IIS restriction enzyme and selection marker. In JUMP vectors and some MoClo standards level 1 plasmids can be used as level 3 assembly destination vectors. (**b**) Conventional modular cloning approach with iterative assemblies. The sequence of interest (SOI) is assembled in parallel with the auxiliary factors (AF) needed to test the SOI, then combined in a second assembly tier. (**c**) Orthogonal modular cloning applied in iterative assemblies. The AF is preassembled in a secondary site of the destination vector thus reducing number of assemblies needed per SOI. (**d**) PhytoBricks standard of fusion sites [11]; format of JUMP parts provided in tool kit; fusion sites of level 1 and level 2 JUMP vectors

are compatible with PhytoBricks basic parts (*see* Fig. 1d) as well as BioBricks. The Standard European Vector Architecture (SEVA) [11] is a large collection of origins of replication (OriV) and antibiotic selection markers (AbR) with a standardized format that allows for simple exchange of vector elements (Fig. 2a). JUMP vectors contain a special SEVA "cargo" that includes the main Golden Gate sites for modular cloning (Main Module), BioBricks Prefix and

A

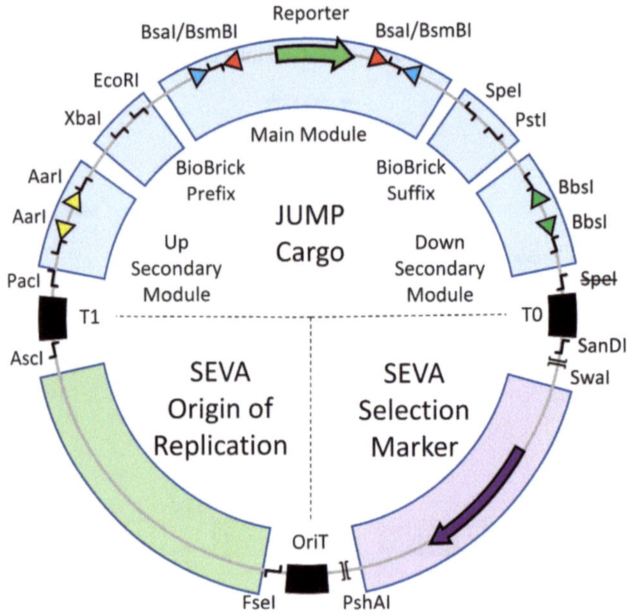

B

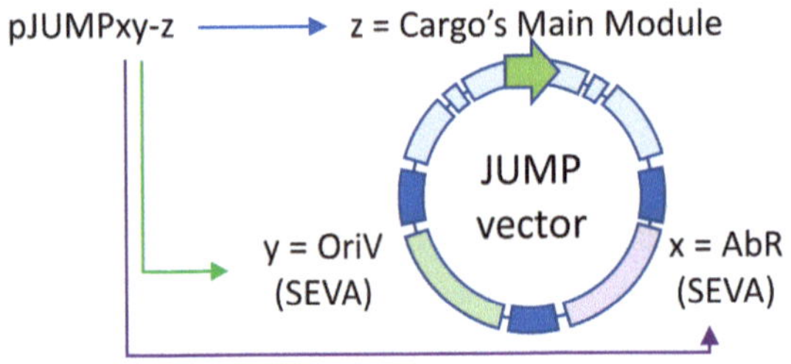

C

AbR	Antibiotic Selection	Notes
1	Ampicillin /Carbenicillin	Default level 0
2	Kanamycin	Default level 1
3	Chloramphenicol	Broad host range marker
4	Spectinomycin /Streptomycin	Default level 2

D

OriV	Origin of replication	Notes
1	R6K	Narrow-host range, (*pir*-dependent)
2x	RK2	Low copy number, broad host range. contains BsmBI, AarI, BtgZI sites
3	pBBR1	Medium-low copy, Broad host range
4	pRO1600 /ColE1	Shuttle E.coli/Pseudomonas *sp*
5	RSF1010	Broad host range
6	p15A	Medium-low copy number
7	pSC101	Low copy number
7ts	pSC101-ts	Thermosensitive replication, low copy
8	pUC	High copy number
9	pBBR322-ROP	Medium-high copy number (Default level 1 and 2)

Fig. 2 Design of joint universal modular plasmids. (**a**) JUMP as a modification of the SEVA design. In SEVA (Standard European Vector Architecture) plasmids, three common short DNA sequences flank three variable regions: origin of replication, marker, and "cargo," which might be any expression cassette. The invariable regions are two transcription terminators (T1 and T0, which flank the cargo) and origin of conjugation (oriT). Invariable regions also contain rare cutting sites, forbidden in the sequence of variable regions. This design means that all variable regions are exchangeable between vectors. JUMP is designed as special cargo of SEVA vectors to allow for compatibility with future OriVs and AbRs of the collection. The cargo contains Upstream (Up) Module (outward AarI); BioBricks prefix (XbaI, EcoRI); Main Module (BsaI and BsmBI outward and inward

Suffix, and the orthogonal Secondary Sites (Upstream and Downstream Module, before and after the Main Module, respectively). JUMP secondary sites are compatible with modular cloning done in the Main Module by having the same acceptor sites as the Main Module in all vectors. These secondary sites are also completely orthogonal due to the use of alternative type IIS restriction sites.

A collection of JUMP plasmids has been constructed [12] and deposited in Addgene for distribution. The collection contains bacterial PhytoBricks basic parts of common use in synthetic biology, vectors to domesticate additional basic parts, and cloning vectors to perform assembly at any level with ten origins of replication and four selection markers. The nomenclature of JUMP vectors is based on that of SEVA (Fig. 2b) for antibiotic selection (Fig. 2c) and origin of replication (Fig. 2d). The full list of parts can be found and ordered here: https://www.addgene.org/browse/article/28203402/. Sequences of the plasmids can be found in the supplementary material of Valenzuela-Ortega and French [12].

2 Materials

Molecular biology enzymes and DNA samples should be stored at −20 °C and kept on ice during use. PCR, digestion, and assembly reactions should be prepared on ice and then incubated as specified. We recommend using bioinformatics software to plan assembly and screening. Examples of software to assist plasmid design include Vector NTI, SnapGene, and Genome Compiler.

2.1 General Materials

1. Waterbath or heat block for heat-shock transformation.
2. Microcentrifuge.
3. Shaking incubator for liquid cultures.
4. Static incubator for agar cultures.
5. Thermocycler for PCR and Golden Gate incubations.
6. 1.5 mL microcentrifuge tubes.
7. 0.2 mL PCR tubes.
8. Ultrapure water.

Fig. 2 (continued) for level 1, respectively, and vice versa for level 2); BioBrick suffix (SpeI, PstI); and Downstream (Down) Module (outward BbsI). SEVA's canonical SpeI site was removed to allow for BioBricks compatibility. (**b**) JUMP nomenclature is conservative with SEVA's style for origins of replication (OriV) and selection marker (AbR) vector regions, while cargo nomenclature is replaced by JUMP's Main Module index. (**c**) Antibiotic selection markers used in JUMP vectors. (**d**) Origins of replication in JUMP vectors

2.2 JUMP Plasmids

1. All necessary parts for the assembly, in level 0 donor vectors. (New parts may be prepared as described below, Subheading 3.2).
2. Destination vector (level 1 for a single transcription unit, or level 2 for multiple transcription units) with suitable replication origin and antibiotic resistance marker for the ultimate host strain to be used.
3. If the final destination vector is level 2 (i.e., will contain multiple level 1 transcription units): core level 1 assembly plasmids. (These are not required if the final construct, to be introduced into the final host, is level 1, that is, contains a single transcription unit.)

2.3 Media and Strains

1. Chemically competent *Escherichia coli* cloning strains such as JM109 (NEB #E4107) or DH5α (NEB #C2987I) (*see* **Note 1**).
2. Helper strain for tri-parental conjugation, if required for introduction of the final construct into the final host (*see* **Note 2**).
3. Autoclaved LB medium (10 g/L tryptone, 5 g/L yeast extract, 10 g/L NaCl) and LB agar (as above plus 15 g/L agar).
4. Antibiotics for selection of plasmids. Level 0: ampicillin or carbenicillin (final concentration 100 μg/mL); level 1, kanamycin (50 μg/mL); level 2, spectinomycin (50 μg/mL) alternative selection vectors, chloramphenicol (18 μg/mL) (*see* **Note 3**).

2.4 Molecular Biology Reagents for Assembly Procedure

1. BsaI-HFv2 (NEB) (*see* **Note 4**).
2. BsmBI (NEB) (*see* **Note 4**).
3. T4 DNA ligase (NEB).
4. 10× T4 DNA ligase buffer (NEB). It is recommended to store this buffer in small aliquots at −20 °C.
5. Shrimp alkaline phosphatase (NEB) (only for assemblies in secondary sites).
6. AarI (ThermoFisher #ER1581) (only for assembly in Upstream Module) (*see* **Note 4**).
7. BbsI-HF (NEB) (only for assembly in Downstream Module) (*see* **Note 4**).
8. T4 polynucleotide kinase (NEB) (for phosphorylation of "dummy" linkers if needed).

2.5 PCR for Amplification of Parts and Clone Screening

1. Oligonucleotide primers from Sigma-Aldrich or a similar supplier.
2. High Fidelity Q5 polymerase (NEB) for amplifying new parts (*see* **Note 5**).
3. GoTaq® G2 Green 2× Master Mix (Promega) for colony screening.

4. Nuclease-free water.
5. TE buffer (QIAGEN).

2.6 Agarose Gel Electrophoresis

1. 50× TAE buffer (Formedium).
2. SYBR Safe (ThermoFisher) or SafeView Nucleic Acid Stain (NBS Biologicals).
3. Safe Imager™ Blue-Light Transilluminator (Invitrogen) with amber filter unit, or similar.

2.7 DNA Purification

1. QIAprep Spin Miniprep Kit (QIAGEN) for plasmid DNA miniprep.
2. QIAquick PCR Purification Kit (QIAGEN) (if purification of PCR reactions is required).
3. QIAquick Gel Extraction Kit (QIAGEN) (if extraction of DNA fragments from agarose gels is required).

3 Methods

3.1 Planning the Assembly

1. Select final destination vector based on the final intended host organism. JUMP vectors, as distributed by Addgene, offer ten different origins of replication for different purposes (*see* Fig. 2d). When only one transcription unit (TU) is needed, it can be built as a level 1 assembly using the destination vector with the OriV of interest. However, to build assemblies with more transcription units, intermediate assemblies should be built using the "core set" of level 1 and level 2 vectors with OriV #9.
2. Decide whether to assemble in main or secondary sites. While all transcription units and genetic elements can be combined in Main Modules (Subheading 3.7), if some of the elements are reused in multiple assemblies they can be introduced in a secondary site to reduce the total number of assemblies (*see* Fig. 1b, c). Figure 3 shows a flowchart to plan assemblies. Briefly, after deciding on the final destination vector type (OriV), elements to be assembled are divided into common elements (shared among multiple assemblies) and unique genetic devices or sequences of interest (SOI). Common elements can be prepared using modular cloning (if assembly is needed) and then introduced in a secondary site via conventional restriction-ligation-transformation cloning or two-step assembly (*see* Subheading 3.7).
3. Plan individual assembly steps. When planning a level 2 assembly with multiple TU, the level 1 assemblies used to prepare each individual TU should be done in different level 1 destination vectors (*see* Fig. 1d), which will allow for assembly in a level

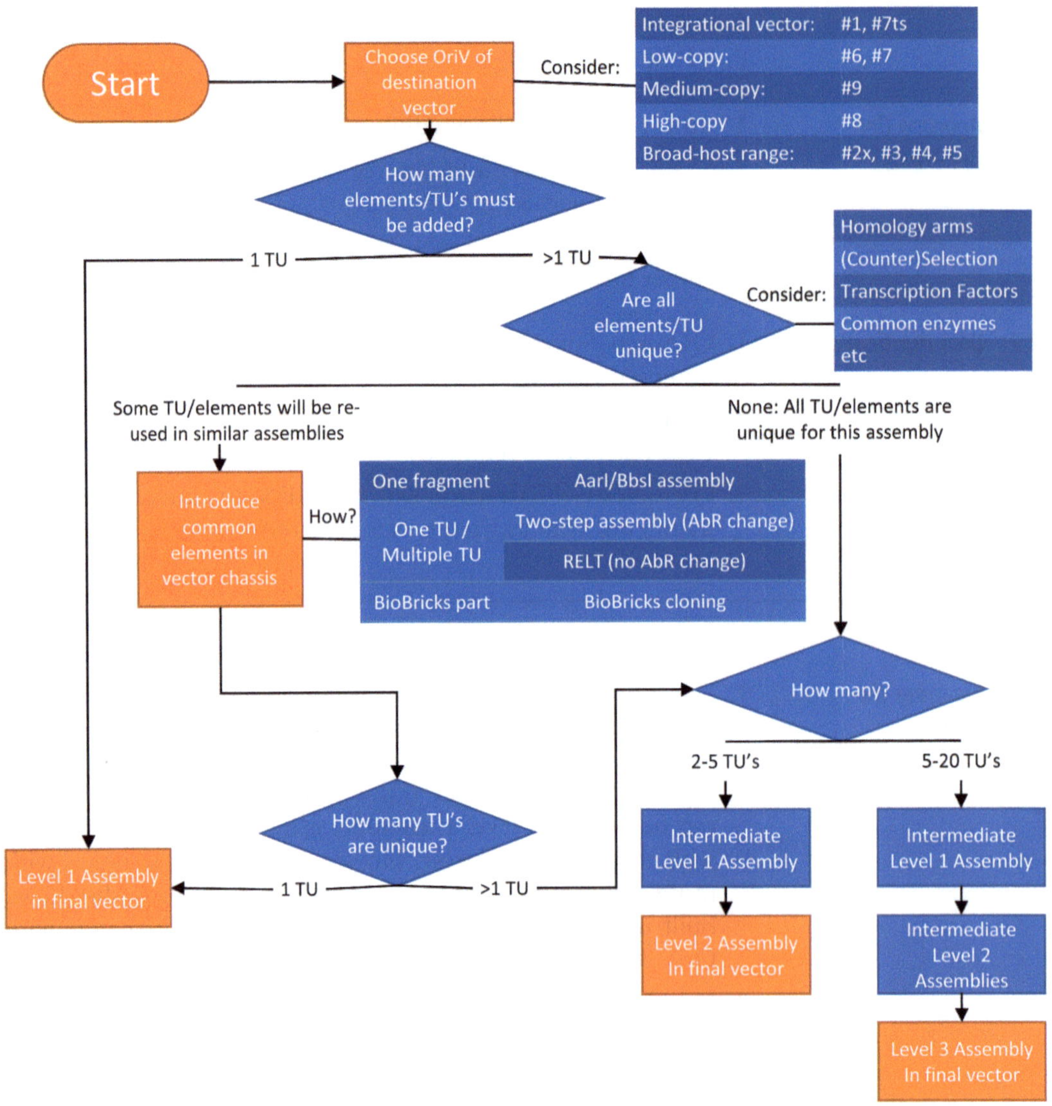

Fig. 3 JUMP assembly-planning guide. After choosing the OriV of the final vector (*see* Fig. 2d for more details), the elements and transcription units (TU) needed in the assembly can considered as "unique" for the assembly or common. The destination level is chosen depending on the number of "unique" TU, and any number of "common" elements can be introduced in secondary sites. Common elements can be introduced in JUMP backbones in multiple ways: if assemblies come from modular cloning they can be assembled in the destination site using the two-step assembly (Subheading 3.8) if the selection markers in the donor and destination vectors are different, or with conventional cloning (restriction endonuclease–ligase–transformation or RELT). If the common elements are available as BioBricks, they can be introduced within JUMP vector's BioBricks Prefix or Suffix. If sequences are designed ad hoc to be introduced in a secondary site, and the inserts are flanked by AarI or BbsI sites, they can be assembled directly via Golden Gate using AarI/BbsI

2 destination vector in the required order. Multiple level 2 assemblies can be combined in a level 1 vector (level 1 vectors are equivalent to a level 3 destination vector, and level 2 vectors are equivalent to level 0).

4. Select the correct parts. Ensure that the parts assembled contain compatible fusion sites. JUMP is designed to use PhytoBricks standardized parts, but researchers should be sure that the selected parts complete a correct assembly (*see* Fig. 1d to find possible alternative part combinations). "Dummy" linkers can be used to replace parts (e.g., when the number of parts assembled is not sufficient to have a complete assembly). Dummy linkers can be prepared with annealed oligonucleotides as explained in Subheading 3.6. Design software may be useful (*see* **Note 6**).
5. Design any new parts required (i.e., parts which are not already in the JUMP kit). Parts must be flanked by sites recognized by the restriction enzyme used during the assembly (BsmBI for insertion into level 0 vectors, and BsaI for assembly in level 1 vectors) and must not have internal sites for these enzymes. Only BsaI and BsmBI sites must be removed from parts to allow assemblies with JUMP vectors (including those into secondary sites). Optionally, AarI, BbsI, and BtgZI sites can be removed to allow direct assembly using those enzymes and allow compatibility with other plasmid tool kits. Generation of new parts from existing sequences is known as domestication. Guidelines for correct part domestication are described in Subheading 3.2.

3.2 Domestication of New Parts (If Required)

DNA molecules used as assembly parts can either have a linear form (e.g., PCR products) or be circular plasmids (Fig. 4a). We recommend domesticating parts into plasmid vectors, such as the Universal Part acceptor pJUMP19-Uac. This allows users to ensure the sequence is correct, store and amplify the part, and share it with other labs. To store the part in a level 0 plasmid, it must be made in a linear form to introduce it into the part acceptor plasmid (level 0 assembly); therefore, it is convenient to generate it in such way that it can be used for both storage in a level 0 vector and direct assembly in a level 1 vector.

1. Design the part sequence. To allow level 0 and level 1 assemblies, parts should be flanked by overlapping BsaI (for level 1 assembly) and BsmBI sites (for level 0 assembly) as shown in Fig. 4a. Digestion with these enzymes should yield the correct 4-base overhang for the particular part type (Fig. 1d).
2. Decide whether to make the part by synthesis, PCR, or annealing oligonucleotides. Small parts (up to ~100 bases) can be generated with oligonucleotides containing the whole part sequence and flanking restriction sites, which should be annealed to each other before assembly (*see* Subheading 3.5). Alternatively, oligonucleotide-based parts can be generated without flanking restriction sites leaving 4 base pair overhangs after annealing, which should correspond to the fusion site of

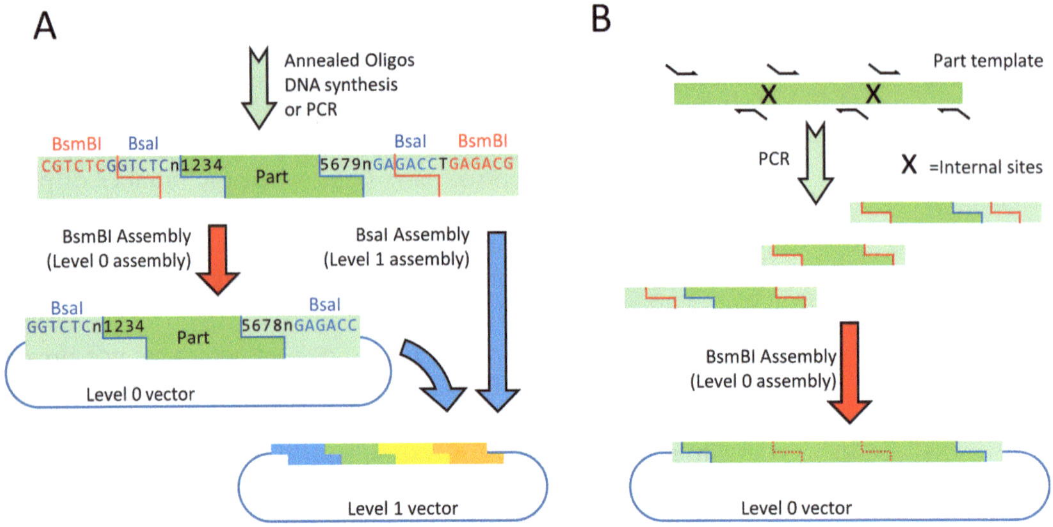

Fig. 4 Domestication of basic parts. (**a**) Domestication of parts using a Universal Acceptor plasmid (such as pJUMP18-Uac). Linear fragments with overlapping BsaI and BsmBI sites can be used for both level 1 assembly and for domestication Golden Gate reaction to introduce the basic part into a level 0 vector. (**b**) PCR-based domestication allows for removal of internal forbidden sites during cloning. Parts are amplified as multiple subpart fragments with primers that introduce silent single-point mutations removing internal sites with flanking BsmBI sites in subpart fragments. BsmBI is used to assemble the subparts in the Universal Acceptor plasmid

the part. Parts made this way must be phosphorylated during the preparation step using T4 polynucleotide kinase. To generate longer parts, users can order the whole sequence as a synthetic DNA fragment (from suppliers such as Integrated DNA Technologies Ltd. or Twist Bioscience) or amplify it via PCR with primers that introduce the flanking restriction sites and fusion sites corresponding to the part type (*see* Fig. 2a). PCR procedure is described in Subheading 3.4.

3. If parts are to be made by PCR, and the natural sequence contains BsaI or BsmBI sites: the domestication PCR can be used to remove internal restriction sites from parts, as described for GoldenBraid 2.0 [5] (Fig. 4b). Briefly, the part template is amplified between the ends and internal sites, generating subpart fragments that end in BsmBI sites. The BsmBI sites generate fusion sites that allow all subparts to be introduced into a part acceptor vector using BsmBI assembly. Single-base mutations are introduced in the fusion sites between subpart fragments, removing internal restriction sites without changing amino acid sequence when the part codes for a protein sequence. Note that this is unnecessary if parts are synthesized or made from oligonucleotides, since the “forbidden” internal restriction sites may be removed in the sequence design stage before ordering.

4. Decide on a suitable level 0 vector for part storage. For general purposes, Universal Part acceptor pJUMP19-Uac should be used. Optionally, researchers who wish to characterize the function of promoters and terminators can domesticate these parts in special level 0 vectors that permit part characterization (*see* **Note** 7).
5. Test the purity of DNA following synthesis and reconstitution or preparation. DNA must be free of contaminants that might inhibit the endonuclease or ligase present in the assembly reaction, and free of unwanted DNA molecules that might interfere with the assembly. Purification kits (for plasmid mini-prep, PCR reaction cleaning or gel-extraction) give a suitable DNA purity to perform assemblies, but it is advisable to check the absorbance spectrum of the purified parts to estimate concentration and ensure that the ratio of absorbance at 260 nm/230 nm and 260 nm/280 nm correspond to pure DNA (2.0–2.2 and 1.8, respectively).

3.3 Preparation of Parts and Vectors by Plasmid DNA Miniprep

1. Grow the plasmid-bearing cells. *E. coli* cultures must be incubated at 37 °C, or 30 °C when a thermosensitive vector is used. Shake liquid cultures appropriately to ensure sufficient aeration and homogeneity of culture.
2. Perform plasmid DNA miniprep. We used the QIAprep Spin Miniprep Kit (QIAGEN), following the manufacturer's protocol, adjusting culture and elution volumes depending on the copy number of the plasmid. For vectors with low-copy number (oriV #2, #3, #5, #6, #7, #9), 10 mL of overnight culture is used for the miniprep and DNA is eluted twice in the same 50 μL of nuclease free water. For medium-copy number vectors (oriV #9), use 10 mL culture and elute DNA in 50 μL of nuclease free water twice (total 100 μL). For high-copy number vectors (oriV #4, #8) use 5 mL culture and elute DNA in 50 μL of nuclease free water twice (total 100 μL).
3. Check DNA concentration and purity. If the DNA concentration is too low for analysis of an assembly clone, DNA can be concentrated (*see* **Note 8**).

3.4 Preparation of Parts by PCR Amplification

1. Perform the PCR reaction. For part amplification, a high-fidelity polymerase should be used to avoid introducing unwanted mutations. We routinely use Q5 High-Fidelity Polymerase according to the manufacturer's protocol.
2. Check the product by agarose gel electrophoresis.
3. Purify DNA. If the only band amplified corresponds to the expected size, the PCR can be cleaned using a PCR-reaction cleaning kit according to the manufacturer's protocol. If multiple bands appear in the PCR amplification (including primer

dimers), the band of the correct size should be purified from an agarose electrophoresis gel using a gel-purification kit. We used the QIAquick Gel Extraction Kit (QIAGEN) repeating the wash step with PE buffer, eluting DNA in 50 μL of nuclease free water and concentrating it in 10 μL (*see* **Note 5**). Note that gel-purification should only be applied when multiple bands appear in a gel. This type of purification will result in lower DNA concentrations and higher level of chemical impurities than with the PCR-purification kit.

3.5 Preparation of Parts by Annealing Oligonucleotides

1. Prepare 100 μM stock solution of each oligonucleotide in TE buffer or nuclease-free water. This should be stored at −20 °C.
2. Combine the two strands. For oligonucleotide parts with restriction sites on both ends, combine 5 μL of each oligonucleotide (100 μM stock solution) and 40 μL of nuclease free water. For oligonucleotides without restriction sites, which are designed to anneal so as to leave the correct four-base overhangs, combine 5 μL of each oligonucleotide (100 μM), 34 μL of nuclease-free water, 5 μL of T4 DNA ligase buffer, and 1 uL T4 DNA polynucleotide kinase (PNK) and incubate for 30 min at 37 °C.
3. Anneal the two strands by heating at 95 °C for 5 min and cool down slowly to room temperature. This can be done using a heat-block and switching it off to allow gradual cooling, or in a thermocycler adjusted to cool at a controlled speed (such as 0.2 °C per second).
4. Prepare working solutions. We recommend diluting 20-fold in TE buffer (to a concentration of 0.5 μM) and storing at −20 °C until used.

3.6 Modular Cloning Assembly in Main Module

Assemblies in the Main Module of JUMP vector work as conventional Golden Gate assemblies, regardless of the level. For an assembly in a JUMP vector of a certain level, the destination vector is combined with insert parts from one level lower: level 1 assembly products are inserts for level 2 vectors; level 2 assembly products (and basic parts) are inserts for level 1 vectors (which also work as acceptors for level 3 assemblies).

1. Prepare assembly reactions. Parts are assembled in equimolar concentrations as shown in Table 1. BsaI is used for level 1 and level 3 assemblies and BsmBI for level 0 and level 2 assemblies. Note that, when calculating the molar concentration of a molecule, the molecular weight is that of the whole molecule, including the backbone carrying the inserts.
2. Carry out the assembly reaction as shown in Table 2. For difficult assemblies (higher number of parts, parts of larger size) or when a higher colony number is desirable (combinatorial assemblies to generate libraries), the incubation protocol can be extended (Table 3).

Table 1
Golden Gate assembly reaction setup

Reagent	Initial Conc.	Vol. (μL)	Final Conc.
Backbone DNA	x	x	20 fmol/20 μL
Insert DNA	x	x	20 fmol/20 μL
T4 ligase buffer	10×	2	1×
T4 DNA ligase (enzyme)	400 U/μL	0.25	100 U/20 μL
BsmBI/BsaI-HF	10/20 U/μL	1 (or 0.5)	10–20 U/20 μL
H_2O (nuclease free)	–	x	–
Total		20	

Table 2
Golden gate assembly incubation conditions

BsaI-HF	BsmBI	Incubation	
37 °C	42 °C	15 min	
37 °C 16 °C	42 °C 16 °C	3 min 3 min	×30 cycles
37 °C	55 °C	15 min	
80 °C	80 °C	5 min	
10 °C	10 °C	Hold	

Table 3
Golden Gate assembly incubation conditions for difficult assemblies

BsaI-HF (overnight)	BsmBI (overnight)	Incubation	
37 °C	42 °C	15 min	
37 °C 16 °C	42 °C 16 °C	5 min 5 min	×60 cycles
37 °C	55 °C	60 min	
80 °C	80 °C	5 min	
10 °C	10 °C	Hold	

3. Transform *E. coli* with 1–2.5 μL of assembly reaction, and plate 10% of the transformation on LB agar containing the antibiotic corresponding to the destination vector.

3.7 Two-Step Assembly in Secondary Modules

All JUMP vectors enable introduction of flanking sequences on either side of the Main Module. Conventional cloning allows introduction of any sequence using AarI (in the Upstream Module) or BbsI (in the Downstream Module). Moreover, because the receiver fusion sites of both secondary sites are the same as the ones used by the Main Module across all JUMP vectors (GGAG and CGCT), the sequences introduced in secondary sites can originate from other JUMP vector Main Modules. To maximize the utility of this design principle, we have developed a two-step assembly method to efficiently assemble TU from basic parts in either secondary site.

1. Assemble inserts, without destination vector, using the methodology described in Subheading 3.6, but increasing concentration to 40 fmol of each part per 20 μL reaction. We recommend using the long overnight protocol to increase efficiency.
2. While the assembly reaction is running, predigest the destination vector (Table 4). Incubate at 37 °C for as long as the half-assembly is running followed by 20 min at 65 °C to inactivate the restriction enzyme. A predigested backbone can be used multiple times with different insert half-assemblies.
3. Both first-step reactions can be stored frozen, but to store them for long periods, reactions should be purified with the PCR-purification kit.
4. Ligate insert-assembly and predigested destination vector as shown in Table 5 and incubate for 1 h at 16 °C.
5. Transform *E. coli* (*see* **Note 1**) with 1–2.5 μL of the assembly reaction, and plate 10% of the transformation on LB agar containing the antibiotic corresponding to the destination vector.

Table 4
Backbone digestion setup conditions

Upstream Predigestion	Downstream Predigestion
200 fmol of destination vectors	
1 μL AarI enzyme	1 μL BbsI-HF enzyme
0.4 μL AarI's oligo[a] 50×	–
2 μL AarI Buffer 10×	2 μL Cutsmart Buffer 10×
0.5 μL Shrimp alkaline phosphatase Nuclease-free water to a final volume of 20 μL	

[a]Correct AarI digestion of DNA substrate requires the addition of an oligonucleotide provided with the enzyme (ThermoFisher)

Table 5
Ligation setup for two-step assembly

20 fmol of assembly product (10 μL of insert half-assembly reaction)
20 fmol of predigested backbone (2 μL of reaction)
2 μL of T4 ligase buffer 10×
1 μL of T4 DNA ligase enzyme
Nuclease-free water to a final volume of 20 μL

3.8 Selection of Correct Assemblies

1. Choose colonies which have grown on the correct antibiotic and in which the marker gene has been inactivated, suggesting possible replacement by the new insert. The default reporter in the Main Module of all JUMP vectors is a constitutive sfGFP cassette, but we have built and added alternative vectors with a *lacZ'α* gene (included in the Addgene distribution) to allow for white–blue screening when assembling GFP-coding genes. Note that secondary modules do not contain a marker gene.
2. Check candidate colonies by colony PCR. We recommend a thorough screening process because, based on our experience, plasmid cotransformation is a persistent problem in all modular cloning systems (*see* **Note 9**). Test between 4 and 8 colonies of correct phenotype (i.e., without sfGFP reporter). For assemblies in secondary modules (where no reporter gene is present), there may be no way of distinguishing undigested destination vector clones, and it may be necessary to analyze more colonies.
3. Resuspend each colony in 40 μL sterile dH_2O. When all colonies to be analyzed have been resuspended, vortex them for a few seconds.
4. Prepare primers. Use either cargo-flanking primers or one primer specific for the insert and one specific for the vector. SEVA primers PS1 (AGGGCGGCGGATTTGTCC) and PS2 (GCGGCAACCGAGCGTTC) bind in all SEVA and JUMP vectors amplifying the cargo region. Flanking primers should only be used when the sequence of the insert is of a different length to that in the parental destination vector.
5. Prepare reaction mixtures. Prepare master-mix with all reagents except template (Table 6), place 8 μL of Mastermix in a PCR tube and then add 2 μL of template (cell suspension).
6. Perform PCR as in Table 7.
7. Analyze the colony PCR product by agarose gel electrophoresis.
8. Grow cultures of 2–4 colonies that gave a positive result on the colony PCR. The colony resuspension can be used to inoculate the overnight culture to perform a plasmid DNA miniprep.

Table 6
Reaction setup for colony PCR

GoTaq Green Master Mix (Promega)			
Reagent	**Initial Conc.**	**Vol. per reaction (μL)**	**Final Conc.**
Gotaq Green 2× MasterMix	2×	5	1×
Primer F	10 μM	0.2	0.2 μM
Primer R	10 μM	0.2	0.2 μM
H_2O (nuclease free)	–	2.6	–
Template	–	2	–

Table 7
Reaction incubation for colony PCR

95 °C	5 min	×1
95 °C 60 °C[a] 68 °C	20 s 30 1 min/kb	×30 cycles
68 °C	5 min	×1
10 °C	Hold	

[a]Adjust annealing temperature for each primer pair. 60 °C is suitable for primers PS1/PS2

9. Test for cotransformation. We recommend confirming that no part-donor plasmid has been cotransformed by inoculating LB agar plates with 2–5 μL (*see* **Note 6**) of the colony resuspension. Cotransformants will show growth with the antibiotic corresponding to the donor vector.
10. The next day, take a sample of the overnight cultures for possible propagation of the clones, and store it at 4 °C.
11. Perform the plasmid DNA miniprep (*see* Subheading 3.3).
12. Restriction test. Verify the plasmids by digesting at least 200 ng of the plasmid DNA with restriction enzymes that will generate a different pattern for the correct assembly from that expected for the parental destination vector. Analyze by gel electrophoresis.
13. Sequence. Sanger sequencing offers definitive confirmation of assemblies. SEVA primers PS1 and PS2 flank cargo and will be sufficient to sequence the ends of assemblies. We only recommend sequencing of the whole assembled sequence for newly domesticated parts, as the PCR-free Golden Gate assembly procedure is not expected to introduce new mutations.

4 Notes

1. Homemade chemically competent *E. coli* cells were prepared based on the protocol of Chung et al. [13]. Briefly, LB medium prewarmed at 37 °C is inoculated with a fresh overnight culture of the strain of interest (inoculum–culture = 1:100), and incubated at 37 °C with shaking at 200 rpm. When the OD reaches 0.5, the culture is transferred to ice for 30–60 min, and then centrifuged in a prechilled centrifuge (at 4 °C) for 10 min at 3900 × *g*. Supernatant is gently decanted and the pellet is resuspended in prechilled sterile TSS (TSS–culture = 1:10). TSS ("Transformation and Storage Solution) is LB broth containing 10% (w/v) polyethylene glycol 4000, 5% (v/v) dimethyl sulfoxide, and 50 mM $MgCl_2$.
2. Our triparental conjugation was based on the methodology published by de Lorenzo and Timmis [14]. We used a helper *E. coli* strain carrying plasmid RK600 [14], which carries *tra* genes for conjugation. Briefly, 0.1 mL of overnight culture of donor, recipient and receiving strain were mixed with 5 mL of fresh LB, incubated for 5 h at 37 °C without agitation (30 °C when using mesophilic bacteria or thermosensitive vectors), then 100 μL were streaked for colony isolation on a selective medium.
3. Chloramphenicol selection is done with 18 μg/mL chloramphenicol, which is about half of the concentration commonly used with *E.coli*. We used the broad-host chloramphenicol resistance gene from pSEVA3b61 [15] and found that higher chloramphenicol concentrations inhibited growth of cells carrying this marker.
4. Isoschizomers of these restriction enzymes are available and may also work, but protocol conditions here have been optimized for the enzymes listed.
5. Other polymerases can be used, but we recommend using High Fidelity polymerases to reduce introduction of mutations in parts. Alternatively, DNA synthesis can be obtained from Integrated DNA Technologies Ltd. among other suppliers.
6. Some bioinformatics software packages (such as SnapGene) allow for in silico construction of assemblies, which helps in confirming that chosen inserts match and provides a simple way of obtaining the sequence map of the newly built plasmid.
7. Promoters and terminators can be domesticated into the universal acceptor or into the specialized acceptors pJUMP19-Pac and pJUMP19-Tac. These vectors have inserts in secondary sites that allow direct visualization of the activity of these two types of parts. Vector pJUMP19-Pac has eGFP without

promoter in the Downstream Module, which will be expressed by any promoter introduced in the Main Module. Vector pJUMP19-Tac has the same eGFP in the Downstream Module plus a constitutive promoter in the Upstream site. The domestication of a terminator in the Main Module of pJUMP19-Tac will be indicated by transcription termination before the eGFP and will be shown by a decrease in fluorescence.

8. We routinely increased the concentration of DNA samples using a vacuum concentrator (Eppendorf™ Concentrator Plus). Alternatively, other common methodologies such as ethanol precipitation can be used to concentrate nucleic acids [16].
9. We have found that plasmid cotransformation occurs very frequently upon transformation with Golden Gate assembly mixes: unselected donor plasmids can be introduced and replicate without the presence of their specific selecting antibiotic. We have found that this also occurs using other MoClo tool kits and, in some assemblies, cotransformant colonies can be more than 10% of the total.

Acknowledgments

This work was supported by the Biotechnology and Biological Sciences Research Council (BBSRC) [grant number BB/J01446X/1].

References

1. Cameron DE, Bashor CJ, Collins JJ (2014) A brief history of synthetic biology. Nat Rev Microbiol 12:381–390
2. Del Vecchio D (2015) Modularity, context-dependence, and insulation in engineered biological circuits. Trends Biotechnol 33:111–119
3. Chao R, Mishra S, Si T, Zhao H (2017) Engineering biological systems using automated biofoundries. Metab Eng 42:98–108
4. Engler C, Gruetzner R, Kandzia R, Marillonnet S (2009) Golden gate shuffling: a one-pot DNA shuffling method based on type IIs restriction enzymes. PLoS One 4:e5553
5. Sarrion-Perdigones A, Vazquez-Vilar M, Palaci J, Castelijns B, Forment J, Ziarsolo P, Blanca J, Granell A, Orzaez D (2013) GoldenBraid 2.0: a comprehensive DNA assembly framework for plant synthetic biology. Plant Physiol 162:1618–1631
6. Engler C, Youles M, Gruetzner R, Ehnert TM, Werner S, Jones JD, Patron NJ, Marillonnet S (2014) A golden gate modular cloning toolbox for plants. ACS Synth Biol 3:839–843
7. Moore SJ, Lai HE, Kelwick RJ, Chee SM, Bell DJ, Polizzi KM, Freemont PS (2016) EcoFlex: a multifunctional MoClo kit for E. coli synthetic biology. ACS Synth Biol 5:1059–1069
8. Andreou AI, Nakayama N (2018) Mobius assembly: a versatile Golden-Gate framework towards universal DNA assembly. PLoS One 13:e0189892
9. Pollak B, Cerda A, Delmans M, Alamos S, Moyano T, West A, Gutierrez RA, Patron NJ, Federici F, Haseloff J (2019) Loop assembly: a simple and open system for recursive fabrication of DNA circuits. New Phytol 222:628–640
10. Vasudevan R, Gale GAR, Schiavon AA, Puzorjov A, Malin J, Gillespie MD, Vavitsas K, Zulkower V, Wang B, Howe CJ et al (2019) CyanoGate: a modular cloning

suite for engineering cyanobacteria based on the plant MoClo Syntax. Plant Physiol 180:39–55

11. Martinez-Garcia E, Aparicio T, Goni-Moreno-A, Fraile S, de Lorenzo V (2015) SEVA 2.0: an update of the Standard European Vector Architecture for de-/re-construction of bacterial functionalities. Nucleic Acids Res 43: D1183–D1189
12. Valenzuela-Ortega M, French C (2019) Joint Universal Modular Plasmids (JUMP): a flexible and comprehensive platform for synthetic biology. bioRxiv 799585
13. Chung CT, Niemela SL, Miller RH (1989) One-step preparation of competent Escherichia coli: transformation and storage of bacterial cells in the same solution. Proc Natl Acad Sci U S A 86:2172–2175
14. de Lorenzo V, Timmis KN (1994) Analysis and construction of stable phenotypes in gram-negative bacteria with Tn5- and Tn10-derived minitransposons. Methods Enzymol 235:386–405
15. Wright O, Delmans M, Stan GB, Ellis T (2015) GeneGuard: a modular plasmid system designed for biosafety. ACS Synth Biol 4:307–316
16. Green MR, Sambrook J (2016) Precipitation of DNA with ethanol. Cold Spring Harb Protoc. https://doi.org/10.1101/pdb.prot093377

Part IV

Combinatorial Methods for Pathway Optimization

Chapter 16

A Step-by-Step Protocol for COMPASS, a Synthetic Biology Tool for Combinatorial Gene Assembly

Gita Naseri and Bernd Mueller-Roeber

Abstract

For industry-scale production of high-value chemicals in microbial cell factories, the elimination of metabolic flux imbalances is a critical aspect. However, a priori knowledge about the genetic design of optimal production pathways is typically not available. COMPASS, COMbinatorial Pathway ASSembly, is a rapid cloning method for the balanced expression of multiple genes in biochemical pathways. The method generates thousands of individual DNA constructs in modular, parallel, and high-throughput manner. COMPASS employs inducible artificial transcription factors derived from plant (*Arabidopsis thaliana*) regulators to control the expression of pathway genes in yeast (*Saccharomyces cerevisiae*). It utilizes homologous recombination for parts assembly and employs a positive selection scheme to identify correctly assembled pathway variants after both in vivo and in vitro recombination. Finally, COMPASS is equipped with a CRISPR/Cas9 genome modification system allowing for the one-step multilocus integration of genes. Although COMPASS was initially developed for pathway engineering, it can equally be employed for balancing gene expression in other synthetic biology projects.

Key words Transcription factor, Combinatorial assembly, Positive selection, Recombination-based cloning, Synthetic biology

1 Introduction

In the last few decades, microorganisms such as the yeast *Saccharomyces cerevisiae* have been used to produce many high-value primary and secondary metabolites, enzymes, and pharmaceutical proteins [1, 2]. Generating an optimal microbial cell factory typically requires changing the expression levels of multiple genes and involves iterative cycles whereby increasingly refined strains are designed and constructed [3]. Metabolic engineering approaches that are based on a priori knowledge about metabolic network topographies and regulation, bottlenecks, and competitive reactions are called rational approaches (Fig. 1a) [3]. A major challenge in such rewiring of cellular metabolism is the time and effort it takes to construct strains with the target production level [4]. To bridge

Sunil Chandran and Kevin W. George (eds.), *DNA Cloning and Assembly: Methods and Protocols*, Methods in Molecular Biology, vol. 2205, https://doi.org/10.1007/978-1-0716-0908-8_16, © Springer Science+Business Media, LLC, part of Springer Nature 2020

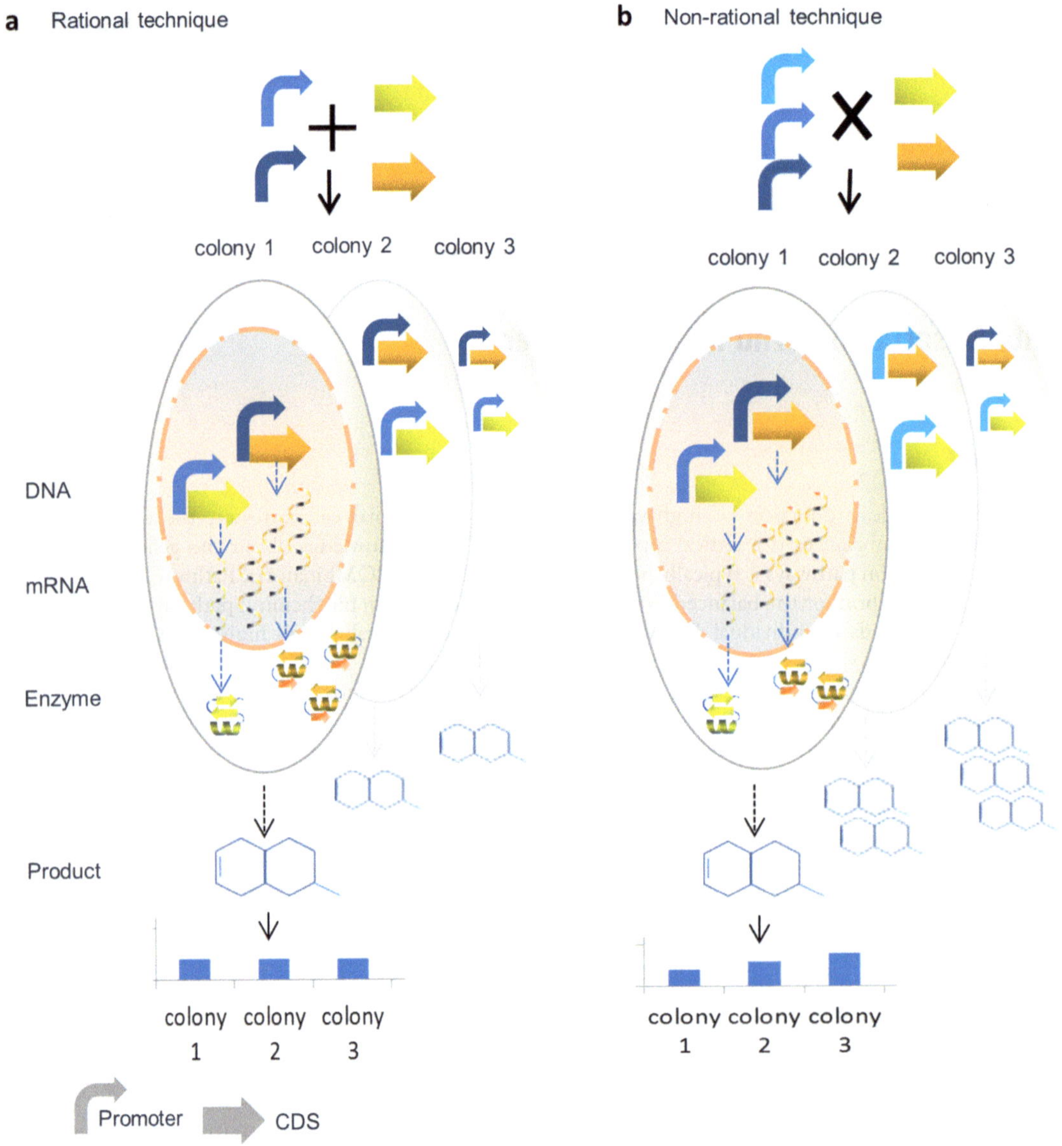

Fig. 1 Rational versus nonrational cloning methods. (**a**) Rational cloning. Promoters are assigned to coding sequences in a predefined manner (highlighted by "+"). Cells use the two-step process of transcription and translation to read each gene and produce a unique protein or enzyme that performs a specialized function. All cells essentially produce the same amount of the final product. (**b**) Nonrational cloning. Promoters are randomly assigned to coding sequences (highlighted by "x") resulting in libraries of cell variants producing different amounts of the final product. This diversity allows for the balanced expression of genes for maximal product output without prior knowledge of the best combination of expression levels of the individual genes. The host's transcription machinery interacts with the promoters upstream of the coding DNA sequences (CDSs). The transcripts are then translated into proteins (enzymes) to produce the final compound

biological knowledge gaps, high-throughput, nonrational techniques that do not rely on a priori knowledge can be employed (Fig. 1b). For example, random or semirandom genetic changes are induced in a population of strains, and the modified strains are then screened for the desired properties [5–7]. The so-called combinatorial engineering approaches are "nonrational" and allow for generating and studying many simultaneous genetic alterations. To this end, several combinatorial approaches for balancing metabolic flux have been developed to adjust gene expression at the DNA, RNA and protein levels [8–18].

Methods established for pathway optimization in yeast, such as VEGAS or CRISPR-AID, rely on the constitutive expression of pathway genes which can be metabolically burdensome for the host cell [15–17, 19]. Moreover, the combinatorial cloning approaches used to generate libraries with these methods require a construction step in *Escherichia coli*. However, large plasmids cannot be easily replicated in *E. coli* and assembly products might be unstable or unclonable [20]. The available methods employ plasmid-based systems (e.g., VEGAS and ePathOptimize) [15, 16] or genomic integration systems (e.g., MAGE and CRISPR-AID) [17, 18] for the engineering of metabolic pathways. As plasmids can be easily manipulated, plasmid-based expression systems are often preferred in synthetic biology projects. However, segregational and/or structural instability of plasmid-based systems are critical aspects that must be considered in such projects [21, 22]. Stable expression levels are typically achieved by integration of pathway genes into chromosomal DNA [23].

COMPASS is a new method that addresses these issues [24]. It relies on overlap-based in vitro cloning methods, like SLiCE [25], Gibson assembly [26], and NEBuilder HiFi assembly (New England Biolabs), and inexpensive in vivo transformation-associated recombination (TAR) [27]. COMPASS is an entirely sequence-independent approach that permits the removal of undesired scar sequences, the weakness of the VEGAS method [15]. COMPASS employs inducible plant artificial transcription factors (ATFs), covering a tenfold range of transcriptional outputs for controlling the expression of pathway genes [23].

Importantly, COMPASS employs only 11 core plasmids to generate combinatorial libraries of thousands to millions of individual constructs. The method allows for fast combinatorial assembly of up to ten pathway genes by passing through three levels of cloning. Level 0 allows for the assembly of individual parts including (1) the combinatorial assembly of ATFs and cognate DNA binding sites (BS) to generate ATF/BS units, and (2) the assembly of coding DNA sequences (CDSs), yeast terminators, and promoters of *E. coli*selection markers to generate CDS units in Entry vector X (Fig. 2a). Level 1 allows for the combinatorial assembly of ATF/BS (regulator) units upstream of up to ten CDS (gene) units (from Level 0) into Set 1 (Destination vector I.1 and Acceptor vectors A, B, C, and D; Fig. 2b) and Set 2 vectors (Destination vector II and Acceptor vectors E, F, G, and H; Fig. 2c) to generate ATF/BS-CDS modules. At Level 2, up to five regulator–gene

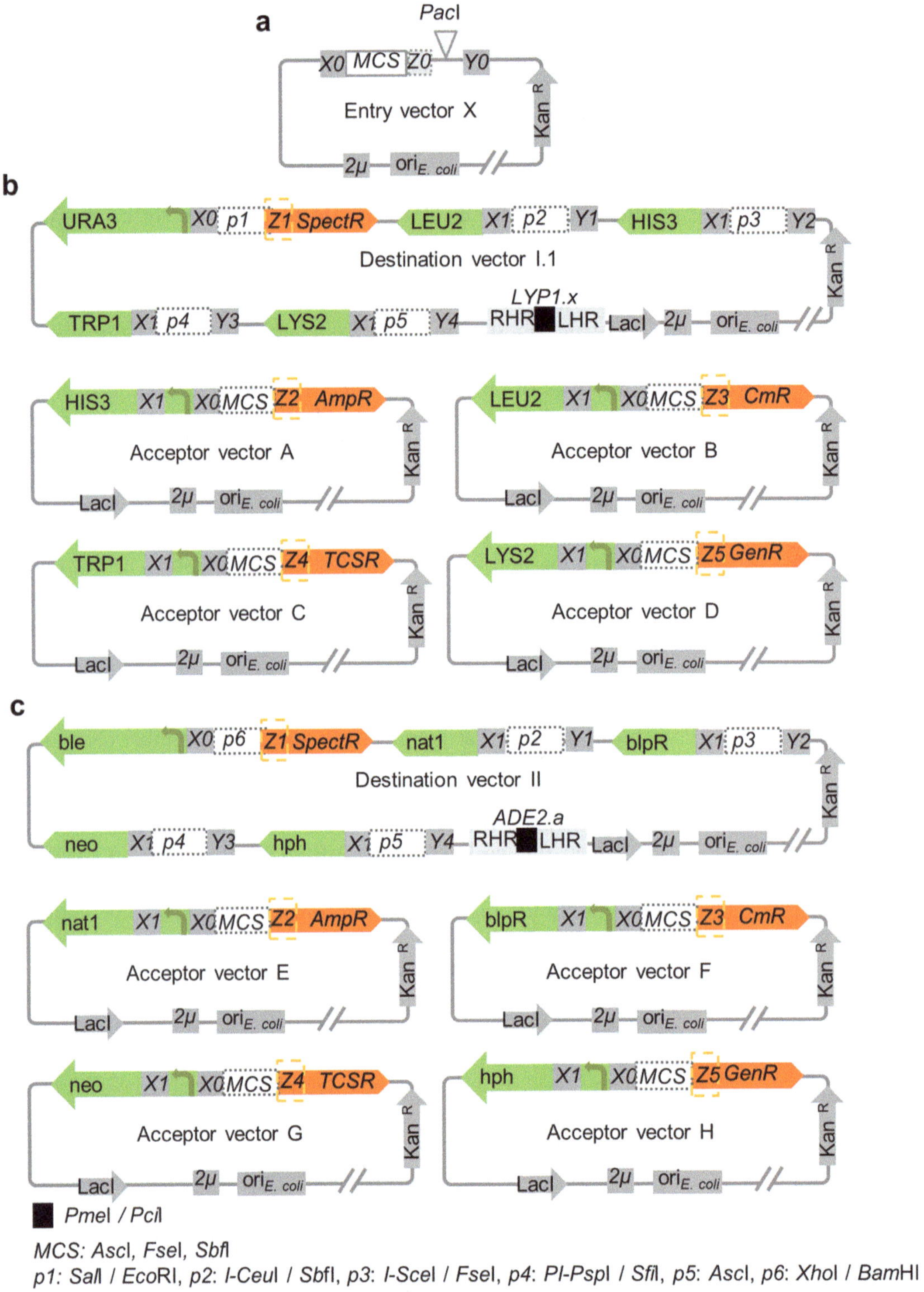

Fig. 2 Design of COMPASS vectors. (**a**) Entry vector X. The vector contains a multicloning site (*MCS*) flanked by *X0* and *Z0* sequences to assemble ATF/BS units (ATF and BS) within the *MCS*, and a PacI site flanked by *Z0* and *Y0* sequences to assemble CDS units. (**b**) Design of Set 1 vectors. Destination vector I.1 has cloning site

modules are assembled into a single vector in a combinatorial manner. COMPASS employs positive selection protocols, which strongly reduces the need for checking individual constructs [21].

At Level 1, by coupling COMPASS with multilocus CRISPR/Cas9-mediated genome modification, it is possible to create libraries of genetically stable yeast isolates with a complexity of thousands to millions of members through just four combinatorial cloning reactions, followed by the decoupled integration of the constructs into ten different sites of the yeast genome. Moreover, the way COMPASS is designed allows for integrating regulator–gene modules located on the Level 2 vectors into the yeast genome [24].

2 Materials

2.1 Media

1. LB: 5 g yeast extract, 10 g Bacto peptone, 15 g Bacto agar (only for making agar plates), 10 g NaCl, pH 7.0, fill to 1 L with ddH_2O.
2. Kanamycin: 1 mL of 50 mg/mL kanamycin stock solution, fill to 1 L with LB.
3. Spectinomycin: 1 mL of 50 mg/mL spectinomycin stock solution, fill to 1 L with LB.

Fig. 2 (continued) p1, flanked by *X0* and *Z1*, whereby a *URA3* marker gene is placed 5′ (left) to *X0*. It also contains sites *p2*, *p3*, *p4*, and *p5* flanked upstream by *Y1*, *Y2*, *Y3*, and *Y4* (right), respectively, and downstream by *X1* (left). The CDSs and terminators of *HIS3*, *LEU2*, *TRP1* and *LYS2* are fused upstream (right) to *X1* of *p2*, *p3*, *p4*, and *p5*, respectively. Acceptor vectors A, B, C, or D have an *MCS* flanked by *X0* and *Z2*, *Z3*, *Z4*, or *Z5*, respectively. In each Acceptor vector, a yeast auxotrophic marker (A: *HIS3*; B: *LEU2*; C: *TRP1; and* D: *LYS2*) is placed downstream (left) of *X0*. (**c**) Design of Set 2 vectors. Destination vector II is similar to Destination vector I.1, except for the following: (1) site *p1* is replaced by site *p6*, (2) the URA3 encoding fragment is replaced by a functional *Ble* dominant marker gene, and (3) the CDSs and terminators of *HIS3*, *LEU2*, *TRP1*, and *LYS2* are replaced by the CDSs and terminators of nat1, blp^R, neo, and hph, respectively. Acceptor vectors E, F, G, and H are similar to Acceptor vectors A, B, C, or D, respectively, except for the yeast auxotrophic markers *HIS3* (A), *LEU2* (B), *TRP1* (C), and *LYS2* (D) that are replaced by yeast dominant selection markers nat1 (E), blp^R (F), neo (G), or hph (H), respectively. All vectors contain the *E. coli pUC19* replication origin, the kanamycin resistance gene *nptII*, and the yeast *2μ* replication origin. *p1*, *p2*, *p3*, *p4*, *p5*, and *p6* encompass SalI/EcoRI, I-CeuI/SbfI, I-SceI/FseI, PI-PspI/SfiI, AscI, and XhoI/BamHI restriction sites, respectively. *X0* represents the sequence 5′ (left) of the *MCS* in Entry vector X. *Z0* represents the last 30 bp of the minimal *CYC1* promoter and overlaps with the HR upstream (left) of the pathway gene. *Y0* represents the sequence 3′ (right) of the PacI cleavage site in Entry vector X. *Z1–Z5* sequences represent the first 30 bp of the CDS of the *E. coli* selection marker genes including $Spect^R$, Amp^R, Cm^R, TCS^R, and Gen^R. *Y1–Y4* sequences provide HRs to the last 30 bp of the terminator of *Amp*^{R}, *Cm*^{R}, *TCS*^{R}, and *Gen*^{R}, respectively. *X1* represents the sequence downstream (left) of the sites *p2–p5* in Destination vectors; it provides an HR to the forward primer amplifying the ATF/BS-CDS module and is required for cloning in Level 2. (The figure was adopted from an original figure published in ref. 24)

4. Ampicillin: 1 mL of 50 mg/mL ampicillin stock solution, fill to 1 L with LB.
5. Chloramphenicol: 1 mL of 25 mg/mL chloramphenicol stock solution, fill to 1 L with LB.
6. Triclosan: 1 mL of 14.5 mg/mL triclosan stock solution, fill to 1 L with LB.
7. Gentamicin: 1 mL of 50 mg/mL gentamicin stock solution, fill to 1 L with LB.
8. SOC medium: 5 g yeast extract, 20 Bacto peptone, 0.5 g NaCl, 20 g glucose, pH 7.0, fill to 1 L with ddH_2O.
9. YPAD medium: 10 g yeast extract, 20 g Bacto peptone, 40 mg adenine hemisulfate, 20 g glucose, 20 g Bacto agar (only for making agar plates), pH 5.8, fill to 1 L with ddH_2O.
10. YPD medium: 10 g yeast extract, 20 g Bacto peptone, 20 g glucose, 20 g Bacto agar (only for making agar plates), pH 5.8, fill to 1 L with ddH_2O.
11. 2× YPAD medium: 10 g yeast extract, 40 g Bacto peptone, 80 mg adenine hemisulfate, 40 g glucose, pH 5.8, fill to 1 L with ddH_2O.
12. SC medium: 6.7 g yeast nitrogen base without amino acids and without ammonium sulfate, purchased SC medium, 20 g glucose, 20 g Bacto agar (only for making agar plates), pH 5.8, fill to 1 L with ddH_2O.
13. SC-thialysine medium: SC-Lys, filter-sterilized (0.2 μm filter) 50 mg/L thialysine hydrochloride.
14. YNB medium: 1.7 g yeast nitrogen base with amino acids and without ammonium sulfate, 5 g NH_3SO_4, 20 g glucose, 20 g bacto agar (only for making agar plates), pH 5.8, fill to 1 L with ddH_2O.
15. G418 medium: 1 mL of 200 mg/mL, fill to 1 L with YNB.
16. Hygromycin B medium: 1 mL of 200 mg/mL, fill to 1 L with YNB.
17. Phleomycin medium: 1 mL of 20 mg/mL, fill to 1 L with YNB.
18. Bleomycin medium: 1 mL of 100 mg/mL, fill to 1 L with YNB.
19. Nourseothricin medium: 1 mL of 100 mg/mL, fill to 1 L with YNB.

2.2 Plasmids

1. Nine plant-derived ATF expression plasmids: pFM003B-NLS-JUB1-GAL4AD, pFM003B-NLS-DBD_{JUB1}-GAL4AD, pGN004B-NLS-JUB1-EDLLAD-EDLLAD, pFM003B-NLS-ATAF1-GAL4AD, pFM004B-NLS-GAL4AD-ATAF1, pFM004B-NLS-GAL4AD-RAV1, pFM004B-NLS-GAL4AD-

Table 1
Expression and reporter plasmids

Number	*Expression plasmid*	*Reference*
1	pFM003B-NLS-JUB1-GAL4AD	[23]
2	pFM003B-NLS-ATAF1-GAL4AD	[23]
3	pFM004B-NLS-GAL4AD-ATAF1	[23]
4	pGN004B-NLS-JUB1-EDLLAD-EDLLAD	[23]
5	pFM004B-NLS-GAL4AD-ANAC102	[23]
6	pFM004B-NLS-GAL4AD-GRF7	[23]
7	pFM0031-ANAC102-NLS-VP64AD	[23]
8	pFM003B-NLS-DBD_{JUB1}-GAL4AD	[23]
9	pFM004B-NLS-GAL4AD-RAV1	[23]
Number	*Reporter plasmid*	*Reference*
1	pGN005B-JUB1_2X	[23]
2	pGN005B-JUB1_4X	[23]
3	pGN005B-ATAF1_2X	[23]
4	pGN005B-ANAC102_4X	[23]
5	pGN005B-RAV1_4X	[23]
6	pGN005B-GRF7_4X	[23]

GRF7, pFM004B-NLS-GAL4AD-ANAC102, and pFM0031-ANAC102-NLS-VP64AD (*see* Table 1).

2. Six BS reporter plasmids: pGN005B-JUB1-2X, pGN005B-JUB1-4X, pGN005B-ATAF1-2X, pGN005B-ANAC102-4X, pGN005B-GRF7-4X, and pGN005B-RAV1-4X (*see* Table 1).
3. Eleven COMPASSvectors: Entry vector X (Addgene #126939), Acceptor vector A–H (Addgene ID; A: #126924, B: #126930, C: #126931, D: #126932, E: #126933, F: #126934, G: #126935, H: #126936), Destination vector I.1 (Addgene #127542), and II (Addgene #126938) (*see* Fig. 2 and ref. 24).
4. Single-guide RNA (sgRNA) plasmids (*see* Table 2).

2.3 Restriction Enzymes

1. AscI, BamHI, FseI, SalI, EcoRI, I-CeuI, SbfI, I-SceI, PI-PspI, SfiI, XhoI, AsiSI, and PI-SceI. Enzymes are used at a concentration at 1 U/50 μL digestion reaction.

2.4 Kits

1. NEBuilder® HiFi DNA Assembly Master Mix.
2. Column-based DNA cleanup kit.

Table 2
sgRNA-encoding plasmids

Plasmid	Target site	Cas9	Selection marker	Reference
pCOM001	*leu2.a, his3.a, lys2.a, trp1.a*	iCas9	URA3	[24]
pCOM002	*leu2.a, his3.a, lys2.a, trp1.a, ura3.a*	iCas9	neo	[24]
pCOM004	*X-3, XI-2, XII-5*	iCas9	LYS2	[24]
pCOM005	*X-4, XI-3, XII-5*	iCas9	LYS2	[24]
pCOM006	*ADE2, URA3, HIS3*	iCas9	LEU2, LYS2, Hph^R	[24]
pCOM007	*X-3, XI-2, XII-2*	iCas9	LYS2	[24]
pCOM008	*X-2, XI-5, XII-4*	iCas9	LYS2	[24]
pCOM009	*neo.a, ble.a, blpR.a, nat1.a, hph.a*	iCas9	URA3	[24]

3. Plasmid miniprep (e.g., NucleoSpin Plasmid QuickPure kit, Macherey-Nagel).
4. Plasmid maxiprep (e.g., Qiagen large-construct kit, Qiagen).

2.5 Reagents for SLiCE Cloning

1. 10× SLiCE buffer: 500 mM Tris–HCl (pH 7.5 at 25 °C), 100 mM $MgCl_2$, 10 mM ATP, 10 mM DTT, store at −20 °C.
2. PPY SLiCE extract (*see* ref. 25).

2.6 PCRs

1. A thermocycler.
2. 0.2-mL PCR tubes.
3. A NanoDrop spectrophotometer for DNA quantification.
4. 1% and 2% agarose gels, equipment and supplies; DNA ladders.
5. PhusionDNA polymerase (New England Biolabs).
6. PrimeSTAR GXL DNA polymerase (Takara Bio).

2.7 *Escherichia coli* Transformation

1. NEB 5-alpha, NEB 10-beta, and NEB 5-alpha electrocompetent *E. coli* cells.
2. 37 °C incubator and shaking incubator.

2.8 Yeast Transformation

1. PEG MW 3350 (50% w/v): 50 g PEG 3350, up to 100 mL ddH_2O.
2. TE: 10 mM Tris–HCl, 1 mM Na_2EDTA, pH 8.0.
3. 2 mg/mL single-stranded carrier DNA: 200 mg of salmon sperm DNA, 100 mL TE.
4. 1 M lithium acetate: 10.2 g lithium acetate dehydrate, fill to 100 mL with ddH_2O.
5. 1.5 mL Eppendorf tubes.
6. Heating orbital shaker.

2.9 Extraction of Plasmid Libraries from Yeast

1. 10 mg/mL zymolyase solution: 200 mg Zymolyase 20T, 1 mL Tris–HCl (pH 7.5), 10 mL 50% (w/v) glycerol, fill to 9 mL with ddH_2O.
2. 1 M sorbitol solution: 182 g sorbitol, fill to 1 L with ddH_2O.
3. SPE solution: 182 g sorbitol, 2.08 g $Na_2HPO_4 \cdot 7H_2O$, 0.32 g $NaH_2PO_4 \cdot H_2O$, 20 mL 0.5 M EDTA (pH 7.5), fill to 1 L with ddH_2O.
4. Lysis buffer: 10 g SDS, 1 M Tris–HCl (pH 7.5), 40 mL 0.5 M EDTA (pH 8), pH 12.8, fill to 1 L with ddH_2O.
5. 100 μg/mL RNase A: add 1 vial 100 mg/mL RNase A stock per bottle Buffer P1 (both included in the Qiagen large-construct kit).
6. 100 mM ATP: 1.1 g ATP, 6 mL 1 M Tris–HCl (pH 8), fill to 20 mL with ddH_2O.
7. 10× TAE buffer: 48.4 g Tris base, 11.42 g glacial acetic acid, 20 mL 0.5 M EDTA (pH 8.0), fill to 1 L with ddH_2O.
8. 10× TAE–3 M sodium acetate: 40.8 g sodium acetate trihydrate, 80 mL 10× TAE buffer, pH 8.0 with glacial acetic acid, fill to 100 mL with ddH_2O.
9. TE: 1 mL 1 M Tris–HCl (pH 8), 0.2 mL 0.5 M EDTA (pH 8), pH 8, fill to 100 mL with ddH_2O.
10. Cooling centrifuge at 4 °C.
11. 1.5 mL Eppendorf tubes.

2.10 Inducing Medium

1. SC inducing medium: 1 L SC medium without glucose, 20 mM IPTG, 20 g galactose, 10 g raffinose.
2. YPDA inducing medium: 1 L YPDA medium without glucose, 20 mM IPTG, 20 g galactose.
3. 1.5 mL Eppendorf tubes.

2.11 Other Solutions and Buffers

1. TBE: 10.8 g Tris base, 5.5 g boric acid, 4 mL 0.5 M EDTA, pH 8.3, fill to 1 L with ddH_2O.
2. 24 mM β-mercaptoethanol.
3. 99.5% (v/v) ethanol.
4. Zymolyase 20T.
5. Isopropanol.
6. 0.1 mg/mL ethidium bromide: 0.1 mg ethidium bromide, fill to 1 mL with ddH_2O.

3 Methods

Digested plasmids and PCR-amplified DNA parts required to be gel-purified, unless mentioned otherwise. Multiplex PCR-amplified parts are not gel-purified. All primer sequences are listed in Table 3.

3.1 Level 0: Assembly of ATFs and BSs in Entry Vector X

ATFs and BSs are assembled in Entry vector X to generate ATF/BS (regulator) units (Fig. 3a) in 1 week through the following steps:

1. Mix plant-derived artificial transcription factor (ATF) expression plasmids (Table 1) [23] in equal molar ratio. Amplify DNA fragments containing the IPTG-inducible *GAL1* promoter, the CDSs of plant derived ATFs, and the *CYC1* terminator ($Pro_{mGAL1\text{-}LacI}$-ATF-Ter_{CYC1}) by multiplex-PCR using primers ATF_for and ATF_rev (*see* **Note 1**).
   (a) Primer ATF_for overlaps with the *X0* sequence, sequence 5′ (left) of the *MCS* in Entry vector X.
   (b) Primer ATF_rev overlaps with primer BS_for.
2. Mix the corresponding binding site (BS) reporter plasmids (Table 1) [23] in equal molar ratios. Amplify DNA fragments containing one, two, or four copies of the BSs upstream of the minimal *CYC1* promoter ($Pro_{CYC1min\text{-}BS}$) by multiplex-PCR using primers BS_for and BS_rev (*see* **Note 1**).
   (a) Primer BS-rev overlaps with the *Z0* sequence, the last 30 bp of the minimal *CYC1* promoter of Entry vector X.
3. Digest 1 μg of Entry vector X with SbfI, AscI, and/or FseI.
4. Mix PCR-amplified plant-derived ATF(s), the corresponding BS(s), and digested Entry vector X at a molar ratio of 2:10:1 in a single cloning reaction tube and perform in vitro cloning (e.g., SLiCE [25], Gibson assembly [26], NEBuilder HiFi DNA assembly (New England Biolabs)).
5. Transform the reaction cocktail into *E. coli* cells followed by plating onto LB agar medium containing kanamycin.
6. To identify the particular combinations of ATFs and BSs, check colonies by using primers EXSEQ-for and EXSEQ-rev and primers PROSEQ-for and PROSEQ-rev, respectively.
7. Inoculate appropriate colonies into LB liquid medium containing kanamycin for selection, confirm the particular combinations of ATFs and BSs by sequencing.

3.2 Level 0: Assembly of CDS Units in Entry Vector X

The CDS, a yeast terminator and a promoter of an *E. coli* selection marker are assembled in PacI-digested Entry vector X to generate CDS (gene) units (Fig. 3b) in 1 week through the following steps:

Table 3
Primer sequences

Name	Sequence (5′–3′)	Reference
ATF-for	ACAATCTGCTCTGATGCCGCATAGTTAAGCGAAGTACGGA TTAGAAGCCGCCGAGCGGG	[24]
ATF-rev	GAATTCCTGTAGAGACCACATCATCCACGGGCCC TCACCAAGCGCCCTCACCAAGCTCTTAAAACGGGAA	[24]
BS-for	CCGTGGATGATGTGGTCTCTACAGGAATTC	[24]
BS-rev	TATTAATTTAGTGTGTGTATTTGTGTTTGT	[24]
EXSEQ-for	CTCAGTACAATCTGCTCTGA	[24]
EXSEQ-rev	GCCCTCACCAAGCGCCCTCA	[24]
PROSEQ-for	TGAGGGCGCTTGGTGAGGGC	[24]
PROSEQ-rev	TATTAATTTAGTGTGTGTAT	[24]
Z0_for	CTATAGACACACAAACACAAATACACACACTAAATTAATA	[24]
Y0_rev	TATTAATTTAGTGTGTGTATTTGTGTTTGT	[24]
X0_for	ACAATCTGCTCTGATGCCGCATAGTTAAGC	[24]
Z0_rev	TATTAATTTAGTGTGTGTATTTGTGTTTGT	[24]
Z0_for	CTATAGACACACAAACACAAATACACACACTAAATTAATA	[24]
Z1_rev	AGGCATAGACTGTACCCCAAAAAAACAGTCTTTGAAAGATGATACTC TTT	[24]
Z2_rev	AGGGAATAAGGGCGACACGGAAATGTTGAATACTCATCGAAACGAT	[24]
Z3_rev	ATACCGCATCAGGAAATTGTAAACGTTAATAAATCATACCTGACC TCCAT	[24]
Z4_rev	GCTTACCGGAAAGAAATAAATTTTTTCAAAAACCCAT	[24]
Z5_rev	GTTTTAGGGCGACTGCCCTGCTGCGTAACATCGTTGCTGCTGCG TAACAT	[24]
X1_for	GACGTGATTAAGCACACAAAGGCAGCTTGGAGT	[24]
Y1_rev	GGGTGGGCGAAGAACTCCAGCATGAGATCC	[24]
Y2_rev	GGCAACCGAGCGTTCTGAACAAATCCAGAT	[24]
Y3_rev	GAAGATCCTTTGATCTTTTCTACGGGGTCT	[24]
Y4_rev	GAGAGCGTTCACCGACAAACAACAGACC	[24]

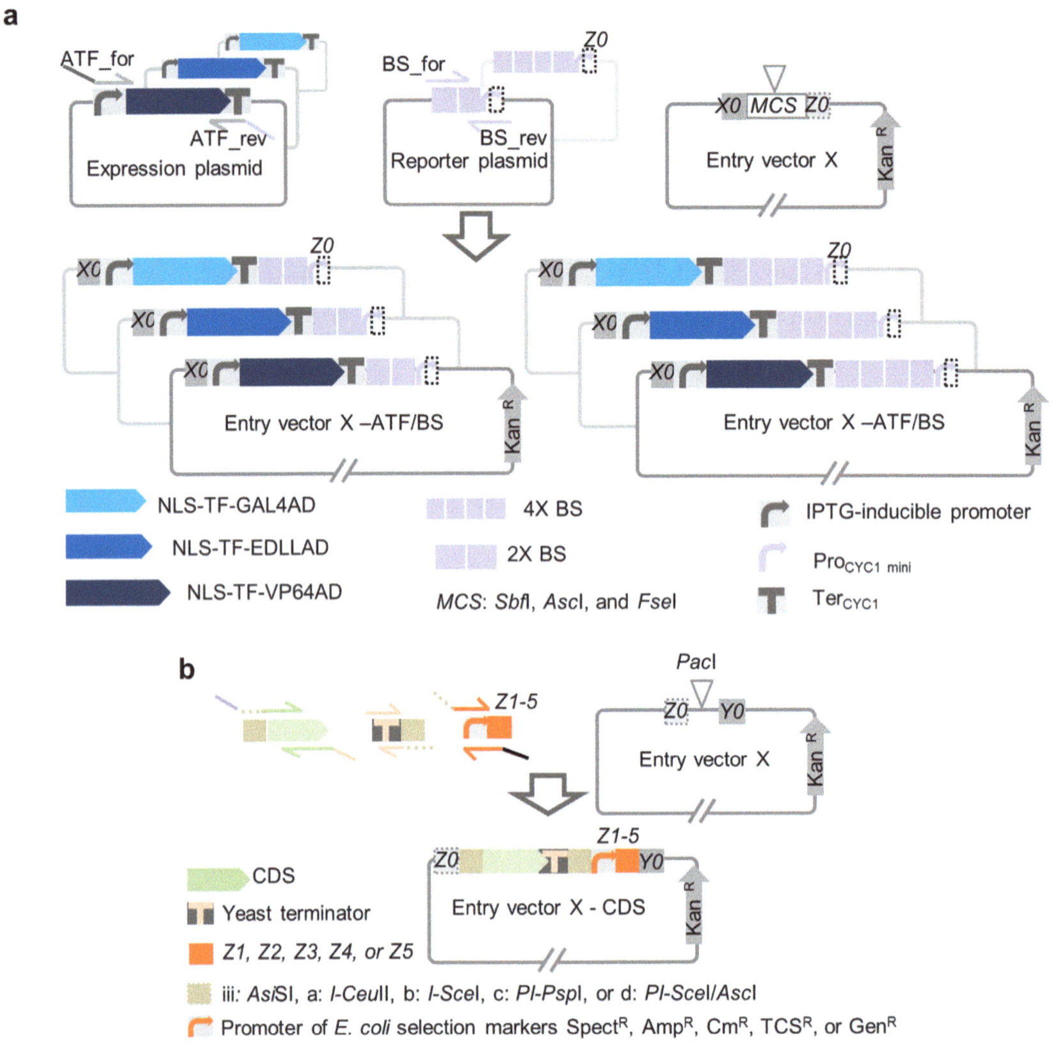

Fig. 3 Construction of assembly units. (**a**) Combinatorial assembly of ATF and corresponding BS units in Entry vector X. Tripartite fragments, harboring an inducible promoter, the ATF (combination of NLS, TF, and various ADs), and the yeast terminator are PCR-amplified from expression plasmids using primers ATF_for and ATF_rev. Synthetic promoters, harboring one to multiple copies of BSs (corresponding to the ATF) upstream of the minimal *CYC1* promoter, are PCR-amplified from reporter plasmids using primers BS_for and BS_rev. ATF and BS amplicons are mixed in 1:10 molar ratio and cloned into FseI/AscI-linearized Entry vector X. The diagram shows six possible outcomes, if a TF is fused to a GAL4-, EDLL-, or VP64-AD, and two synthetic promoters harboring two or four copies of the corresponding BS. (**b**) Cloning of CDS units into Entry vector X. CDSs with rare RE recognition sites (iii, a, b, c, d) in their 5′ regions, a yeast terminator carrying similar RE sites in the 3′ regions, promoters, and (downstream of the promoters) the first 30 bp of the CDS of the *E. coli* selection marker genes (*Z1*, *Z2*, *Z3*, *Z4*, or *Z5*) are cloned into PacI-digested Entry vector X. The *Z* and RE sites are defined based on Level 1 vectors (Destination vectors I/II: *Z1* and iii; Acceptor vectors A/E: *Z2* and a; B/F: *Z3* and b; C/G: *Z4* and c; and D/H: *Z5* and d). *X0* represents the sequence at the 5′ end (left) of the *MCS* in Entry vector X. *Z0* represents the sequence of the last 30 bp of the minimal *CYC1* promoter which is placed downstream (right) of the *MCS* in Entry vector X. *Z1–Z5* represent the first 30 bp of the CDS of the *E. coli* selection marker gene including Spect^R, Amp^R, Cm^R, TCS^R, and Gen^R. *Z0* represents the sequence at the 3′ end (right) of the PacI cleavage site in Entry vector X. (The figure was adopted from an original figure published in ref. 24)

1. Amplify coding sequence of the gene of interest by PCR (*see* **Note 1**).
   (a) Forward primer: The 5′ (left) region of linearized vector (called *Z0*) overlaps with the 5′ (left) homology region (HR) of the primer. A rare restriction enzyme (RE) cleavage site (Destination vectors I/II: AsiSI; Acceptor vectors A/E: I-CeuI, B/F: I-SceI, C/G: PI-PspI, and D/H: PI-SceI/AscI) is required to be introduced upstream the gene's translation start codon. This is an optional possibility (*see* **Note 2**).
   (b) Reverse primer: The 3′ (right) HR of the primer overlaps with the 5′ sequence of the yeast terminator.
2. Amplify a yeast terminator of interest by PCR [28] (*see* **Note 1**).
   (a) Forward primer: The primer overlaps with the 5′ region of the yeast terminator.
   (b) Reverse primer: The same rare RE recognition site as above is introduced at the 3′ end of the terminator. This is an optional possibility (*see* **Note 2**). The 5′ (left) HR of the primer overlaps with the yeast terminator.
3. Amplify promoter of an *E. coli* selection marker by PCR [28, 29] (*see* **Note 1**).
   (a) Forward primer: The 5′ (left) HR of the primer overlaps with the revers primer amplifying the yeast terminator and is defined based on the upstream region of the chosen *E. coli* selection marker.
   (b) Reverse primer: The 3′ (right) HR of the primer overlaps with the *Y0* sequence of the vector and introduces *Z1*, *Z2*, *Z3*, *Z4*, or *Z5* sequences, the first 30 bp of the *E. coli*selection markers CDSs (see Level 1).
4. Digest Entry vector X with PacI.
5. Mix PCR-amplified CDS, yeast terminator and promoter of *E. coli* selection marker, and digested Entry vector X in a single cloning reaction tube for in vitro overlap-based cloning (e.g., SLiCE [25], Gibson assembly [26], NEBuilder HiFi DNA assembly (New England Biolabs)).
6. Transform the cloning reaction into *E. coli* cells followed by plating onto LB agar medium containing kanamycin.
7. Perform colony-PCR using primers Z0_for/Y0_rev.
8. Inoculate appropriate colonies into LB liquid medium containing kanamycin followed by sequencing to identify the correct assemblies.

Up to five gene units can be combinatorially assembled in Level 1 vectors (*see* **Note 3**).

3.3 Level 1: Combinatorial Assembly of Regulator and Gene Units in Regulator–Gene Modules

Regulator and gene units are combinatorially assembled in Level 1 vectors to generate a library of regulator–gene modules (Fig. 4) in 1 week through the following steps:

1. Mix an equal molar ratio of nine Entry vectors X-regulators to perform multiplex PCR-amplification of fragments containing IPTG inducible promoter, ATF, *CYC1* terminator and the *CYC1* minimal promoter harboring ATF biding site (regulator units) using primers X0_for/Z0_rev (*see* **Note 1**).
   (a) Primer X0_for is defined based on the *X0* sequence in the Level 1 vectors.
   (b) Primer Z0_rev is defined based on the last 30 bp of the *CYC1* minimal promoter.
2. Prepare an equimolar ratio of the five Entry vectors X-CDS to perform multiplex PCR-amplification of five CDSs, yeast terminators and promoters of *E. coli*selection markers (five gene units) using primers Z0_for, Z1_rev, Z2_rev, Z3_rev, Z4_rev, and Z5_rev at a molar ratio of 5:1:1:1:1:1 (*see* **Note 1**).
   (a) Primer Z0_for overlaps with primer Z0_rev.
   (b) Primers Z1_rev, Z2_rev, Z3_rev, Z4_rev, and Z5_rev are defined based on the *Z1* (Destination vector I.1), *Z2* (Acceptor vectors A), *Z3* (Acceptor vectors B), *Z4* (Acceptor vectors C), and *Z5* (Acceptor vectors D), and represent the first 30 bp of the CDSs of the *E. coli*selection marker genes spectinomycin ($Spect^R$), ampicillin (Amp^R), chloramphenicol (Cm^R), triclosan (TCS^R), or gentamicin (Gen^R), respectively.
3. Digest Destination vector I.1 with the SalI and EcoRI. Acceptor vectors A, B, C, and D are digested with FseI and AscI. Digested plasmids are mixed in equimolar ratio

   PCR-amplified (nine) ATF/BS units, (five) CDS units, and (five) digested vectors are mixed at a molar ratio of 2:2:1 in a single reaction tube for NEBuilder HiFi DNA assembly.
4. Transform the cloning reaction cocktail into *E. coli* cells (*see* **Note 4**).
5. Plate each one-fifth of the *E. coli* cells onto LB agar media containing either spectinomycin, ampicillin, chloramphenicol, triclosan, or gentamicin to select for Destination vectors I-ATF/BS-CDS1, and Acceptor vectors A-ATF/BS-CDS2, B-ATF/BS-CDS3, C-ATF/BS-CDS4, or D-ATF/BS-CDS5.
6. Check single colonies using primers ATF-for and BS-rev.
7. Inoculate appropriate colonies into LB liquid medium containing proper antibiotics (*see* **step 5**) followed by sequencing to detect the regulator sequences upstream of CDS1 to CDS5.

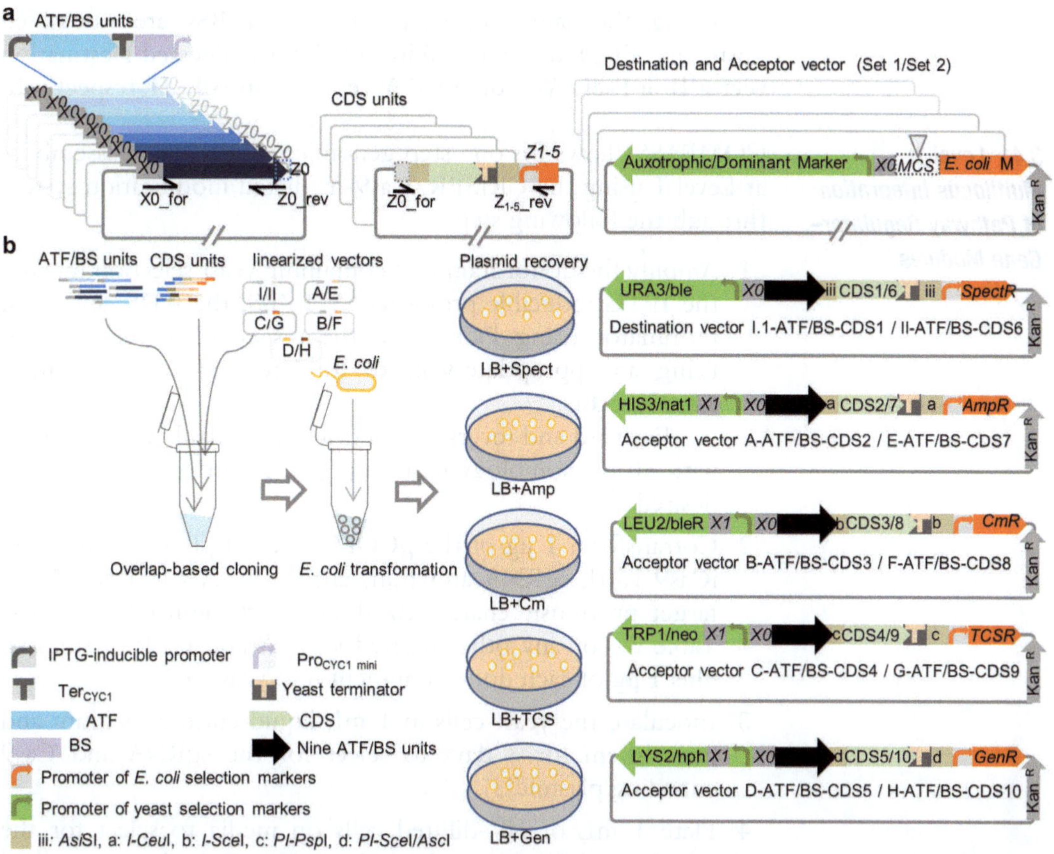

Fig. 4 Combinatorial assembly of regulators and gene units in the modules. (**a**) Combinatorial assembly of regulator units upstream of gene units. Nine regulator units are PCR-amplified from corresponding Entry vectors X using primers X0_for and Z0_rev. Five gene units are PCR-amplified from corresponding Entry vectors X using primers Z0_for, Z1_rev, Z2_rev, Z3_rev Z4_rev, and Z5_rev mixed at a molar ratio of 5:1:1:1:1:1. The *X0* sequence of the linearized Level 1 vector overlaps with the 5′ regions of the regulator units, and the 3′ regions of the regulator units, called *Z0*, overlap with the 5′ regions of the forward primers amplifying the CDS units. The 3′ regions of the CDS units overlap with *Z1–Z5* of the five linearized vectors. Five Set 1 or Set 2 vectors of Level 1 are digested (SalI/EcoRI-digested Destination vector I.1 or XhoI/BamHI-digested Destination vector II, FseI/AscI-digested Acceptor vectors A or E, B or F, C or G, and D or H). (**b**) The COMPASS workflow to generate regulator–gene modules in either Set 1 or Set 2 vectors. PCR-amplified ATF/BS unites (primers X0_for and Z0_rev, on Entry vectors X-ATF/BS), five CDS units (primers Z0_for, Z1_rev, Z2_rev, Z3_rev, Z4_rev, and Z5_rev at a molar ratio of 5:1:1:1:1:1, on mixed Entry vectors X-CDS), and five linearized Destination vectors I/II and Acceptor vectors A/E, B/F, C/G, and D/H are mixed at a molar ratio of 2:2:1 to perform in vitro overlap-based cloning in a single tube to generate different ATF/BS-CDS modules in the five vectors by providing the missing promoter sequences of the *E. coli* selection markers within the assembly units. Libraries of Destination vectors I-CDS1/II-CDS6, Acceptor vectors A-CDS2/E-CDS7, B-CDS3/F-CDS8, C-CDS4/G-CDS9, or D-CDS5/H-CDS10 are generated. X0_for overlaps with *X0*, Z0_rev overlaps with Z0_for, while primers Z1_rev–Z5_rev overlap with *Z1–Z5* sequences of the linearized vector. For simplicity, IPTG-inducible promoters and terminators are not included in the figure. *X0* represents the sequence 5′ (left) of the *MCS* in Entry vector X. *X1* represents the sequence upstream (left) of the sites *p2–p5* in Destination vectors which is required for cloning in Level 2. *Z0* represents the last 30 bp of the minimal *CYC1* promoter and overlaps with the HR upstream (left) of the pathway gene. *Z1–Z5* sequences represent the first 30 bp of the CDS of the *E. coli* selection marker genes including Spect^R, Amp^R, Cm^R, TCS^R, and Gen^R. (The figure was adopted from an original figure published in ref. 24)

Using the same approach, ATFs and BSs are assembled upstream of CDS6 to CDS10 in SalI/EcoRI-digested Destination vector II or FseI/AscI-digested Acceptor vectors E–H, respectively.

3.4 Level 1: Multilocus Integration of Pathway Regulator–Gene Modules

COMPASS allows for one-step genomic integration of assemblies at Level 1 using the CRISPR/Cas9-mediated modification system through the following steps:

1. Amplify the donor fragment containing yeast selection marker, the IPTG-inducible promoter, ATF/BS, the CDS, and yeast terminator using PCR from the respective Level 1 vector using an appropriate pair of forward and reverse primers (*see* **Note 1**).

   Forward and reverse primers allow for flanking the donors with 50 bp homology regions (HRs) for integration into the desired loci.
2. Cotransform 1 μg of the pCOM series of plasmids encoding iCas9 [30], sgRNA and trans-encoded RNA (tracrRNA) to target previously characterized sites with high efficiency (*see* Table 3), or any other desired sgRNA and Cas9 plasmid(s), plus 1 μg of each donor fragment into yeast cells.
3. Inoculate the yeast cells in 1 mL liquid culture medium and grow them for 3 days to select for the sgRNA and Cas9 encoding plasmid(s).
4. Plate 1 mL of 10^2-diluted cells on media to select for the sgRNA and Cas9 encoding plasmid(s). Colonies are washed off the plate(s).
5. Plate the yeast cells on selective medium to screen for colonies with integrated selection markers. Colonies are washed off.
6. Perform multiplex-PCR on single colonies using primers amplifying regulatory units upstream of each CDS (*see* **Note 5**).
7. Plate the cells on nonselective induction plates to eliminate iCas9 and sgRNA encoding plasmid(s).
8. To disrupt the coding sequences of yeast selection markers within the regulator–gene modules, cotransform 1 μg of the pCOM001, pCOM002 and pCOM009 plasmids encoding iCas9, tracrRNA [29] and sgRNAs (pCOM001: *leu2.a*, *his3.a*, *lys2.a*, and *trp1.a*; pCOM002: *leu2.a*, *his3.a*, *lys2.a*, *trp1.a*, and *ura3.a*; pCOM009: *ble.a*, *nat1.a*, *blpR.a*, *neo.a*, and *hph.a*) (*see* Table 3) into yeast cells. Inoculate the yeast cells in 1 mL liquid culture medium for 3 days followed by plating 1 mL of 10^2-fold diluted cell culture on media that select for the presence of pCOM001 (SC-Ura), pCOM002 (YNB containing G418), and pCOM009 (SC-Ura).
9. Streak out each single colony onto different selection plates (i.e., SC-Leu, SC-His, SC-Lys, SC-Trp, SC-Ura, YPDA

containing G418, hygromycin B, phleomycin, bleomycin, nourseothricin medium). After 2 days, cells of a colony that did not grow on the selective plates are streaked out on nonselective YPDA agar medium to eliminate pCOM001, pCOM002, and pCOM009 plasmid(s).

3.5 Level 2: Combinatorial Assembly of Regulator–Gene Modules in Destination Vector I.1

Up to five regulator–gene modules are combinatorially assembled in Destination vector I.1 (Fig. 5) in 1 week through the following steps:

1. Prepare and digest an equimolar ratio of Destination vectors I.1-ATF/BS-CDS1 with I-CeuI or/and SbfI.
2. Mix an equal molar ratio of the library of Acceptor vectors A-ATF/BS-CDS2 to perform multiplex PCR amplification of ATF/BS-CDS2 modules, containing *HIS3* promoter, *ATF/BS*, *CDS 2*, a yeast terminator, and promoter, ORF and terminator of Amp^R, using primers X0_for and Y1_rev (*see* **Notes 1** and **5**).
3. Assemble the PCR-amplified fragments in the library of Destination vectors I.1-ATF/BS-CDS1 using transformation associated recombination (TAR). Thereby, a library of Destination vectors I.1-ATF/BS-CDS1-ATF/BS-CDS2 is generated.
4. Plate the transformed cells on SC-Ura/-His medium.
5. Scrape the yeast cells from the plates after 4 days to extract the plasmid library according to the method described by Noskov et al. [20].
6. Transform the plasmid library extracted from yeast into *E. coli* cells by electroporation (*see* **Notes 4** and **6**), followed by growth on LB agar plates containing ampicillin.
7. Scrape the transformed *E. coli* cells from the plates to recover the plasmid library.
8. Use the recovered library of Acceptor vectors B-ATF/BS-CDS3 for multiplex PCR amplification of ATF/BS-CDS3 modules containing *LEU2* promoter, *ATF/BS*, *CDS 3*, a yeast terminator, and promoter, ORF and terminator of Cm^R using primers X0_for and Y2_rev (*see* **Note 1**).
9. Assemble the PCR-amplified fragments in the library of I-SceI or/and FseI-digested Destination vectors-ATF/BS-CDS1-ATF/BS-CDS2 using TAR. Thereby, a library of Destination vectors I.1-ATF/BS-CDS1-ATF/BS-CDS2-ATF/BS-CDS3 is generated.
10. Plate the transformed cells on SC-Ura/-His/-Leu medium.
11. Scrape the cells from the plates after 4 days to extract the plasmid library according to the method described by Noskov et al. [20].

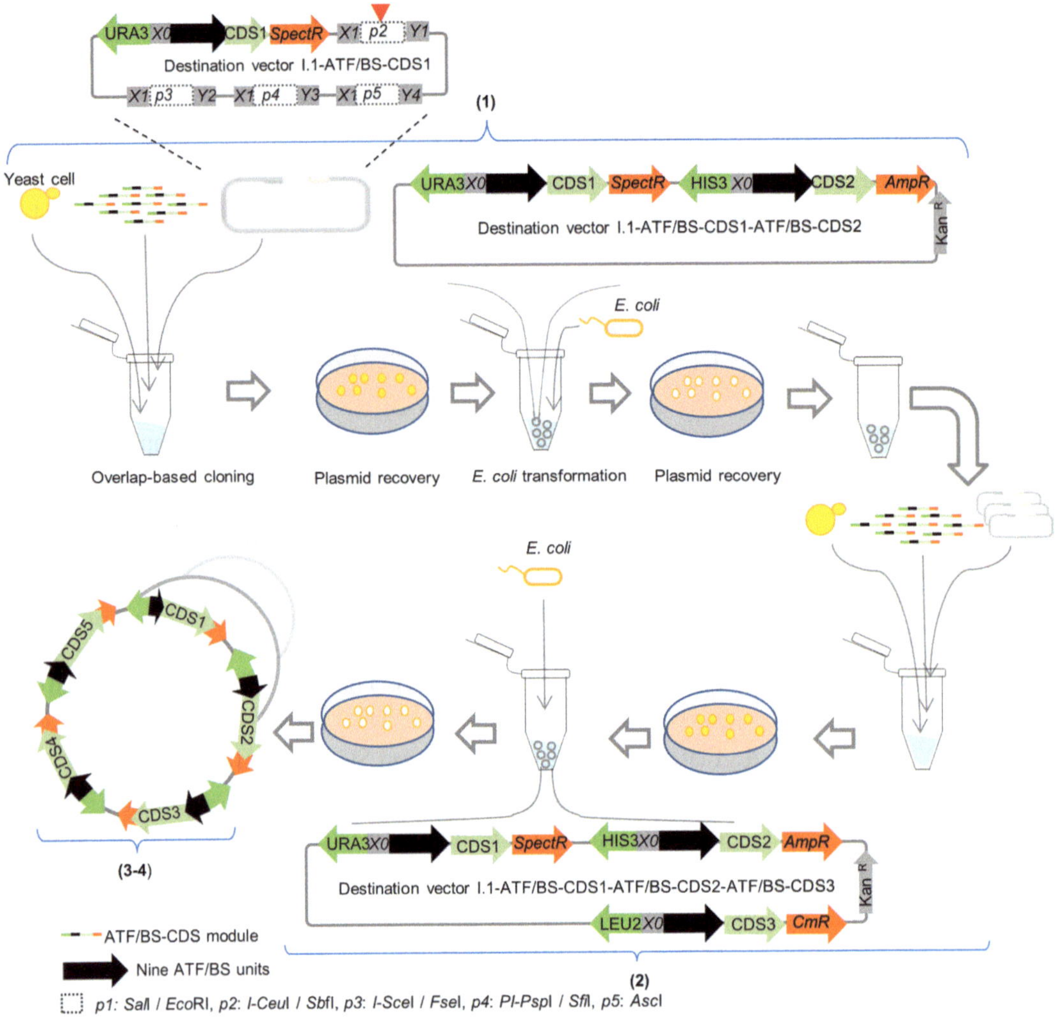

Fig. 5 COMPASS workflow for the combinatorial assembly of regulator–gene modules in Destination vector I.1. Multiplex PCR-amplified ATF/BS-CDS2 modules (*Pro*$_{HIS3}$*-ATF/BS-CDS2-Pro*$_{AmpR}$*-AmpR-Ter*$_{AmpR}$, with primers X1_for and Y1_rev on Acceptor vector A) are assembled *via* TAR in Destination vectors I.1-CDS1 digested at site *p2* (red arrowhead). Successful assemblies lead yeast cells to grow on SC-Ura/-His medium. The colonies are scraped from the plates and the plasmid library is extracted. The pool of all randomized members is then transformed into *E. coli*. The successful assemblies allow cells to grow on LB plates containing ampicillin. Cells are scraped from the plates and the plasmid library is extracted (1). Multiplex PCR-amplified ATF/BS-CDS3 modules (*Pro*$_{LEU2}$*-ATF/BS-CDS3-Pro*$_{CmR}$*-CmR-Ter*$_{CmR}$, with primers X1_for and Y2_rev on Acceptor vector B) modules are assembled in the Destination vectors I.1-CDS1-CDS2 library digested at site *p3*. Yeast cells with successful constructs grow on yeast SC-Ura/-His/-Leu medium. The recovered plasmid library is transformed into *E. coli* cells. The *E. coli* cells with successful constructs grow on LB plates containing chloramphenicol (2). Using the same procedure, the libraries of ATF/BS-CDS4 (*Pro*$_{TRP1}$*-ATF/BS-CDS4-Pro*$_{TCSR}$*-TCSR-Ter*$_{TCSR}$, with primers X1_for and Y3_rev on Acceptor vector C) and ATF/BS-CDS5 (*Pro*$_{LYS2}$*-ATF/BS-CDS5-Pro*$_{GenR}$*-GenR-Ter*$_{GenR}$, with primers X1_for and Y4_rev on Acceptor vector D) modules are assembled into sites *p4* and *p5*, respectively (3–4). For simplicity, the IPTG-inducible promoters and terminators are not included in the figure. *Y1–Y4* overlap with the last 30 bp of terminators of the Amp^R, Cm^R, TCS^R, and Gen^R genes, respectively. *X0* represents the sequence 5′ (left) of the *MCS* in Entry vector X. *X1*

12. Transform the extracted plasmid library into *E. coli* cells by electroporation. The transformed cells are then grown on LB agar plates containing chloramphenicol.
13. Recover the plasmid library from the *E. coli* cells scraped from the plates.
14. Amplify ATF/BS-CDS4 modules (*TRP1* promoter, ATF/BS-CDS4, yeast terminator, promoter, ORF and terminator of TCS^R) using multiplex PCR amplification (primers X0_for and Y3_rev, on the library of Acceptor vectors C-ATF/BS-CDS4) (*see* **Note 1**).
15. Assemble the PCR-amplified fragments in the library of PI-PspI or/and SfiII-digested Destination vectors-ATF/BS-CDS1-ATF/BS-CDS2-ATF/BS-CDS3 using TAR. Therefore, a library of Destination vectors I.1-ATF/BS-CDS1-ATF/BS-CDS2-ATF/BS-CDS3-ATF/BS-CDS4 is generated.
16. Plate the transformed cells on SC-Ura/-His/-Leu/-Trp medium.
17. Scrape the yeast cells from the plates after 4 days to extract the plasmid library according to the method described by Noskov et al. [20].
18. Transform the extracted plasmid library into *E. coli* by electroporation. The transformed cells are then grown on LB agar plates containing triclosan.
19. Recover the plasmid library from the *E. coli* scraped from the plates.
20. Use the library of Acceptor vectors D-ATF/BS-CDS5 for multiplex PCR amplification of ATF/BS-CDS5 modules containing *LYS2* promoter, ATF/BS, CDS 5, a yeast terminator, and promoter, ORF and terminator of Gen^R using primers X0_for and Y4_rev.
21. Assemble the PCR-amplified fragments in the library of Destination vectors-ATF/BS-CDS1-ATF/BS-CDS2-ATF/BS-CDS3-ATF/BS-CDS4 digested with AscI (without gel-purification) using TAR. Thereby, a library of Destination vectors I.1-ATF/BS-CDS1-ATF/BS-CDS2-ATF/BS-CDS3-ATF/BS-CDS4-ATF/BS-CDS5 is generated.
22. Plate the transformed yeast cells on SC-Ura/-His/-Leu/-Trp/-Lys medium.

Fig. 5 (continued) represents the sequence downstream (left) of the sites *p2–p5* in Destination vectors which is required for cloning at Level 2. *Y1–Y4* represent the last 30 bp of the terminators of the *E. coli* selection marker genes Amp^R, Cm^R, TCS^R, and Gen^R. (The figure was adopted from an original figure published in ref. 24)

23. Scrape the yeast cells from the plates to extract the plasmid library according to the method described by Noskov et al. [20].
24. Transform the extracted plasmid library into *E. coli* by electroporation. The cells are then grown on LB agar plates containing gentamicin.
25. Extract the plasmid library from *E. coli* cells scraped from the plates.

    Sequences *Y1*, *Y2*, *Y3*, and *Y4* represent to the last 30 bp of the terminator sequences of the Amp^R, Cm^R, TCS^R, or Gen^R gene, respectively.

Using the same approach, libraries of ATF/BS-CDS6 to ATF/BS-CDS10 modules are assembled into, respectively, sites *p6*, *p2*, *p3*, *p4*, and *p5* of Destination vector II (Fig. 2c) to generate a library of Destination vectors II-ATF/BS-CDS6-ATF/BS-CDS7-ATF/BS-CDS8-ATF/BS-CDS9-ATF/BS-CDS10 (*see* **Note** 7). While NEBuilder HiFi is preferred for generating constructs at Level 1, TAR is the desired approach to generate constructs at Level 2.

3.6 Level 2: Single-Locus Integration of Destination Vectors

The library of Destination vector I.1 (Fig. 2a, b) is integrated into the *LYP1.x* site (79 bp downstream of the start codon of LYP1) [30] through the following steps:

1. Cotransform 5 μg of the PmeI/PciI-digested Destination vector I.1 library, and 1 μg of pCRCT-array1 (encoding Cas9 and sgRNA allowing for integration of the library into the *LYP1.x* site) [30] into yeast cells.
2. Inoculate yeast cells in 1 mL liquid culture medium for 3 days to select for the presence of the sgRNA and Cas9 encoding plasmid(s).
3. Plate 1 mL of 10^2-diluted cells on media that select for the presence of the sgRNA and Cas9 encoding plasmid(s). Colonies are washed off the plate(s).
4. Plate the yeast cells on selective medium with thialysine (*see* **Notes 8** and **9**) to screen for colonies with integrated selection markers. Colonies are washed off.
5. Perform multiplex PCR on single colonies; primers are designed to amplify ATF/BS regulatory units upstream of each CDS (*see* **Notes 1** and **4**).
6. Plate the yeast cells on nonselective induction plates.
7. For disruption of the auxotrophic selection markers implemented in regulator–gene modules in engineered strains, transform 1 μg of pCOM001 (Fig. 4) or pCOM002 (Fig. 4) into the cells. The yeast cells are inoculated in 1 mL liquid culture medium for 3 days followed by plating 1 mL of 10^2-fold

diluted cell culture on media that select for the presence of pCOM001 (SC-Ura) and pCOM002 (YNB containing G418).

8. When colonies appeared, streak out each single colony onto different selective plates (i.e., SC-Leu, SC-His, SC-Lys, SC-Trp, SC-Ura). After 2 more days, streak out cells of a colony that did not grow on either of the selective plates on nonselective YPDA agar medium to remove the pCOM001 or pCOM002 plasmids.

Using the same approach, the library of Destination vector II and pCRCT-array1 (encoding Cas9 and crRNA allowing for integration of the library into the *ADE2.a* site) [29] are transformed into the yeast cell to integrate Destination vector II into the *ADE2.a* site (154 bp downstream of the start codon of ADE2) (*see* **Note 10**). For the disruption of the dominant selection markers included in Destination vector II, plasmid pCOM009 (Table 2) is used.

4 Notes

1. To PCR-amplify the DNA fragment required for cloning, 50 μL PCR reactions are set up as recommended by the supplier of PhusionDNA polymerase. *See* Tables 4 and 5 for cycling parameters with short and long primer pairs, respectively. To PCR-amplify the DNA donor fragments required for genomic integration, 50-μL PCR reactions are set up as recommended by the supplier of the PrimeSTAR GXL DNA polymerase. *See* Table 6 for cycling parameters with PrimeSTAR GXL DNA polymerase. To colony-PCR the DNA fragment, 5-μL PCR reactions are set up as recommended by the supplier of the *Taq*DNA polymerase. *See* Table 7 for cycling parameters of colony-PCR.
2. We suggest introducing rare restriction enzyme cutting sites through primer sequences upstream of the CDS and downstream of the yeast terminator which allows for replacement of the CDS(s) and yeast terminator(s) with other CDS(s) and terminator(s) at a later stage when alternative COMPASSvectors are needed.
3. Combinatorial assembly of gene units in Level 1 vectors. To generate CDS units, Level 1 vectors can be used as an alternative to Entry vector X (Level 0). The positive selection of truncated plasmid markers is rendered active by providing promoter sequences during the assembly process. Therefore, Destination vector I-CDS1 or II-CDS6, Acceptor vectors A-CDS2 or E-CDS7, B-CDS3 or F-CDS8, C-CDS4 or G-CDS9, and D-CDS5 or H-CDS10 are generated. The CDS, a yeast terminator and a promoter of an *E. coli* selection marker are combinatorially assembled in either Set 1 or Set 2 of

Table 4
Cycling parameters with short primer pair (≤30 bp)

Step	Temperature (°C)	Time
Initial denaturation	98	30 s
35 cycles	98 55 72	10 s 25 s 30 s/kb
Final extension	72	5 min
Storage	4	Hold

Table 5
Cycling parameters with long primer pair (≥30 bp)

Step	Temperature (°C)	Time
Initial denaturation	98	30 s
10 cycles	98 55 72	10 s 25 s 30 s/kb
25 cycles	98 72 72	10 s 25 s 30 s/kb
Final extension	72	5 min
Storage	4	Hold

Table 6
Cycling parameters with PrimeSTAR GXL DNA polymerase

Step	Temperature (°C)	Time
10 cycles	98 60 68	10 s 25 s 1 min/kb
25 cycles	98 68 68	10 s 25 s 1 min/kb
Final extension	68	5 min
Storage	4	Hold

Table 7
Cycling parameters for colony-PCR

Step	Temperature	Time
Initial denaturation	95	30 s
35 cycles	95 55 72	10 s 25 s 1 min/kb
Final extension	72	5 min
Storage	4	Hold

Level 1 vectors to generate CDS (gene) units (Fig. 6) in 1 week through the following steps:

(a) Amplify coding sequences of the five genes of interest by PCR.

Forward primer: The 5′ (left) homology region (HR) of the primer is defined based on the 5′ (left) region of linearized Level 1 vector (called *X0*). Rare RE cleavage sites (Destination vectors I.1 (or II): iii, Acceptor vectors A (or E): a, B (or F): b, C (or G): c, and D (or H): d), compatible to the Level 1 and Level 2 vectors, are introduced before the gene's translation start codon through primer sequences. For optional possibilities (*see* **Note 2**).

Reverse primer: The 3′ (right) HR of the primer is defined based on the 5′ region of the chosen yeast terminator.

(b) Amplify five yeast terminators of interest by PCR [28].

Forward primer: The primer is defined based on the 5′ region of the chosen yeast terminator.

Reverse primer: The same rare RE recognition site as above is introduced at the 3′ end of the terminator. For optional possibilities (*see* **Note 2**).

The 5′ (left) HR of the primer overlaps with the yeast terminator.

(c) Amplify promoters of the five *E. coli* selection markers-$Spect^R$, Amp^R, Cm^R, TCS^R, and Gen^R by PCR [28, 30].

Forward primer: The upstream HR of the promoter of the antibiotic resistance gene is defined by the yeast terminator; the same rare RE recognition site downstream of the terminator is introduced, while the downstream (right) HR of the *E. coli* promoter is defined based on Z sequences (*Z1*: $Spect^R$, *Z2*: Amp^R, *Z3*: Cm^R, *Z4*: TCS^R, or *Z5*: Gen^R).

Reverse primer: The 3′ (right) HR of the primer overlaps with the *Y0* region of the vector and introduces Z

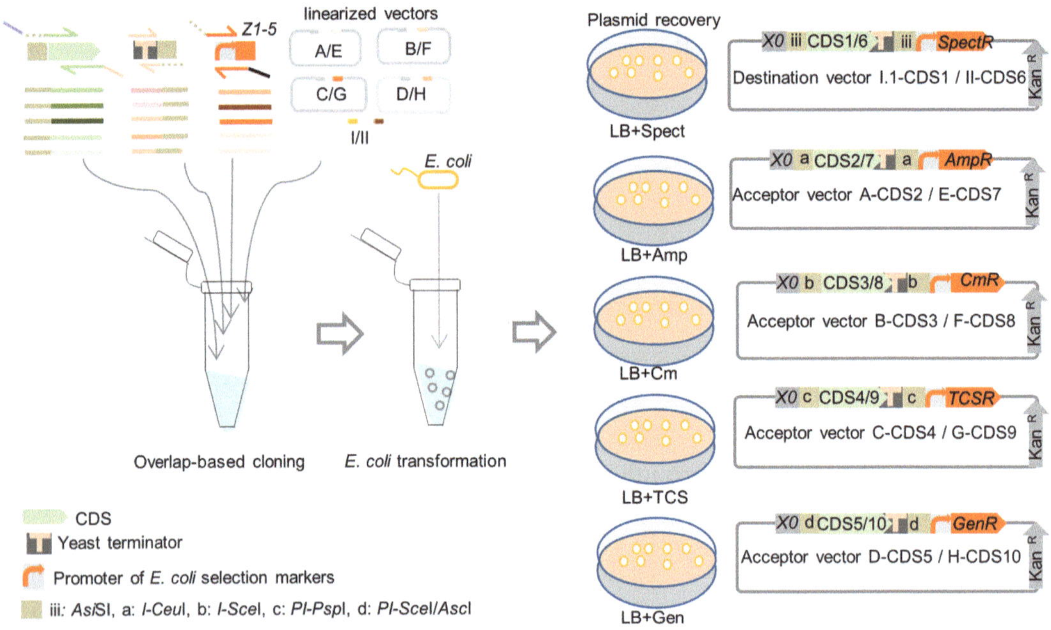

Fig. 6 Combinatorial assembly of gene units in Level 1 vectors. Five Set 1 or Set 2 vectors of Level 1 are digested (SalI/EcoRI-digested Destination vector I.1 or XhoI/BamHI-digested Destination vector II, FseI/AscI-digested Acceptor vectors A or E, B or F, C or G, and D or H). The five linearized vectors, five selected CDSs, five yeast terminators, and five promoters of *E. coli* selection markers (Spect^R, Amp^R, Cm^R, TCS^R, and Gen^R) are mixed in a single tube to perform overlap-based cloning. The *X0* sequence of the Level 1 vectors overlaps with the upstream (left) HR of the CDSs. The terminator of yeast defines the upstream HR of the promoter of the *E. coli* selection marker genes. Rare RE cleavage sites are introduced before the translation start codon of the gene and downstream of the yeast terminator through primer sequences (Destination vectors I.1 or II: *iii*, Acceptor vectors A or E: *a*, B or F: *b*, C or G: *c*, and D or H: *d*). The downstream (right) HR of the promoter of *E. coli* selection marker is defined based on the *Z* sequences (Destination vectors I.1 or II: *Z1*, Acceptor vectors A or E: *Z2*, B or F: *Z3*, C or G: *Z4*, and D or H: *Z5*). By providing promoter sequences of the five *E. coli* selection markers Spect^R, Amp^R, Cm^R, TCS^R, and Gen^R, the truncated markers on the plasmids are activated allowing for positive selection. This allows for screening five Destination vectors I.1-CDS1 or II-CDS6, Acceptor vectors A-CDS2 or E-CDS7, B-CDS3 or F-CDS8, C-CDS4 or G-CDS9, and D-CDS5 or H-CDS10 generated in a single cloning reaction plated after transformation into *E. coli* on LB agar media containing either spectinomycin, ampicillin, chloramphenicol, triclosan, or gentamicin. *X0* represents the sequence 5′ (left) of the *MCS* in Entry vector X. *Z1–Z5* sequences represent the first 30 bp of the CDS of the *E. coli* selection marker genes including Spect^R, Amp^R, Cm^R, TCS^R, and Gen^R. (The figure was adopted from an original figure published in ref. 24)

sequences, representing the first 30 bp of the CDSs of the *E. coli*selection markers encoded by corresponding Level 1 vectors: *Z1* (Destination vector I.1 or II), *Z2* (Acceptor vectors A or E), *Z3* (Acceptor vectors B or F), *Z4* (Acceptor vectors C or G), and *Z5* (Acceptor vectors D or H), representing the first 30 bp of the CDSs of *Spect^R*, *Amp^R*, *Cm^R*, *TCS^R*, or *Gen^R*, respectively.

(d) Mix an equal molar ratio of PCR-amplified (five) CDS units, (five) yeast terminators, (five) promoters of *E. coli* selection markers, and (five) linearized vectors at a molar ratio of 2:2:10:1 to perform NEBuilder HiFi DNA assembly in a single reaction tube.

(e) Transform the cloning reaction cocktail into *E. coli* cells.

(f) Plate each one-fifth of the *E. coli* cells onto LB agar media containing either spectinomycin, ampicillin, chloramphenicol, triclosan, or gentamicin allowing for screening for the presence of Destination vectors I-CDS1, or II-CDS6, Acceptor vectors A-CDS2 or E-CDS7, B-CDS3 or F-CDS8, C-CDS4 or G-CDS9, and D-CDS5 or H-CDS10.

(g) Inoculate appropriate colonies into LB liquid medium containing proper antibiotics (*see* **step 6**) followed by sequencing.

4. We recommend using *E. coli* NEB 5-alpha and NEB 10-beta chemical competent cells for Level 0 and Level 1 assemblies, while *E. coli* NEB 10-beta electrocompetent cells are recommended for Level 2 assemblies.

5. Sequencing analysis of library integrated into the genome:

   (a) Plate yeast strains harboring the pathway genes on noninducing YPDA medium (2% glucose). Cells are grown at 30 °C for 3–4 days.

   (b) Perform colony PCRs using primers to amplify plant-derived ATF/BSs located upstream of each CDS.

   (c) Use the PCR-amplified fragments for sequencing to identify ATF/BS sequences. Moreover, the PCR-amplified fragments can be cloned into the blunt-end cloning vector pJET1.2 (Thermo Scientific CloneJET PCRCloning kit) according to the manufacturer's protocol. The extracted plasmids are subsequently sequenced to identify ATF/BS.

6. To PCR-amplify long DNA fragments (larger than 5 kb), we suggest to use PrimeSTAR GXL DNA Polymerase (Takara Bio, Saint-Germain-en-Laye, France) according to the manufacturer's recommendations.

7. Addition of β-mercaptoethanol at a final concentration of 24 mM increases the transformation efficiency of NEB 5-alpha by 140% (New England Biolabs). This is helpful to increase the retransformation efficiency of libraries with large plasmids.

8. Yeast cells need to be plated on YNB media with the appropriate dominant selection marker.

9. Yeast cells with a disrupted *LYP1* gene can grow in the presence of thialysine, a toxic lysine analog taken up by the cells via a permease encoded by the *LYP1* locus [31].
10. The *ADE2* gene encodes a phosphoribosyl amino imidazole carboxylase required for adenine biosynthesis. Mutations in *ADE2* result in the accumulation of red pigment on YPD medium (without adenine hemisulfate) [32].

Acknowledgments

COMPASS was previously developed through funding by the Federal Ministry of Education and Research of Germany (BMBF; grant numbers FKZ 031A172 and FKZ 031B0223). G.N. received a fellowship from the Potsdam Graduate School, University of Potsdam.

References

1. Nielsen J, Fussenegger M, Keasling J, Lee SY, Liao JC, Prather K, Palsson B (2014) Engineering synergy in biotechnology. Nat Chem Biol 10(5):319–322. https://doi.org/10.1038/nchembio.1519
2. Nielsen J (2013) Production of biopharmaceutical proteins by yeast: advances through metabolic engineering. Bioengineered 4 (4):207–211. https://doi.org/10.4161/bioe.22856
3. Pinto JP, Pereira R, Cardoso J, Rocha I, Rocha M (2013) TNA4OptFlux – a software tool for the analysis of strain optimization strategies. BMC Res Notes 6:175–188
4. Du J, Yuan Y, Si T, Lian J, Zhao H (2012) Customized optimization of metabolic pathways by combinatorial transcriptional engineering. Nucleic Acids Res 40(18):e142. https://doi.org/10.1093/nar/gks549
5. Rodrigues R (2015) Improving microbial chemical production strains through transcriptional regulatory network rewiring. PhD Dissertaion, Imperial College London, Department of Life Sciences
6. Subtil T, Boles E (2012) Competition between pentoses and glucose during uptake and catabolism in recombinant *Saccharomyces cerevisiae*. Biotechnol Biofuels 5:14. https://doi.org/10.1186/1754-6834-5-14
7. Li YJ, Wang MM, Chen YW, Wang M, Fan LH, Tan TW (2017) Engineered yeast with a CO_2-fixation pathway to improve the bio-ethanol production from xylose-mixed sugars. Sci Rep 7:43875. https://doi.org/10.1038/srep43875
8. Thodey K, Galanie S, Smolke CD (2014) A microbial biomanufacturing platform for natural and semisynthetic opioids. Nat Chem Biol 10(10):837–844. https://doi.org/10.1038/nchembio.1613
9. Pfleger BF, Pitera DJ, Smolke CD, Keasling JD (2006) Combinatorial engineering of intergenic regions in operons tunes expression of multiple genes. Nat Biotechnol 24 (8):1027–1032. https://doi.org/10.1038/nbt1226
10. Bond-Watts BB, Bellerose RJ, Chang MC (2011) Enzyme mechanism as a kinetic control element for designing synthetic biofuel pathways. Nat Chem Biol 7(4):222–227. https://doi.org/10.1038/nchembio.537
11. Shen CR, Lan EI, Dekishima Y, Baez A, Cho KM, Liao JC (2011) Driving forces enable high-titer anaerobic 1-butanol synthesis in *Escherichia coli*. Appl Environ Microbiol 77 (9):2905–2915. https://doi.org/10.1128/AEM.03034-10
12. Bujara M, Panke S (2010) Engineering in complex systems. Curr Opin Biotechnol 21:586–591
13. Holtz WJ, Keasling JD (2010) Engineering static and dynamic control of synthetic pathways. Cell 140(1):19–23. https://doi.org/10.1016/j.cell.2009.12.029
14. DiCarlo JE, Norville JE, Mali P, Rios X, Aach J, Church GM (2013) Genome engineering in *Saccharomyces cerevisiae* using CRISPR-Cas systems. Nucleic Acids Res 41(7):4336–4343. https://doi.org/10.1093/nar/gkt135

15. Mitchell LA, Chuang J, Agmon N, Khunsriraksakul C, Phillips NA, Cai Y, Truong DM, Veerakumar A, Wang Y, Mayorga M, Blomquist P, Sadda P, Trueheart J, Boeke JD (2015) Versatile genetic assembly system (VEGAS) to assemble pathways for expression in *S. cerevisiae*. Nucleic Acids Res 43 (13):6620–6630. https://doi.org/10.1093/nar/gkv466
16. Jones JA, Vernacchio VR, Lachance DM, Lebovich M, Fu L, Shirke AN, Schultz VL, Cress B, Linhardt RJ, Koffas MA (2015) ePathOptimize: a combinatorial approach for transcriptional balancing of metabolic pathways. Sci Rep 5:11301. https://doi.org/10.1038/srep11301
17. Wang HH, Isaacs FJ, Carr PA, Sun ZZ, Xu G, Forest CR, Church GM (2009) Programming cells by multiplex genome engineering and accelerated evolution. Nature 460 (7257):894–898. https://doi.org/10.1038/nature08187
18. Lian J, HamediRad M, Hu S, Zhao H (2017) Combinatorial metabolic engineering using an orthogonal tri-functional CRISPR system. Nat Commun 8(1):1688. https://doi.org/10.1038/s41467-017-01695-x
19. Zhou YJ, Gao W, Rong Q, Jin G, Chu H, Liu W, Yang W, Zhu Z, Li G, Zhu G, Huang L, Zhao ZK (2012) Modular pathway engineering of diterpenoid synthases and the mevalonic acid pathway for miltiradiene production. J Am Chem Soc 134(6):3234–3241. https://doi.org/10.1021/ja2114486
20. Noskov VN, Chuang R-Y, Gibson DG, Leem S-H, Larionov V, Kouprina N (2010) Isolation of circular yeast artificial chromosomes for synthetic biology and functional genomics studies. Nat Protoc 6(1):89–96
21. de Boer CG, Hughes TR (2012) YeTFaSCo: a database of evaluated yeast transcription factor sequence specificities. Nucleic Acids Res 40 (Database issue):D169–D179. https://doi.org/10.1093/nar/gkr993
22. Karim AS, Curran KA, Alper HS (2013) Characterization of plasmid burden and copy number in *Saccharomyces cerevisiae* for optimization of metabolic engineering applications. FEMS Yeast Res 13(1):107–116. https://doi.org/10.1111/1567-1364.12016
23. Naseri G, Balazadeh S, Machens F, Kamranfar I, Messerschmidt K, Mueller-Roeber B (2017) Plant-derived transcription factors for orthologous regulation of gene expression in the yeast *Saccharomyces cerevisiae*. ACS Synth Biol 6(9):1742–1756. https://doi.org/10.1021/acssynbio.7b00094
24. Naseri G, Behrend J, Rieper L, Mueller-Roeber B (2019) COMPASS for rapid combinatorial optimization of biochemical pathways based on artificial transcription factors. Nat Commun 10 (1):2615. https://doi.org/10.1038/s41467-019-10224-x
25. Zhang Y, Werling U, Edelmann W (2012) SLiCE: a novel bacterial cell extract-based DNA cloning method. Nucleic Acids Res 40 (8):e55. https://doi.org/10.1093/nar/gkr1288
26. Gibson DG (2011) Enzymatic assembly of overlapping DNA fragments. Methods Enzymol 498:349–361. https://doi.org/10.1016/B978-0-12-385120-8.00015-2
27. Kouprina N, Larionov V (2016) Transformation-associated recombination (TAR) cloning for genomics studies and synthetic biology. Chromosoma 125(4):621–632. https://doi.org/10.1007/s00412-016-0588-3
28. Hochrein L, Machens F, Gremmels J, Schulz K, Messerschmidt K, Mueller-Roeber B (2017) AssemblX: a user-friendly toolkit for rapid and reliable multi-gene assemblies. Nucleic Acids Res 45(10):e80. https://doi.org/10.1093/nar/gkx034
29. Jang CW, Magnuson T (2013) A novel selection marker for efficient DNA cloning and recombineering in *E. coli*. PLoS One 8(2): e57075. https://doi.org/10.1371/10.1371/journal.pone.0057075.t001
30. Bao Z, Xiao H, Liang J, Zhang L, Xiong X, Sun N, Si T, Zhao H (2015) Homology-integrated CRISPR-Cas (HI-CRISPR) system for one-step multigene disruption in *Saccharomyces cerevisiae*. ACS Synth Biol 4(5):585–594. https://doi.org/10.1021/sb500255k
31. Zwolshen JH, Bhattacharjee JK (1981) Genetic and biochemical properties of thialysine-resistant mutants of *Saccharomyces cerevisiae*. J Gen Microbiol 122:281–287
32. Dorfman BZ (1969) The isolation of adenylosuccinate synthetase mutants in yeast by selection for constitutive behavior in pigmented strains. Genetics 61:377–389

Chapter 17

SCRaMbLE-in: A Fast and Efficient Method to Diversify and Improve the Yields of Heterologous Pathways in Synthetic Yeast

Reem Swidah, Jamie Auxillos, Wei Liu, Sally Jones, Ting-Fung Chan, Junbiao Dai, and Yizhi Cai

Abstract

The synthetic chromosome rearrangement and modification by LoxP-mediated evolution (SCRaMbLE) system is a key component of the synthetic yeast genome (Sc2.0) project, an international effort to construct an entire synthetic genome in yeast. SCRaMbLE involves the introduction of thousands of symmetrical LoxP (LoxPsym) recombination sites downstream of every nonessential gene in all 16 chromosomes, enabling numerous genome rearrangements in the form of deletions, inversions, duplications, and translocations by the Cre-LoxPsym recombination system. We highlight a two-step protocol for SCRaMbLE-in (Liu, Nat Commun 9(1):1936, 2018), a recombinase-based combinatorial method to expedite genetic engineering and exogenous pathway optimization, using a synthetic β-carotene pathway as an example. First, an in vitro phase uses a recombinase toolkit to diversify gene expression by integrating various regulatory elements into the target pathway. This combinatorial pathway library can be transformed directly into yeast for traditional screening. Once an optimized pathway which is flanked by LoxPsym sites is identified, it is transformed into Sc2.0 yeast for the in vivo SCRaMbLE phase, where LoxPsym sites in the synthetic yeast genome and Cre recombinase catalyze massive genome rearrangements. We describe all the conditions necessary to perform SCRaMbLE and post-SCRaMbLE experiments including screening, spot test analysis, and PCRTag analysis to elucidate genotype-phenotype relationships.

Key words Metabolic engineering, Yeast, Synthetic biology, SCRaMbLE-in, SCRaMbLE

1 Introduction

Genomic and structural variation lead to phenotypic diversification and drive evolutionary changes within the population of different species [1]. In nature, genomic evolution through single-nucleotide polymorphisms [2], insertion–deletion [2], and structural variation [3] occurs on extremely long time scales. Various techniques to manipulate DNA and accelerate diversification have been developed by genetic engineers. Traditionally, insertion and deletion events are introduced into the genome in vivo or in vitro

Sunil Chandran and Kevin W. George (eds.), *DNA Cloning and Assembly: Methods and Protocols*, Methods in Molecular Biology, vol. 2205, https://doi.org/10.1007/978-1-0716-0908-8_17,

via homologous recombination [4], while mutations are created via site directed/random mutagenesis or error-prone PCR at the level of single nucleotides [5]. More recently, advanced genome-editing technologies such as CRISPR-Cas9 allow the introduction of multiple mutations precisely or randomly [6]. While these techniques are proficient at introducing small-scale genetic modifications, very few techniques that drive large-scale structural evolution and genomic rearrangement have been reported. SCRaMbLE, a recently developed system for diversifying gene expression through genome shuffling, provides a potential solution. Such large-scale rearrangements could enable rapid exploration of genotypic and phenotypic diversity to optimize microbial strains for the production of medicines, biofuels, high-value chemicals, and antibiotics [7].

SCRaMbLE is derived from the synthetic yeast genome project (Sc2.0). The Sc2.0 project is designed to encode an inducible genomic rearrangement system in the synthetic yeast strain. Three main principles were applied in the design of each chromosome (Fig. 1). First, the synthetic strain bearing the synthetic chromosomes must maintain a wild-type phenotype and fitness. Second, all elements that cause genome instability such as tRNA genes, transposons, subtelomeric repeat regions, and introns are removed. All tRNA genes are relocated to a tRNA neochromosome, and the telomeric regions were replaced with a universal telomere cap. Third, genomic flexibility is extended by introducing LoxPsym sites at the 3′ untranslated region of nonessential genes, which enables a large-scale genomic inducible evolution system, known as (SCRaMbLE) (Fig. 2) [8, 9]. The Sc2.0 genome is also designed to encode further modifications, for example all UAG stop codons were swapped for UAA. Short recoded sequences called PCRTag sequences are integrated within the open reading frames (ORFs) in the synthetic genome. The PCRTags facilitate the differentiation between the native and the synthetic chromosomes using simple techniques such as qPCR [8].

Here, we describe a simple protocol to induce SCRaMbLE in a synthetic strain SynII bearing a heterologous β-carotene production pathway on the SCRaMbLE-in vector as an example (Fig. 3) [10]. After induction with Cre in vivo, the synthetic pathway is randomly integrated into the synthetic genome. Concomitantly, Cre will rearrange the synthetic genome and diversify the genetic background of the host via different deletion, inversion, and duplication events (Figs. 4 and 5). This allows for fast selection of the best production strain bearing the desired synthetic pathway. All SCRaMbLEd strains are subjected to PCRTag analysis to identify which genes have been deleted, and whole genome sequencing is used to identify all genomic rearrangements including deletions, inversions, duplications, and insertions. Strains with the desired phenotype can be obtained after one round of SCRaMbLE or can be developed further with multiple independent rounds of SCRaMbLE experiments [11].

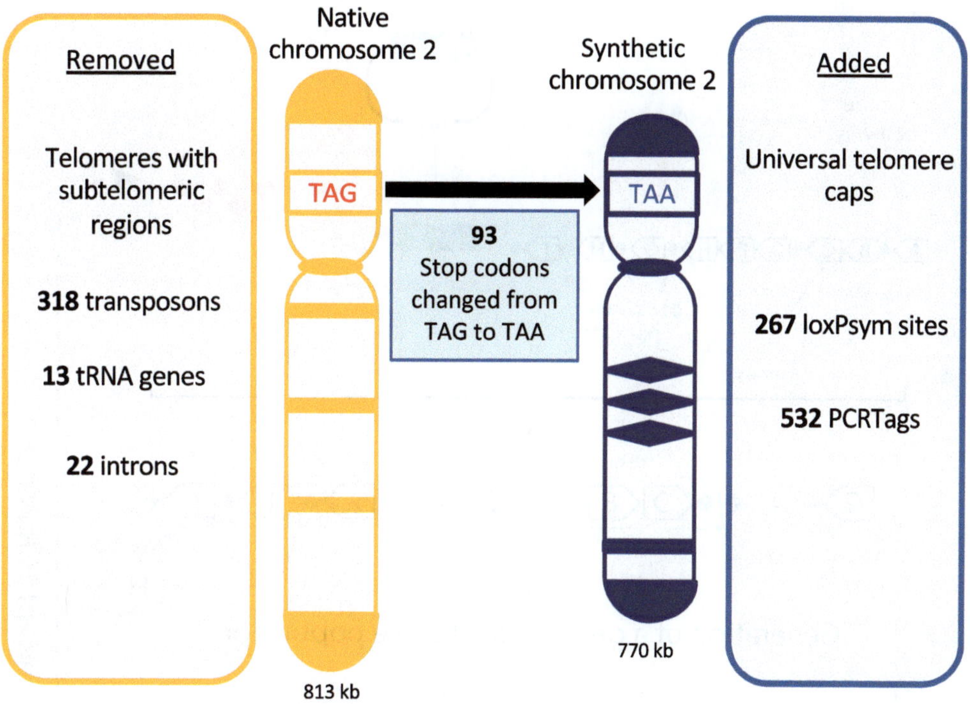

Fig. 1 The principles of Synthetic Yeast Genome Project (Sc2.0). The design principles for the construction of the synthetic chromosome 2 involved the removal of the elements that cause genome instability such as transposons, subtelomeric repeats. and introns, while the telomeric regions were replaced with the universal telomere caps. All tRNA genes were relocated to a separate neochromosome and all TAG stop codons were changed to TAA. The LoxPsym recombination sites were introduced downstream of every nonessential gene to allow structural rearrangements upon induction with Cre recombinase. PCRTags were also introduced as a watermark to allow the discrimination between a synthetic chromosome and a wild-type yeast chromosome

2 Materials

2.1 Strains, Plasmids, and Oligonucleotides

Yeast strains used in this study are listed in Table 1 (*see* **Note 1**). Plasmids used in this study are listed in Table 2. PCRTag verification oligos for the wild-type BY4741, and SynII are listed in Table 3.

2.2 PCR Components

1. 2× GoTaq® Green Master Mix (Promega): GoTaq® DNA Polymerase is supplied in 2× Green GoTaq® Reaction Buffer (pH 8.5): 400 μM dATP, 400 μM dGTP, 400 μM dCTP, 400 μM dTTP, and 3 mM $MgCl_2$. Used for traditional PCR.
2. FastStart Essential DNA Green Master (Roche): Contains FastStart Taq DNA Polymerase and double-stranded DNA specific SYBR Green I dye. Used for LightCycler 480 PCR.

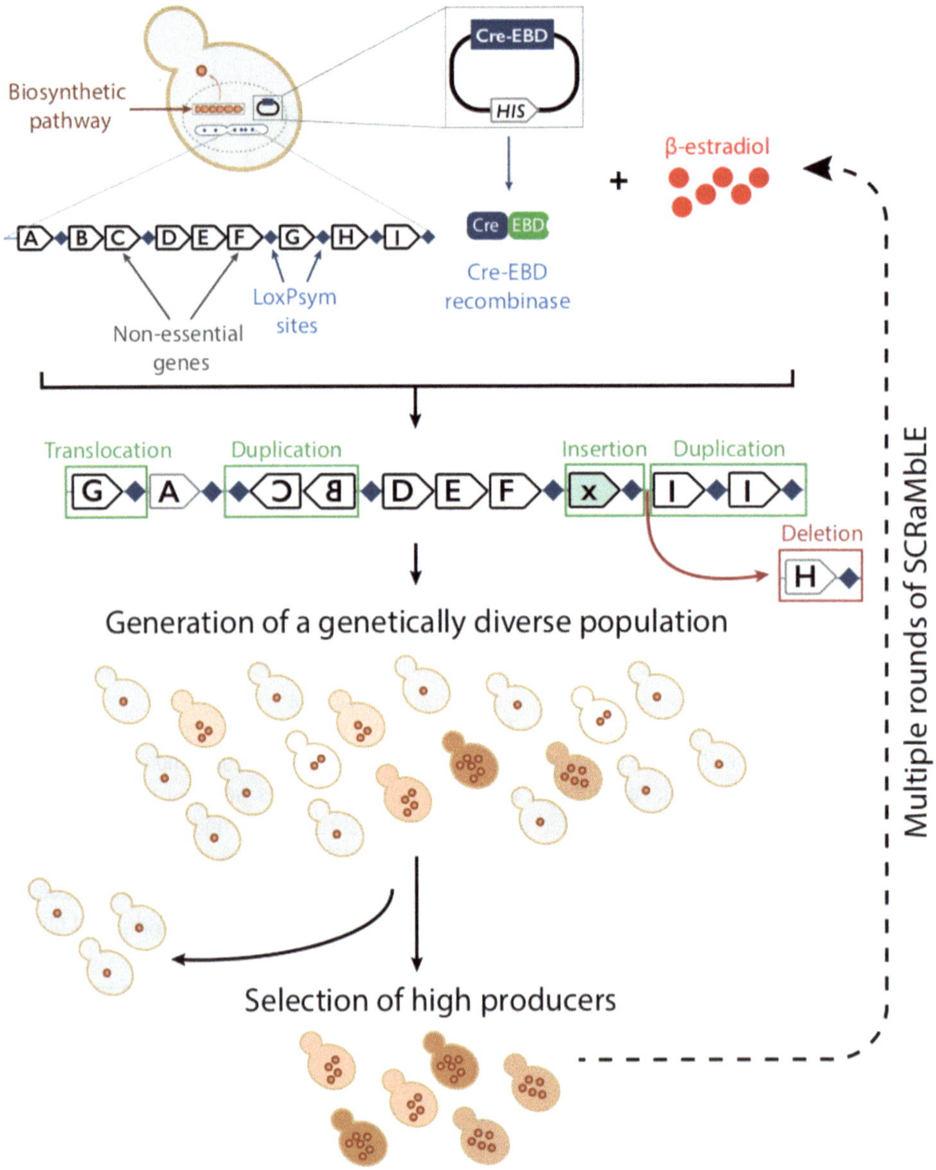

Fig. 2 Overview of the SCRaMbLE method. The biosynthetic pathway for the production of a compound is expressed in a synthetic yeast strain along with a plasmid containing the gene for the Cre recombinase. The expression of Cre recombinase is induced by β-estradiol and triggers genome-wide rearrangements. Various rearrangement events such as translocations, duplications, insertions, inversions, and deletions would occur within the synthetic chromosomes resulting in the generation of a genetically diverse population of SCRaMbLEd yeast strains. High producers are selected and can be further subjected to multiple rounds of SCRaMbLE in order to further improve the production of the compound

2.3 Transformation Components

1. 1 M lithium acetate (LiAc) (lithium acetate dihydrate; Sigma-Aldrich): Dissolve 6.6 g lithium acetate (LiAc) in sterile ddH_2O and complete the volume to 100 mL; filter-sterilize with a 0.22 μm filter. Store at room temperature.

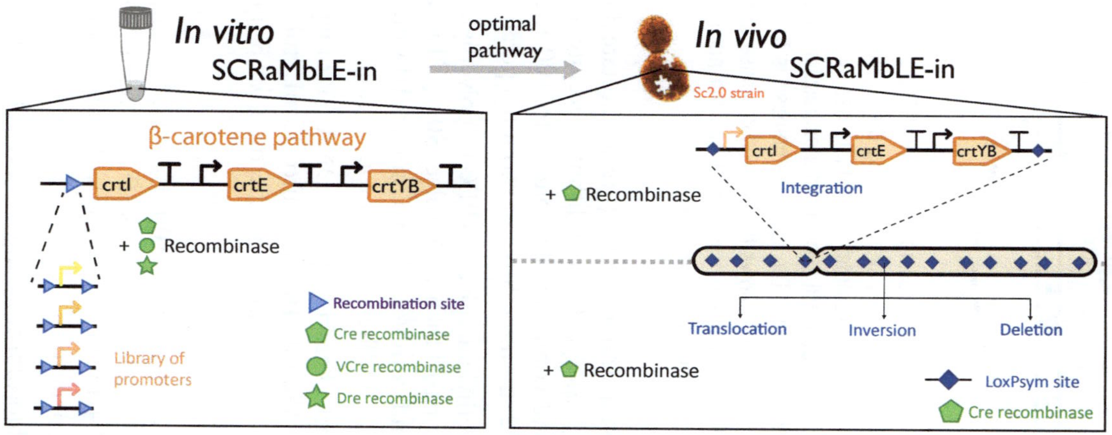

Fig. 3 The SCRaMbLE-in tool kit. An overview of the SCRaMbLE-in toolkit for optimizing production of a compound in a synthetic yeast. In vitro SCRaMbLE-in is a rapid prototyping tool for optimizing pathway flux. A recombination site is placed upstream of "promoter-less" a gene in the heterologous pathway and using a recombination system (three available: Cre-loxP, VCre-Vlox, and Dre-rox), promoters of different activities can be introduced upstream of the "promoter-less" gene. The library of assemblies are transformed into yeast and screened for highly productive strains. This optimized pathway assembly can be carried over to in vivo SCRaMbLE-in, where the heterologous pathway is flanked by LoxPsym sites and introduced into a synthetic yeast strain. As the synthetic strain contains LoxPsym sites downstream of every nonessential gene of a synthetic chromosome, this allows for the integration of the pathway to different locations in the synthetic chromosome. In addition, the synthetic chromosome can undergo multiple rearrangement events including translocations, insertions, deletions, and duplications to optimize the chassis strain for optimal compound production

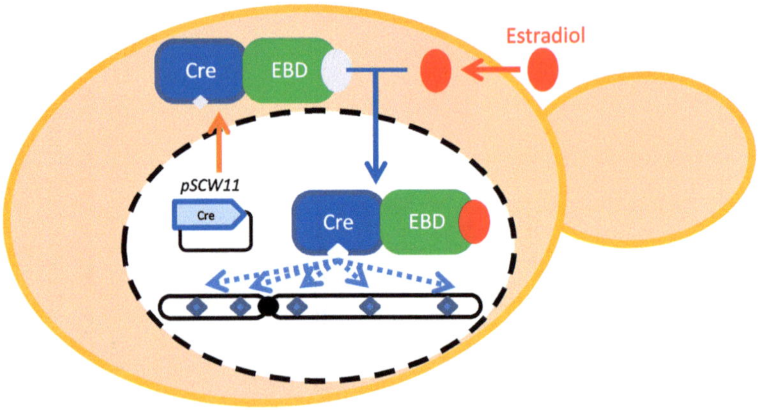

Fig. 4 Cre recombinase system using the *pSCW11*-Cre-EBD construct. Estradiol-dependent SCRaMbLE induction system using Cre-recombinase under the control of a daughter-specific promoter pSCW11 to use in *Saccharomyces cerevisiae*. The *pSCW*11 promoter will allow for the expression of Cre to express only once in the daughter cell life. The Cre recombinase is fused to the estrogen-binding domain (EBD) which allows the Cre recombinase to be expressed upon the induction with estradiol and translocate into the nucleus where it causes different genomic rearrangement events

2. 0.1 M lithium acetate (LiAc) (lithium acetate dihydrate; Sigma-Aldrich): Dissolve 0.66 g lithium acetate (LiAc) in sterile ddH_2O and complete the volume to 100 mL; filter-sterilize with a 0.22 μm filter. Store at room temperature.
3. 50% polyethylene glycol (PEG) (polyethylene glycol MW 3350; Sigma-Aldrich): Dissolve 50 g PEG-3350 in sterile ddH_2O and complete the volume to 100 mL. Filter-sterilize with a 0.22 μm filter. Store at room temperature.
4. 10 mg/mL herring sperm DNA (Promega UK Limited D1815): Aliquot 100 μL into 1 mL tubes and store at −20 °C. Before using, boil the tube for 5 min and place on ice immediately.

2.4 Genomic Extraction Components

1. Solution A: 2% Triton X-100, 1% SDS, 100 mM NaCl, 10 mM Tris–HCl (pH 8), 1 mM Na_2EDTA. Prepare the solution using sterile ddH_2O; filter-sterilize with a 0.22 μm filter. Store at room temperature.
2. Phenol-chloroform-isoamyl alcohol (25:24:1) (Sigma-Aldrich).
3. TE buffer pH 8 (Thermo Fisher Scientific).
4. 100% ethanol.
5. RNase A (100 mg/mL) (Qiagen).

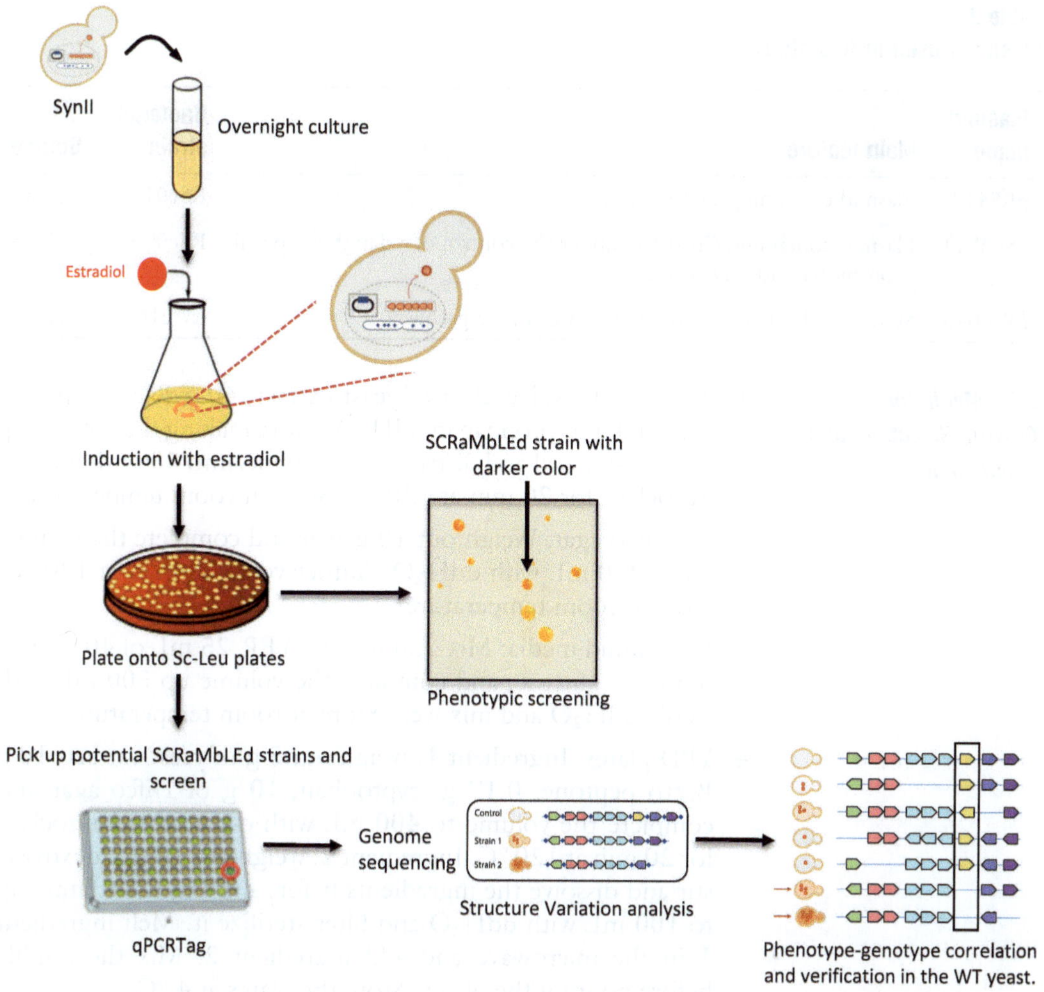

Fig. 5 The workflow using SCRaMbLE to improve and analyze β-carotene production in SCRaMbLEd strains. The workflow for the SCRaMbLE method begins with the growth of the synthetic strain containing the optimal *β-carotene pathway* overnight to saturation. The following day, the culture is back diluted to the desired OD_{600} (0.1), the Cre recombinase is induced with estradiol and incubated for 24 h. Cells are harvested, back diluted to 0.1 OD_{600} and plated onto the selective media and grown for 2–3 days. Cells with the darker orange color are further screened, analyzed by qPCRTag analysis and sequenced to identify the genomic rearrangements associated with the desired phenotype

Table 1
Yeast strains used in this study

Strain	Abbreviation	Genotype	Source
LWy212	BY4741 + pWL045 (SCRaMbLE-in vector containing β-carotene pathway)	*MATA*, *leu2Δ*, *met15Δ*, *LYS2*, *ura3Δ*, *his3Δ1*+ pWL045	Cai lab
LWy033	SynII + pWL045 (SCRaMbLE-in vector containing β-carotene pathway)	*MATA*, *his3Δ1*, *leu2Δ*, *LYS2*, *met15Δ*, *ura3Δ*+ pWL045	Cai lab

Table 2
Plasmids used in this study

Plasmid name	Main feature	Bacterial strain	Source
pRS413	Plasmid containing *HIS3* marker	RSe016	Cai lab
pSCW11	Plasmid containing Cre-EBD under the control of a daughter-specific promoter and a *HIS3* marker	RSe096	Cai lab
pWL045	SCRaMbLE-in vector containing β-carotene pathway	LWe168	Cai lab

2.5 Media for Growth, Selection, and Maintenance

1. 2× YEP: Dissolve 20 g of yeast extract, 40 g Bacto peptone, and 0.64 g tryptophan in ddH_2O before making the volume up to 1 L. Mix well and distribute as 250 mL into 500 mL bottles. Autoclave for 20 min at 120 °C. Store at room temperature.
2. 4% Difco agar: Weigh out 10 g agar and complete the volume up to 250 mL with ddH_2O. Autoclave for 20 min at 120 °C. Store at room temperature.
3. YPD liquid media: Mix 250 mL 2× YEP, 25 mL of 40% filter-sterilized glucose, and complete the volume up 500 mL with sterile ddH_2O and mix well. Store at room temperature.
4. YPD plates: Ingredient 1: weigh out 5 g of yeast extract, 10 g Bacto peptone, 0.17 g tryptophan, 10 g of Difco agar and complete the volume to 400 mL with ddH_2O and autoclave for 20 min at 120 °C. Ingredient 2: weigh out 10 g of dextrose, stir and dissolve the ingredients before making the volume up to 100 mL with ddH_2O and filter-sterilize it. Melt ingredient 1 in the microwave and add ingredient 2. Mix thoroughly before pouring the plates. Store the plates at 4 °C.
5. 2× SC -His, -Leu liquid media: Add 8.5 g of yeast nitrogenous base without ammonium sulfate and amino acids (SLS), 8.5 g ammonium sulfate (Fisher Scientific), 10 g of SC -His, -Leu dropout mix (Formedium), 1.3 g leucine and complete the volume up to 2.5 L with ddH_2O. Autoclave for 20 min at 120 °C. Store at room temperature.
6. 2× SC -Leu liquid media: Mix 250 mL 2× SC -His, -Leu liquid media, 1.5 mL of 100 mM L-histidine hydrochloride, 25 mL of 40% filter-sterilized glucose, and complete the volume up 500 mL with sterile ddH_2O and mix well. Store at room temperature.
7. 1× SC -His, -Leu liquid media: Mix 250 mL 2× SC -His, -Leu, 25 mL of 40% filter-sterilized glucose and complete the volume up to 500 mL with sterile ddH_2O. Store at room temperature.
8. SC -His, -Leu plates: Ingredient 1: weigh out 8.5 g yeast nitrogenous base without ammonium sulfate and amino acids

Table 3
PCRTag verification oligos for the wild type BY4741 and SynII

Oligo for the SynII PCRTag	*Sequence*
SYN_chr02_A1_YBL106C_241-F	TGATGAACTGCTGATACAACCTCTGGCG
SYN_chr02_A1_YBL106C_508-R	CGGCGATATCGTTATCAGAACCGGTAAG
SYN_chr02_A1_YBL106C_1075-F	ATGGACAGTGCTGACATAGCTTGAGCCG
SYN_chr02_A1_YBL106C_1363-R	CTATGGTCAATTGCCAAAGTCAGACGCT
SYN_chr02_A1_YBL106C_2341-F	ATTGGTACTCAAGTCGCCACCTGGG
SYN_chr02_A1_YBL106C_2626-R	TACCTGCATCGAGACCGACCCAAGTTTA
SYN_chr02_A2_YBL105C_1435-F	GGTCAAACCTAAACTTCTACGGTTGCTG
SYN_chr02_A2_YBL105C_1690-R	GTTGGCCGAAAGAGCTAATGCCCCATTA
SYN_chr02_A2_YBL105C_1720-F	ACTAGGACTCAAGTCTGAGCCG
SYN_chr02_A2_YBL105C_1963-R	TAGAGGCACCAAGTGGTGTTGCCAT
SYN_chr02_A2_YBL105C_2989-F	AGCAGCGCTTGATGATCTTTGGTCG
SYN_chr02_A2_YBL105C_3235-R	TGCCCAACAATCACAGGGCGAAAAC
SYN_chr02_A2_YBL104C_403-F	ACCGGCTGAGGTGCTAACAGCTGAT
SYN_chr02_A2_YBL104C_670-R	TTCAGACGTCCAGTCAGCTGCTTTAATC
SYN_chr02_A2_YBL104C_1174-F	GTCGCTACGACTCAAATCCCAACCGCTA
SYN_chr02_A2_YBL104C_1456-R	CTGGAGATGGATTGCTATCGCTAAAGCC
SYN_chr02_A2_YBL104C_2218-F	TCTGCTAGCAGCGCCACTACCG
SYN_chr02_A3_YBL104C_2461-R	TGACACTTCAGTTTTAGCCGCTTCAACC
SYN_chr02_A3_YBL103C_793-F	AGGACTGTAGGTACTGCTAACACTAGCA
SYN_chr02_A3_YBL103C_1015-R	TGATAGCCCAGTTGACCGTGCTATGAAC
SYN_chr02_A3_YBL102W_111-F	CTTATTTAGCTCATGGGCTGACAGCTTG
SYN_chr02_A3_YBL102W_351-R	AGCCAAGACGAACAACAAACTACCCATG
SYN_chr02_A3_YBL101C_52-F	ATTGCCGTGGCTCAACAAGCTGTCAGCT
SYN_chr02_A3_YBL101C_256-R	GTTGTTATCAACCCCTAGCCCAGTTAAC
SYN_chr02_A3_YBL101C_349-F	ACCTGAGTTACCGCTGACGAAAGCACTG
SYN_chr02_A3_YBL101C_781-R	TTCAAAGCCAGACCCAGAGTGCCCAAGT
SYN_chr02_A4_YBL101C_2827-F	TGAGCTAACAACGGCGCTGCTATTGTGA
SYN_chr02_A4_YBL101C_3115-R	TACCACCAAGGGTGGTAACAACAATGGT
SYN_chr02_A4_YBL099W_330-F	TGTCTTGTTTGGCAGTGACCGTTTAGTC
SYN_chr02_A4_YBL099W_630-R	AGTATCCAAAGCAACGGCGGTTTTGCCG
SYN_chr02_A4_YBL099W_1212-F	CGGCTTAAGTGTCAGCAGAGTTGGCAGT

(continued)

Table 3
(continued)

SYN_chr02_A4_YBL099W_1413-R	AGGGACCTGTTCTTCGGTGGCTAATGGT
SYN_chr02_A4_YBL098W_165-F	AGCCAGAGGTATCGACGCCTTAAAGAGC
SYN_chr02_A5_YBL098W_489-R	GCTATAAGCACCATCGCAGCCGATAACG
SYN_chr02_A5_YBL097W_327-F	CAACGGTGCTAGCAATGGTGACGATTCA
SYN_chr02_A5_YBL097W_582-R	CAATGATTTAGCACCGCCTTCGTCAAAG
SYN_chr02_A5_YBL097W_1587-F	TCGTGGTGATGTCTCAAGCGGTTTGTTC
SYN_chr02_A5_YBL097W_1776-R	ATCGACGCCATCCATATCGTCTTCGAAA
Oligo for the WT PCRTag	*Sequence*
WT_chr02_A1_YBL106C_241-F	ACTACTGGAAGATATGCATCCTCGAGCA
WT_chr02_A1_YBL106C_508-R	TGGTGACATTGTCATTCGCACAGGGAAA
WT_chr02_A1_YBL106C_1075-F	GTGAACGGTTGAAACGTAAGAGCTACCA
WT_chr02_A1_YBL106C_1363-R	TTACGGACAACTTCCCAAAAGCGATGCA
WT_chr02_A1_YBL106C_2341-F	GTTTGTGGATAGATCACCGCCCGGA
WT_chr02_A1_YBL106C_2626-R	CACTTGTATTGAGACTGATCCGTCCTTG
WT_chr02_A2_YBL105C_1435-F	AGTGAGTCCCAAGGAACGTCTATTCGAC
WT_chr02_A2_YBL105C_1690-R	ATTAGCTGAAAGGGCAAACGCTCCTTTG
WT_chr02_A2_YBL105C_1720-F	GGATGGGGAAAGATCACTACCA
WT_chr02_A2_YBL105C_1963-R	CCGTGGTACTAAATGGTGCTGTCAC
WT_chr02_A2_YBL105C_2989-F	TGCCGCAGAACTACTACGTTGATCA
WT_chr02_A2_YBL105C_3235-R	CGCTCAACAAAGTCAGGGTGAAAAT
WT_chr02_A2_YBL104C_403-F	TCCAGCACTAGTAGAGACGGCGCTA
WT_chr02_A2_YBL104C_670-R	CAGCGATGTTCAGAGTGCCGCACTGATA
WT_chr02_A2_YBL104C_1174-F	ATCAGATCTGGAAAGGTCCCATCCTGAG
WT_chr02_A2_YBL104C_1456-R	ATGGAGGTGGATCGCGATTGCAAAGGCT
WT_chr02_A2_YBL104C_2218-F	CCTTGACGCTGCACCGGATCCA
WT_chr02_A3_YBL104C_2461-R	CGATACCAGTGTATTGGCTGCAAGTACG
WT_chr02_A3_YBL103C_793-F	TGGGGAATATGTGGAAGATACGGATGCT
WT_chr02_A3_YBL103C_1015-R	CGACTCACCTGTGGATAGAGCAATGAAT
WT_chr02_A3_YBL102W_111-F	ATTGTTCTCAAGCTGGGCGGATTCTCTC
WT_chr02_A3_YBL102W_351-R	CGCAAGAACAAATAGTAGGGACCCCATT
WT_chr02_A3_YBL101C_52-F	GTTACCATGAGAGAGTAGCGAATCTGCG
WT_chr02_A3_YBL101C_256-R	ATTATTGAGCACTCCATCGCCCGTCAAT
WT_chr02_A3_YBL101C_349-F	GCCACTATTGCCCGAAACAAATGCGGAA

(continued)

Table 3 (continued)

WT_chr02_A3_YBL101C_781-R	CAGTAAACCTGATCCTGAGTGTCCGTCC
WT_chr02_A4_YBL101C_2827-F	GCTTGACACTACAGCTGATGAGTTATGG
WT_chr02_A4_YBL101C_3115-R	CACAACGAAAGGCGGCAATAATAACGGC
WT_chr02_A4_YBL099W_330-F	CGTTCTTTTCGGTTCCGATAGACTGGTT
WT_chr02_A4_YBL099W_630-R	GGTGTCTAAGGCGACAGCAGTCTTACCT
WT_chr02_A4_YBL099W_1212-F	TGGTTTGTCCGTTTCTCGTGTCGGTTCC
WT_chr02_A4_YBL099W_1413-R	TGGAACCTGTTCTTCTGTAGCCAAAGGA
WT_chr02_A4_YBL098W_165-F	TGCTCGTGGCATTGATGCTCTGAAATCA
WT_chr02_A5_YBL098W_489-R	AGAGTATGCTCCGTCACAACCTATGACA
WT_chr02_A5_YBL097W_327-F	GAATGGGGCATCTAACGGAGATGACAGT
WT_chr02_A5_YBL097W_582-R	TAAACTCTTGGCGCCACCTTCATCGAAA
WT_chr02_A5_YBL097W_1587-F	CAGAGGCGACGTTAGTTCGGGGCTTTTT
WT_chr02_A5_YBL097W_1776-R	GTCAACACCGTCCATGTCATCTTCAAAC
WT_chr02_A5_YBL093C_100-F	GTTACCTGAATTTGAGCCGGATTGGCTT
WT_chr02_A5_YBL093C_340-R	CATGATGATACAACCACCTCAGCAGGGT
WT_chr02_A5_YBL091C_757-F	ATGGTTGAGAGAGAGACCCGTTGGAAAC
WT_chr02_A5_YBL091C_1039-R	CCCAGAAGGTGCGTGGATGGACTATCAT
WT_chr02_B1_YBL089W_189-F	ATCTGAGAACGCCTCGTTTGCTAAACTC
WT_chr02_B1_YBL089W_540-R	CCCTCTCTCTAGCTGATGCCGATTTACA
WT_chr02_B1_YBL088C_247-F	TGGTGACATTACCCAGGAATAAAGGGGA
WT_chr02_B1_YBL088C_466-R	AGGGGAGCCTATTCATATAGACCTGGGA

(SLS), 25 g ammonium sulfate (Fisher Scientific), 10 g SC -His, -Leu dropout mix (Formedium). Stir and dissolve and before completing the volume up to 2 L with ddH_2O. Ingredient 2: weigh out 100 g dextrose, stir and dissolve before completing the volume up to 500 mL with ddH_2O and filter it. Ingredient 3: weigh out 100 g Difco agar complete the volume up to 2.5 L with ddH_2O. Autoclave ingredients 1, 2, and 3 separately for 20 min at 120 °C. Mix all the ingredients together after melting the agar using the microwave, and pour plates with ~25 mL volume per plate. Store at room temperature or 4 °C. For SC -Leu plates, 0.3 g of L-histidine hydrochloride should be added to ingredient 1 and the same recipe should be followed.

9. 50% glycerol: Dilute glycerol v/v with sterile ddH_2O and filter-sterilize it. Store at room temperature.

2.6 Equipment and Consumables

1. 12-channel multichannel pipettor.
2. Certus-Flex (Fritz Gyger AG).
3. LightCycler 480 (LC480) (Roche Life Science).
4. Echo 550 machine (Labcyte).
5. MP Biomedicals™ FastPrep™-96 Instrument.
6. Pipette tips: 1000 μL, 200 μL, and 10 μL.
7. Reagent reservoirs (Matrix Technologies Corp Ltd).
8. Nunc™ OmniTray™ Single-Well Plate (Thermo Fisher Scientific).
9. 96-well plates (Nunc).
10. Replica blocks: Round replica blocks (Cora Styles).
11. 250 mL flasks. Sterilized by autoclaving.
12. Zirconia/Silica Bead, 0.5 mm diameter (BioSpec).
13. LightCycler® 480 Multiwell 384-well plate, white (Roche).
14. LightCycler® 480 SYBR Green I master (Roche).

3 Methods

3.1 Transforming pRS413/pSCW11 Vectors into the SynII + pWL045 (SCRaMbLE-in Vector Containing β-Carotene Pathway) Synthetic Strain

This section describes the introduction of the pSCW11 vector bearing Cre-EBD into the appropriate strains, which will be used in SCRaMbLE including the wild type strain and the semisynthetic strain. pRS413 is used as a control in both strains.

1. Set up a 5 mL culture from a fresh colony for BY4741 + pWL045 (SCRaMbLE-in vector containing β-carotene pathway) and SynII + pWL045 (SCRaMbLE-in vector containing β-carotene pathway) and grow them overnight in SC -Leu liquid media.
2. Incubate at 30 °C in a shaking incubator (180 rpm) overnight.
3. The following morning, measure the OD_{600} for each strain.
4. Back-dilute to OD_{600} 0.15 in 25 mL SC -Leu.

$$0.15/OD_{600} \times 25 = x\ \text{mL}.$$

5. Incubate at 30 °C until $OD_{600} \approx 0.8$ (approx. 3–4 h).
6. Transfer the culture into a 50 mL falcon tube.
7. Centrifuge (3750 × *g*) for 5 min.
8. Remove the supernatant.
9. Resuspend in 12.5 mL H_2O (1/2 original vol.).
10. Repeat **steps 7** and **8**.
11. Resuspend the pellet in 1 mL 100 mM lithium acetate and transfer to a 1.5 mL microcentrifuge tube.

Table 4
The reagents of the transformation mix for each sample

Volume	Reagent
266.67 µL	50% PEG
40 µL	1 M lithium acetate
12.5 µL	10 mg/mL herring sperm DNA (boiled for 5 min and chilled on ice prior to use)
2 µL	Plasmid (100 ng) or sterile ddH_2O (negative control)
78.83 µL	Sterile ddH_2O

12. Centrifuge (16,200 × *g*) for 15 s.
13. Remove supernatant.
14. Resuspend the pellet in 250 µL of 100 mM lithium acetate.
15. Transfer 2 × 50 µL into 1.5 mL microcentrifuge tube (one for the sample and one for the −ve control).
16. Prepare the transformation mix shown in Table 4. Add the reagents in order in the order they are listed.
17. Vortex the tubes vigorously for 1 min.
18. Incubate the tubes on rotation at 30 °C for 1 h.
19. Heat shock in a 42 °C water bath at for 20 min.
20. Centrifuge (3750 × *g*) for 15 s.
21. Remove transformation mix with a pipette.
22. Resuspend in 200 µL sterile ddH_2O.
23. Plate out 200 µL onto the selection media SC -His, -Leu.
24. Incubate the plates at 30 °C for 2–3 days and check the transformants (*see* **Note 2**).

3.2 SCRaMbLE Procedure

This section describes induction of SCRaMbLE in the β-carotene pathway strains via the addition of estradiol.

1. Set up a 5 mL culture from a fresh colony for BY4741 + pWL045 (SCRaMbLE-in vector containing the β-carotene pathway) + pRS413, SynII + pWL045 (SCRaMbLE-in vector containing the β-carotene pathway) + pRS413, and SynII + pWL045 (SCRaMbLE-in vector containing the β-carotene pathway) + pSCW11 bearing Cre-EBD and grow them overnight in SC -His, -Leu liquid media using glucose as a carbon source.
2. The following morning, back-dilute the overnight culture to OD_{600} 0.1 in 25 mL with the same media (SC -His, -Leu) in a 250 mL flask.

3. To check whether SCRaMbLE worked in the synthetic strain, back-dilute the overnight culture of SynII + pWL045 (SCRaMbLE-in vector containing β-carotene pathway) to OD_{600} 0.1 in 50 mL with the same media (SC -His, -Leu) as two 25 mL culture samples in 250 mL flasks.
4. To sample 1, add estradiol (1 mM prepared in 100% ethanol) to a final concentration of 1 μM. To sample 2, add nothing. Grow the samples at 30 °C for 12–24 h using the shaking incubator at 180 rpm and measure the OD_{600} after 2, 4, 6, 12, and 24 h (*see* **Note 4**).
5. Back-dilute all of the strains to OD_{600} 0.1 in 1 mL sterile ddH_2O.
6. Spread 50 μL of the OD_{600} 0.1 culture from each strain after SCRaMbLE onto an individual SC -Leu plate.
7. Incubate the plate for 2–4 days at 30 °C (*see* **Note 3**).

3.3 Selecting for SCRaMbLEd Mutants Which Have the Desired Phenotype

1. Select for SCRaMbLEd mutants which have the desired phenotype and pick up the healthy colonies which have a darker color of β-carotene and restreak them out to acquire single colonies on fresh SC -Leu plates.

3.4 Enforce the Strains with Desired Phenotypes After SCRaMbLE to Lose the pSCW11 Vector

1. Incubate the strains with the desired phenotype after SCRaMbLE on SC -Leu plates for 1 day at 30 °C.
2. Replica plate colonies onto SC -Leu and SC -His plates.
3. Pick up the strains from the Leu plate which could not grow on the SC -His plates (*see* **Note 5**).

3.5 Checking the Phenotype of the SCRaMbLEd Mutants

1. Restreak all the strains with the desired phenotype (after SCRaMbLE) to single colonies on SC -Leu plates and incubate at 30 °C overnight.
2. Use an SC -Leu plate as a primary plate and replica plate colonies onto different dropout plates (SC -His, SC -Ura, SC -Leu, SC -Lys, SC -Trp, SC -Ade, and SC -Met) and a control YPD plate.
3. Incubate the replica plates at 30 °C overnight.
4. Assess the phenotype of the strains and take a photo of the plates.

3.6 Preparing the Glycerol Stock for All the SCRaMbLEd Mutants

1. Set up a 5 mL culture from a fresh colony of the wild-type strain BY4741 + pWL045 (SCRaMbLE-in vector containing β-carotene pathway) and SynII + pWL045 (SCRaMbLE-in vector containing β-carotene pathway) as controls in addition to all potential SCRaMbLEd candidates and grow them overnight in YPD liquid media.
2. Add 900 μL of the fresh overnight culture + 900 μL of 50% glycerol to a cryotube and store it at −80 °C.

3.7 Performing a Spot Test Assay for the Selective Candidates

Perform a spot test assay in order to check the desired phenotype for the SCRaMbLEd mutants which have already lost the Cre plasmid and compare it with the WT strain and the synthetic strain as controls using a tenfold series of dilutions starting from OD_{600} 0.1.

1. Set up a 5 mL culture from a fresh colony for all the strains which were mentioned in Subheading 3.6 and grow them overnight in YPD liquid media.
2. Back-dilute each of the overnight cultures to OD_{600} 0.1 in 1 mL sterile ddH_2O.
3. Prepare tenfold dilutions for each strain (1/10, 1/100, 1/1000, 1/10,000).
4. Spot 5 μL for all the dilutions from each strain onto an SC -Leu plate.
5. Incubate the SC -Leu plate for 1–2 days at 30 °C.
6. Take a photo of the plates of the spot test assay.

3.8 Genomic DNA Extraction

Extract the yeast genomic DNA prior to verification qPCRTag analysis (from Hoffman and Winston [12]).

1. Set up a 10 mL yeast culture from a fresh colony for each strain and grow them to the saturation phase (overnight is best) in SC -Leu liquid at 30 °C.
2. Collect the cells by centrifugation (3750 × *g* for 5 min).
3. Resuspend the pellet in 0.5 mL of sterile ddH_2O.
4. Transfer the cells to a 1.5 mL Eppendorf tube with a screw top and collect the cells by centrifugation (16,200 × *g* for 10 s).
5. Decant the supernatant and resuspend the cell pellet in the residual liquid by vortexing.
6. Add 0.3 mL of Solution A.
7. Add 0.3 mL of phenol-chloroform-isoamyl alcohol (25:24:1).
8. Add 0.3 g of Zirconia/Silica Beads, 0.5 mm Diameter (BioSpec). Break cells for two rounds of 45 s using a Fast-prep machine, placing the samples on ice in between. Add 0.2 mL of TE buffer (pH 8) and vortex.
9. Microfuge at 16,200 × *g* for 5 min.
10. Transfer the top aqueous layer to new tube (dispose of the old tube in a Waste Disposal Bins for Hazardous Waste—be careful with the remaining phenol in the tube).
11. To the aqueous fraction in the new tube, add 1 mL of 100% ethanol and mix by inversion.
12. Place the tubes at −20 °C for 30 min.
13. Microfuge at 16,200 × *g* for 10 min.

Table 5
The components for the PCRTag master mix using traditional PCR

Master mix	For one reaction (μL)	For 96 reactions + 4 μL as a dead volume (μL)
2× GoTaq Green mix	6.25	625
10 μM Forward verification primer	0.75	75
10 μM reverse verification primer	0.75	75
Sterile ddH_2O	2.25	225
(20 ng/μL DNA template)	2.50	25
Final volume	12.5	125

14. Discard the supernatant and resuspend the pellet in 0.4 mL of TE buffer plus 3 μL of a 10 mg/mL solution of RNase A (0.3 μL if 100 mg/mL).
15. Incubate for 5 min at 37 °C. Add 40 μL of 3 M sodium acetate and 1 mL of 100% ethanol. Mix by inversion.
16. Microfuge at 16,200 × *g* for 10 min.
17. Discard the supernatant.
18. Air-dry the remaining pellet.
19. Wash the pellet with 1 mL 70% ethanol, respin (16,200 × *g* for 5 min) and remove the ethanol.
20. Air-dry again and resuspend pellet in 50 μL of water (this represents approximately 2–4 μg/μL of DNA).

3.9 Verification qPCRTag Using the Traditional Method

The verification forward and reverse WT/SynII PCRTag primers will aid in differentiating between the native chromosome and the synthetic chromosome (*see* **Notes 6–8**).

1. Prepare GoTaq Green PCR master mix for the desired number of PCRs (*see* Table 5).
2. Aliquot 10 μL GoTaq Green PCR master mix into PCR tubes or 96 well plate.
3. Add 2.5 μL of the genomic DNA to each tube or well.
4. Run the PCR reactions using the PCR thermocycling program (*see* Table 6).
5. Load 5–12 μL of each PCR product on an agarose gel and run the samples for 35 min by electrophoresis in order to visualize and assess the presence/absence of the amplicons in the WT, SynII, and SCRaMbLEd strains (Fig. 6a, b).

Table 6
The PCR thermocycling program

Step	Program
Step 1.	98 °C 2 min
Step 2.	98 °C 10 s
Step 3.	52 °C 1 min 30 s
Step 4.	72 °C 1 min/kb
Step 5.	Return to step 2 31 times (32 cycles total)
Step 6.	72 °C 5 min
Step 7.	4 °C HOLD

3.10 Verification qPCRTag Using the LightCycler 480 (LC480)

This method is adapted and modified from Mitchell et al. [13].

1. Prepare the genomic DNA extraction (*see* Subheading 3.8).
2. Prepare 384 μL of qPCR master mix, such as LightCycler® 480 SYBR Green I master (*see* Table 7) in a 1.5 mL microfuge tube. Vortex to mix and add around 50 μL of qPCR master mix as a dead volume which will be lost during dispensing with the Certus-Flex.
3. Dispense the premixed forward and reverse WT/SynII PCRTag verification primers using an acoustic liquid transfer system, such as the Echo 550 (*see* **Note 9**).
4. Dispense the master mix into the LightCycler® 480 Multiwell 384-well plate using any bulk dispenser (Certus-Flex) (*see* **Notes 10–12**).
5. Set up your program using the LightCycler software as shown in Table 8. Insert the LightCycler® 480 Multiwell 384-well plate into the LightCycler® 480 instrument and run the program.
6. Use the qPCR software to analyze the results as it shows the amplification, melt curves and there is an automated crossing point (Cp) calling algorithm built into the software which considers the shape, slope, and time for the amplification curve while determining positive/negative result. The outcome from the PCR reactions will be a red signal for the negative result indicating the absence of the amplicon and green signal for the positive result relating to the presence of the amplicon (Fig. 6c).
7. Export data from the Software as a .txt file.

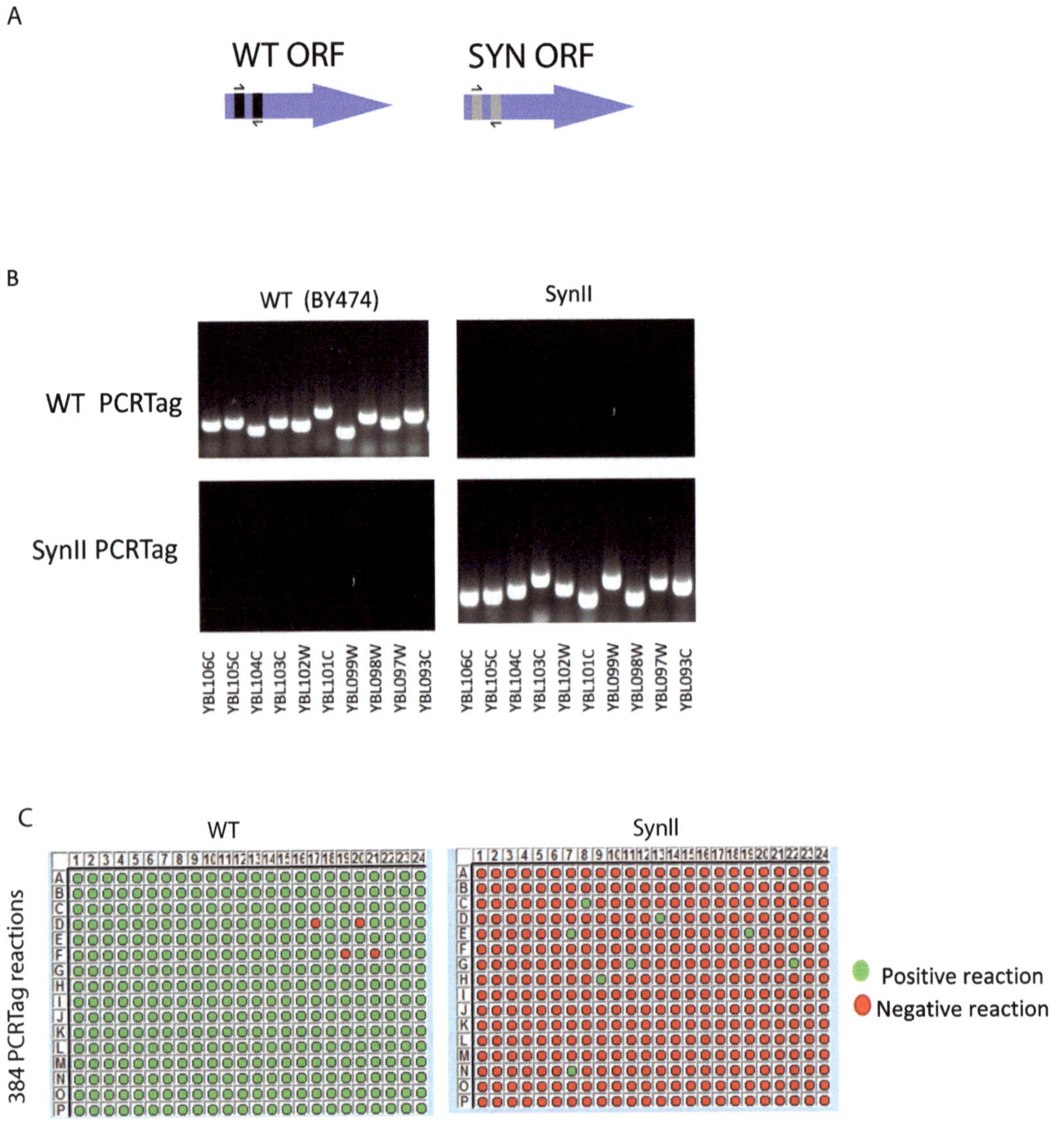

Fig. 6 PCRTag analysis using (**a**) traditional method, (**b**) LC480 method. (**a**) PCRTags are short, unique, and recoded sequences within the open reading (ORF) frame for each gene and which differentiate between the WT and synthetic gene in *S. cerevisiae.* (**b**) PCRTag verification of each gene in the synthetic chromosome and the WT chromosome is carried out traditionally by PCR amplification and the presence or absence of each PCR product is visualized on an agarose gel. (**c**) The q-PCRTag verification is carried out using the LC480 method for 384 reactions. The green color represents a positive result while the red color is a negative result

Table 7
The components for GoTaq Green PCR master mix using the LC480

Master mix	One reaction (μL)	384 reactions + 50 μL as a dead volume (μL)
2× LightCycler® 480 SYBR Greenl master	0.5	217
Sterile ddH_2O	0.4	173.6
20 ng/μL DNA template	0.1	43.4
Final volume	1	434

Table 8
The PCRTag program for the LightCycler 480 (LC480)

Stage	*T*	Time	Ramp rate C/s	
1. Preincubation	95	5 min	4.8	No acquisition
2. Amplification	95	10 s	4.8	No acquisition
Annealing	64	15 s	2.5	Single acquisition
Extinction Note step 2 × 30 cycles	72	1 min	4.8	
3. Melt curve	95	5 s	4.8	No acquisition
	65	1 min	2.5	No acquisition
	97	–	0.11	5 acquisition
4. Cool	40	30 s	30 s	

4 Notes

1. All of the strains and vectors used in this protocol are available as individual strains, which can be obtained from the lab of Prof. Patrick Cai (https://www.cailab.org). Any project details can also be provided upon request.
2. For transformant selection, pick four to eight healthy transformants and streak them out to single colonies onto fresh SC -His, -Leu plates. There are often colonies of differing sizes on a transformation plate; avoid tiny colonies as these usually do not grow after streaking to the next round of selection.
3. You should see a lower OD_{600} for the yeast induced with estradiol (sample 1) than the yeast without estradiol (sample 2; ten times lower). This will be a sign that SCRaMbLE has worked.

4. Failure of the SCRaMbLE experiment could relate to the absence of the estradiol in the media as it induces the expression of Cre recombinase, or the absence of the pSCW11 vector containing Cre-EBD, or a mutation that makes Cre dysfunctional. In order to explore this, restriction analysis with different restriction enzymes should be carried out on the vector.
5. It is very important to be sure that the promising candidates after screening have lost the pSCW11 vector bearing the Cre-EBD recombinase—otherwise there is a potential chance to lose the desired phenotype as one of the disadvantages of Cre-EBD recombinase is that a small amount is liable to leak into the nucleus even without the induction by estradiol.
6. The verification forward and reverse WT/SynII PCRTag primers will aid in differentiating between the native chromosome and the synthetic chromosome. However, missing the particular amplicon for a particular gene from the SCRaMbLEd strains would show a knockout event within the synthetic chromosome.
7. Premixed PCRTag verification primers (forward and reverse) were synthesized by Integrated DNA Technologies and distributed into 96-well and 384-well plates using an Echo 550 acoustic dispenser.
8. For the PCRTag analysis using the traditional PCR reaction, when the confirmation PCRTag analysis relating to the wild type is done, the genomic DNA of the wild type strain should be used as a positive control, while the genomic DNA of SynII is the negative control. On the other hand, when the confirmation PCRTag analysis relating to the SCRaMbLEd strains is done, the genomic DNA of the wild-type strain should be used as a negative control, while the positive control is the genomic DNA of the SynII strain.
9. Prior to using the Echo, defrost the Echo® Qualified 384-well plate containing 10 μM of the premixed forward and reverse PCRTag verification primers and centrifuge at 3750 × *g* for 1 min. Ensure that the LightCycler® 480 Multiwell 384-well plate (Roche) is defined in the Echo software. In Echo plate reformat, select tools > labware definitions . . . and add/import the plate definition if required. For the liquid transfer system, use the Plate Reformat software or alternatively provide a custom a .csv file to program the transfer. Open Echo plate reformat and create a new protocol. In Source plate, in plate format choose 384PP (polypropylene). In default plate type choose 384PP_AQ_BP2 (AQ_BP2 is the calibration for the specific type of liquid: buffer without surfactants). In Destination plate, Choose the defined LightCycler® 480 Multiwell 384-well plate. In Mapping mode, from custom, select the

wells you want to dispense from the source plate to the destination plate. Create a dispensing program with the appropriate amount that you need to dispense. From region information adjust the dispensing volume to dispense 100 nL of 10 μM premixed forward and reverse PCRTag verification primers into the desired wells of the LightCycler® 480 Multiwell 384 plate. Remove the seal from the source plate and run the protocol (optionally, first run a simulation). As instructed by the UI, insert the source plate (384PP) and then insert the destination plate (LightCycler® 480 Multiwell 384-well plate) into the Echo® 550/555 Liquid Handler. Once the protocol is complete, remove the LightCycler® 480 Multiwell 384-well destination plate from the Echo 550 and spin down. Remove the source plate and seal with an appropriate seal, applying a roller to remove any bubbles if necessary. Store the plate at −20 °C.

10. When using a Certus-Flex, it is important to clean the syringe (use water first then 70% ethanol, then water) and calibrate the instrument prior to use. When adding the master mix solution to the syringe, do not fill it up to the top as this can risk leaking and contamination with various solutions. Close the lid tightly, press air on, check the right volume you want to dispense by clicking "prime." Click "flush." The syringe should be washed by the cleaning system using water by flushing. Apply the cleaning cycle 2–3 times to make sure it is entirely clean. During use, click on the Certus software and create a dispensing program set up, go to "new experiment," go to "serial," select the wells that you want to dispense from by highlighting the region, right-click on the plate, from edit region select "fluids," then from add fluids select the fluid you want to dispense and choose the appropriate volume (e.g., to a final volume of 1000 nL). Press "run" to dispense the master mix into the appropriate wells. Once you have finished, remove the master mix by flushing, and wash with water and 70% ethanol. After you finish dispensing by Certus-Flex, seal the plate immediately in order to avoid evaporation. Spin down the plate by centrifugation (3750 × *g* for 2 min) and keep it away from light. Put the plate in the LightCycler® 480 in order to run.

11. If the PCR reaction fails, it may be necessary to optimize PCR parameters. These parameters include the number of the steps for the PCR reactions, including two steps or three steps, melting temperature and the number of cycles. Users should also check all of the reagents for the PCR reaction such as 2× GoTaq Green, primers, or dNTPs. Minimizing false positive results in the PCRTag analysis can be achieved by decreasing the number of PCR cycles and by omitting the positive signals at the end of the cycles. By considering the difference

between the amplification time for the real amplicons and the amplifications which happen at the end of the cycle, one can determine which amplicons are more likely to be related to primer dimers and should therefore be considered false positives. Nevertheless, the quality and the concentration of the genomic DNA for each strain should be detected and optimized very carefully as using very low/high concentrations of the DNA could increase the chances of PCR failure. The ideal concentration for the genomic DNA is 7.5 ng per reaction when the q-PCRTag is performed by the traditional method and <5 ng per reaction when the q-PCRTag is performed by LC480. If nonspecific PCR products are obtained: All PCR conditions need to be optimized in order to increase the efficiency of the PCRTag reaction producing the correct PCRTag product and decrease the chance of getting nonspecific bands. For example, it may be necessary to optimize the annealing temperature and the number of PCR cycles.

12. Using the traditional method, all PCRTag product sizes are supposed to be between 200 and 500 bp, so any band with a larger size should be considered a nonspecific band.

Acknowledgments

Special thanks to Dr. Eva Garcia-Ruiz for the invaluable advice and discussion for using the Echo® 550/555 Liquid Handler and Certus-Flex and Dr. Daniel Schindler for his help with the technical issues, and BBSRC for funding BB/P02114X/1 (to Y.C.) and Royal Society Newton Advanced Fellowship R123288 (to J.D. and Y.C.).

References

1. Peter J et al (2018) Genome evolution across 1,011 Saccharomyces cerevisiae isolates. Nature 556(7701):339–344
2. Mills RE et al (2006) An initial map of insertion and deletion (INDEL) variation in the human genome. Genome Res 16 (9):1182–1190
3. Radke DW, Lee C (2015) Adaptive potential of genomic structural variation in human and mammalian evolution. Brief Funct Genomics 14(5):358–368
4. Thomason LC et al (2014) Recombineering: genetic engineering in bacteria using homologous recombination. Curr Protoc Mol Biol 106:1.16.1–1.1639
5. Ellis HM et al (2001) High efficiency mutagenesis, repair, and engineering of chromosomal DNA using single-stranded oligonucleotides. Proc Natl Acad Sci U S A 98(12):6742–6746
6. Li W et al (2013) Simultaneous generation and germline transmission of multiple gene mutations in rat using CRISPR-Cas systems. Nat Biotechnol 31(8):684–686
7. Yadav VG et al (2012) The future of metabolic engineering and synthetic biology: towards a systematic practice. Metab Eng 14(3):233–241
8. Dymond JS et al (2011) Synthetic chromosome arms function in yeast and generate phenotypic diversity by design. Nature 477:471

9. Richardson SM et al (2017) Design of a synthetic yeast genome. Science 355 (6329):1040–1044
10. Liu W et al (2018) Rapid pathway prototyping and engineering using in vitro and in vivo synthetic genome SCRaMbLE-in methods. Nat Commun 9(1):1936
11. Jia B et al (2018) Precise control of SCRaMbLE in synthetic haploid and diploid yeast. Nat Commun 9(1):1933
12. Hoffman CS, Winston F (1987) A ten-minute DNA preparation from yeast efficiently releases autonomous plasmids for transformation of Escherichia coli. Gene 57(2–3):267–272
13. Mitchell LA et al (2015) qPCRTag analysis - a high throughput, real time PCR assay for Sc2.0 genotyping. J Vis Exp 99:e52941

Index

A

B

C

Sunil Chandran and Kevin W. George (eds.), *DNA Cloning and Assembly: Methods and Protocols*, Methods in Molecular Biology, vol. 2205, https://doi.org/10.1007/978-1-0716-0908-8, © Springer Science+Business Media, LLC, part of Springer Nature 2020

T

U

V

W

X

MIX
Papier aus verantwortungsvollen Quellen
Paper from responsible sources
FSC® C105338

If you have any concerns about our products,
you can contact us on
ProductSafety@springernature.com

In case Publisher is established outside the EU,
the EU authorized representative is:
Springer Nature Customer Service Center GmbH
Europaplatz 3, 69115 Heidelberg, Germany

Printed by Libri Plureos GmbH
in Hamburg, Germany